Viscoelasticity and Rheology

Publication No. 53
of the Mathematics Research Center
The University of Wisconsin–Madison

Academic Press Rapid Manuscript Reproduction

Viscoelasticity and Rheology

Edited by

Arthur S. Lodge Michael Renardy John A. Nohel

Mathematics Research Center
The University of Wisconsin–Madison
Madison, Wisconsin

Proceedings of a Symposium
Conducted by the Mathematics Research Center
The University of Wisconsin–Madison
October 16–18, 1984

1985

ACADEMIC PRESS, INC.
Harcourt Brace Jovanovich, Publishers
Orlando San Diego New York Austin
London Montreal Sydney Tokyo Toronto

ACADEMIC PRESS, INC.
Orlando, Florida 32887

United Kingdom Edition published by
ACADEMIC PRESS INC. (LONDON) LTD.
24–28 Oval Road, London NW1 7DX

LIBRARY OF CONGRESS CATALOGING-IN-PUBLICATION DATA

Main entry under title:

Viscoelasticity and rheology.

"Proceedings of a symposium conducted by the Mathematics Research Center, The University of Wisconsin-Madison, October 16–18, 1984."
Includes index.
1. Viscoelasticity—Congresses. 2. Rheology—Congresses. I. Lodge, Arthur S. II. Renardy, Michael. III. Nohel, John A. IV. University of Wisconsin—Madison. Mathematics Research Center.
QA931.V55 1985 532'.053 85-48018
ISBN 0-12-454940-3 (alk. paper)

PRINTED IN THE UNITED STATES OF AMERICA

85 86 87 88 9 8 7 6 5 4 3 2 1

Contents

Contributors

Numbers in parentheses indicate the pages on which the authors' contributions begin.

Robert C. Armstrong (361), *Department of Chemical Engineering, Massachusetts Institute of Technology, Cambridge, Massachusetts 02139*

A. N. Beris (361), *Department of Chemical Engineering, Massachusetts Institute of Technology, Cambridge, Massachusetts 02139*

R. Byron Bird (105), *Department of Chemical Engineering and Rheology Research Center, University of Wisconsin–Madison, Madison, Wisconsin 53706*

R. A. Brown (361), *Department of Chemical Engineering, Massachusetts Institute of Technology, Cambridge, Massachusetts 02139*

Bernard D. Coleman (125), *Department of Mathematics, Carnegie-Mellon University, Pittsburgh, Pennsylvania 15213*

J. M. Crissman (81), *Polymers Division, National Bureau of Standards, Gaithersburg, Maryland 20899*

Constantine Dafermos (221), *Lefschetz Center for Dynamical Systems, Division of Applied Mathematics, Brown University, Providence, Rhode Island 02912*

Hanswalter Giesekus (157), *Department of Chemical Engineering, University of Dortmund, D-4600 Dortmund 50, West Germany*

Ole Hassager (1), *Danmarks Tekniske Højskole, Lyngby, Denmark, and Mathematics Research Center and Chemical Engineering Department, University of Wisconsin–Madison, Madison, Wisconsin 53706*

Daniel D. Joseph (235), *Department of Aerospace Engineering and Mechanics, University of Minnesota, Minneapolis, Minnesota 55455*

J. V. Lawler (361), *Department of Chemical Engineering, Massachusetts Institute of Technology, Cambridge, Massachusetts 02139*

Arthur S. Lodge (181), *Department of Engineering Mechanics, Engineering Experiment Station and Mathematics Research Center, University of Wisconsin–Madison, Madison, Wisconsin 53705*

R. C. MacCamy (323), *Department of Mathematics, Carnegie–Mellon University, Pittsburgh, Pennsylvania 15213*

David S. Malkus (391), *Mathematics Research Center and Engineering Mechanics Department, University of Wisconsin–Madison, Madison, Wisconsin 53706*

S. J. Muller (361), *Department of Chemical Engineering, Massachusetts Institute of Technology, Cambridge, Massachusetts 02139*

Michael Renardy (345), *Mathematics Research Center and Mathematics Department, University of Wisconsin–Madison, Madison, Wisconsin 53706*

Roger I. Tanner (421), *Department of Mechanical Engineering, University of Sydney, Sydney, New South Wales 2006, Australia, and Mathematics Research Center, University of Wisconsin–Madison, Madison, Wisconsin 53706*

E. K. Walsh (13), *Department of Engineering Sciences, University of Florida, Gainesville, Florida 32611*

K. Walters (47), *Department of Applied Mathematics, The University College of Wales, Penglais, Aberystwyth SY23 3BZ, Wales*

W. O. Williams (209), *Department of Mathematics, Carnegie–Mellon University, Pittsburgh, Pennsylvania 15213*

L. J. Zapas (81), *Polymers Division, National Bureau of Standards, Gaithersburg, Maryland 20899*

Preface

Viscoelasticity and rheology, fields of wide-ranging technological importance, have undergone major developments in recent years. In spite of significant advances, the understanding and construction of constitutive models for such materials is still far from complete. Various models, based partially on molecular theories, have been suggested and, as a first step, have provided useful insights, but none is without serious deficiencies. From a continuum viewpoint, the study of special flow problems, such as viscometric and steady elongational flows, has provided much valuable information. Recently, attention has shifted toward more complex flows, such as flows through contractions, flow over a slot, flow between eccentric cylinders, and unsteady motions of viscoelastic fluids and solids. Such problems often reveal important differences between various constitutive theories and sometimes between different materials. The computation and numerical simulation of such problems, as well as an understanding of the underlying mathematical features of the equations used to model them, are important for evaluating the consequences of choosing a particular constitutive theory and for analyzing the resulting performance of the model in relation to experiments.

The symposium on Viscoelasticity and Rheology, organized by the Mathematics Research Center and held at the University of Wisconsin–Madison on 16–18 October 1984, brought together a wide variety of international experts from various disciplines.

The contributions to the symposium are divided into four broad categories: experimental results, constitutive theories, mathematical analysis, and computation. Experimental topics include the motion of bubbles in viscoelastic fluids (Hassager), wave propagation in viscoelastic solids (Walsh), flows through contractions (Walters), and cold-drawing of polymers (Zapas). Topics in constitutive theories include molecular theories for polymer solutions and melts based on statistical mechanics (Bird), the use and limitations of approximate constitutive theories (Coleman), a comparison of constitutive laws based on various molecular theories (Giesekus), network theories and some of their advantages in relation to experiments (Lodge), and models for viscoplasticity (Williams). Topics in mathematical analysis include the existence, regularity, and development of singularities (Dafermos), change of type (Joseph), interface problems in viscoelasticity (MacCamy), existence for initial value problems and steady flows, and propagation and development of singularities (Renardy). Topics in computation include numerical simulation of flow between eccentric cylinders (Armstrong), flow around spheres and bubbles (Hassager), the hole pressure problem (Malkus), and a review of computational problems related to various constitutive laws (Tanner).

We gratefully acknowledge support by the United States Army under Contract No. DAAG29-80-C-0041 and the National Science Foundation under Grant No. DMS-8210950. We thank Gladys Moran for the efficient organization of the symposium and Elaine DuCharme for assembling the proceedings.

Arthur S. Lodge
Michael Renardy
John A. Nohel

THE MOTION OF VISCOELASTIC FLUIDS AROUND SPHERES AND BUBBLES

Ole Hassager

Danmarks Tekniske Højskole
Lyngby, Denmark and
Mathematics Research Center and
Chemical Engineering Department
University of Wisconsin
Madison, Wisconsin

INTRODUCTION

This paper is concerned with the translation of bubbles and solid spheres in a viscoelastic fluid. The motion is assumed to be caused by a gravitational field or an imposed force, and we will consider the situations in which the motion takes place in an unbounded fluid, quiescent far from the object as well as the situation in which the motion takes place in a finite container (a cylinder). First we will review some experimental observations of bubble shapes, velocity fields and friction coefficients. In the next section we then consider perturbation solutions for motions that are so slow that the viscoelastic fluid behaves almost as a Newtonian fluid. In this so-called low Deborah number limit analytical solutions may be obtained for bubble shapes, friction factors and velocity fields by perturbation methods. In the last section we consider some numerical simulations that apply also for intermediate Deborah numbers.

2. EXPERIMENTAL EVIDENCE

One of the most striking features of translating air bubbles in viscoelastic fluids is that they develop a cusp at the rear pole. Often this effect may be observed in an almost full shampoo bottle by rapidly turning it upside down and watching an air bubble rise. More controlled experiments by Astarita and Apuzzo (1965) with a series of bubbles of increasing volume show a transition from a spherical bubble shape to an elongated ellipsoidal shape and the development of the cusp at the rear pole. Astarita and Apuzzo also measured the rise velocity of the bubbles as a function of the volume, and showed that for some viscoelastic fluids there is a critical volume at which the rise velocity appears to have a discontinuous increase when plotted as function of the volume. Similar measurements have been performed by Calderbank, Johnson and Loudon (1970) and Leal, Skoog and Acrivos (1971) who also documented discontinuous jumps in the rise velocity. The actual value of the jump depends on the particular polymer/ solvent system as well as the temperature. The above investigators have reported jumps by factors in the range of 2 - 10, and the effect is in fact quite remarkable. It is possible that the velocity discontinuity is related to the development of the cusp at the rear pole. It is proposed that the "cusped" bubbles are really not closed surfaces, but rather open surfaces in which the "cusps" continue into thin gas filaments that eventually dissolve in the liquid. In some fluids (Hassager (1979)) the "cusps" lose rotational symmetry and take the form of a knife edge. In these circumstances the knife edge appears to continue into a thin sheet of air that supposedly eventually dissolves in the liquid. Both in the situation where the bubbles extend into filaments or into sheets there must be a critical total volume at which the boundary condition at the rear pole changes from one involving a stagnation point into another condition with no stagnation point. This could certainly give a discontinuous change in rise velocity and one may argue qualitatively that at least two mechanisms that would retard the bubble when it has a rear stagnation point will not be present when the bubble does not close at the rear pole. First as long as the bubble has a rear stagnation point there may be an accumulation of surface active impurities near that point that would cause immobilization of the surface. Second with the stagnation point present there will be an elongational flow near the rear pole that could account for much of the drag on the bubble.

More detailed information on the flow around bubbles and spheres may be obtained by laser Doppler measurements of the fluid velocity fields. The technique, unfortunately, is limited to fluid velocities well above those at which the velocity discontinuity takes place. We will refer to this region as the high Deborah number region. In this region the following two phenomena have been found and documented by laser Doppler anemometry:

First, in the wake region behind the bubble the fluid velocity as seen by an observer stationary with respect to the fluid far from the bubble is in the opposite direction to that in which the bubble is moving (Hassager (1979)). This wake flow is strikingly different from wake flow in Newtonian fluids where fluid is always pulled with the bubble, and has been termed "negative wake". It has been demonstrated also by flow visualization by Contanceau and Hajjam (1982).

Second, in the wake flow region the fluid velocity field (referred to an observer on the bubble) is not steady even at Reynolds numbers much less than unity (Bisgaard 1983). This observation however is currently limited to one particular polymer solution.

The above described two phenomena in the wake behind bubbles in viscoelastic fluids have been observed also in wakes behind solid spheres, whereas the velocity discontinuity at low Deborah numbers does not occur for spheres.

3. LOW DEBORAH NUMBER FLOW

For intermediate and high Deborah numbers there is considerable ambiguity as to the correct constitutive equation to be used for incompressible viscoelastic fluids. However in the limit as the Deborah number tends to zero the correct expression for the stress tensor $\underset{\sim}{\tau}$ is the retarded motion expansion, which through terms of third order may be written:

$$\underline{\underline{\tau}} = -b_1[\underline{\underline{\gamma}}_{(1)} + B_2\underline{\underline{\gamma}}_{(2)} + B_{11}\underline{\underline{\gamma}}_{(1)}\cdot\underline{\underline{\gamma}}_{(1)} + B_3\underline{\underline{\gamma}}_{(3)} + B_{12}(\underline{\underline{\gamma}}_{(1)}\cdot\underline{\underline{\gamma}}_{(2)} + \underline{\underline{\gamma}}_{(2)}\cdot\underline{\underline{\gamma}}_{(1)}) + B_{1:11}\underline{\underline{\gamma}}_{(1)}(\underline{\underline{\gamma}}_{(1)}:\underline{\underline{\gamma}}_{(1)}) + \ldots] \quad (3.1)$$

where

$$\underline{\underline{\gamma}}_{(1)} = \underline{\nabla v} + (\underline{\nabla v})^{\dagger} \text{ and}$$

$$\underline{\underline{\gamma}}_{(n+1)} = \frac{D}{Dt}\underline{\underline{\gamma}}_{(n)} - \{(\underline{\nabla v})^{\dagger}\cdot\underline{\underline{\gamma}}_{(n)} + \underline{\underline{\gamma}}_{(n)}\cdot(\underline{\nabla v})\} \quad (3.2)$$

Here b_1 is the zero shear-rate viscosity, B_2 and B_{11} are constants with dimension of time and B_3, B_{12} and $B_{1:11}$ are constants with dimension of time squared. The B_2 and B_{11} are related to the zero-shear-rate first and second normal stress coefficients $\Psi_{1,0}$ and $\Psi_{2,0}$ by $\Psi_{1,0} = -2b_1B_2$ and $\Psi_{2,0} = b_1B_{11}$. Values of the parameters for various molecular models may be obtained from Table 1 of Bird (1984).

In any given flow situation one may define a characteristic strain-rate, κ say. For example for flow around a spherical object of radius R_0 one could choose $\kappa = V/R_0$ where V is the rise velocity of the object. Then one may expand the velocity and pressure fields as:

$$\underline{v} = \underline{v}_1 + \kappa B_2\underline{v}_2 + \kappa B_{11}\underline{v}_{11} + \cdots \tag{3.3}$$

$$p = p_1 + \kappa B_2 p_2 + \kappa B_{11} P_{11} + \cdots \tag{3.4}$$

where κB_2 and κB_{11} are Deborah numbers that meter the importance of the nonlinear terms associated with B_2 and B_{11} respectively. For creeping motion these expansions give:

To zeroth order:

$$(\underline{\nabla}\cdot\underline{v}_1) = 0 \tag{3.5}$$

$$b_1\nabla^2\underline{v}_1 - \underline{\nabla}p_1 + \underline{g} = 0 \tag{3.6}$$

which are the Newtonian creeping flow equations.

To first order in κB_2 :

$$(\underline{\nabla}\cdot\underline{v}_2) = 0 \tag{3.7}$$

$$b_1\nabla^2\underline{v}_2 - \underline{\nabla}p_2 = -(b_1/\kappa)\ \underline{\nabla}\cdot\underline{\underline{\gamma}}_{(2)1} \tag{3.8}$$

where $\underline{\underline{\gamma}}_{(2)1}$ is the $\underline{\underline{\gamma}}_{(2)}$ tensor constructed from $\underline{v}_1$.

To first order in κB_{11} :

$$(\underline{\nabla}\cdot\underline{v}_{11}) = 0 \tag{3.9}$$

$$b_1\nabla^2\underline{v}_{11} - \underline{\nabla}p_{11} = -(b_1/\kappa)\underline{\nabla}\cdot(\underline{\underline{\gamma}}_{(1)1}\cdot\underline{\underline{\gamma}}_{(1)1}) \tag{3.10}$$

where $\underline{\underline{\gamma}}_{(1)1}$ is the $\underline{\underline{\gamma}}_{(1)}$ tensor constructed from $\underline{v}_1$. The perturbation technique is shown here only through terms of first order in κB_2 and κB_{11} but it can be carried to any order. At each order one solves an equation similar to the Stokes equation for creeping motion of a Newtonian fluid, but with inhomogeneous terms given by the solution to the lower order flow problem. The $\underline{v}^{(1)}$ and $p^{(1)}$ must satisfy the physical boundary conditions for the given flow situation. The perturbation fields must satisfy conditions such that $\underline{v}$ and p still satisfy the physical boundary conditions independent of κB_2 and κB_{11}.

The above perturbation scheme for small Deborah numbers has been used to study the translation of a nearly spherical bubble in an unbounded viscoelastic fluid. The bubble rises at velocity V due to the gravitational acceleration g and is characterized by the radius R_0 of a sphere of the same volume as the bubble. In addition the surface of the bubble is characterized by the coefficient of surface tension σ that enters into the dimensionless Capillary number $Ca = b_1 V/\sigma$. The leading term in the deformation of the bubble has been found to be (Hassager (1977)):

$$R(\theta)/R_0 = 1 + Ca \frac{V}{20R_0} (7\Psi_{1,0} + 4\Psi_{2,0}) P_2(\cos\theta) + 0(Ca^n De^m) \quad ; \quad n + m > 1 \qquad (3.11)$$

Here $R(\theta)$ is the radius of the bubble as function of θ in a spherical coordinate system (r,θ,ϕ) with origin at the center of volume of the bubble and axis coincident with the symmetry axis of the bubble. Also the Deborah numbers have been rewritten in terms of the zero shear-rate normal stress coefficients, and P_n is the nth Legendre polynomial. Note that the second normal stress coefficient plays a relatively minor role in this expression, so that the deformation is primarily determined by the first normal stress coefficient. Since $\Psi_{1,0} \geq 0$ the calculation shows that the bubble is deformed into the shape of an ellipsoid elongated at the poles (a "prolate" ellipsoid). Tiefenbruck and Leal (1980) have extended the analysis to terms of first order in the Capillary number but of second order in some Deborah numbers. The calculated deformations are at least qualitatively in agreement with the experiments but no experiments have been performed under sufficiently controlled conditions that a quantitative comparison can be made. Moreover since the calculations are limited to small deformations they cannot give any insight into the cusp at the rear end, the velocity discontinuity or the high Deborah number velocity fields.

4. NUMERICAL SIMULATIONS AT SMALL AND INTERMEDIATE DEBORAH NUMBERS

In the previous section we saw that the perturbation scheme at small Deborah numbers is of limited interest for the purpose of calculating bubble deformations. This limitation is not just associated with the treatment of the bubble deformation, but is inherent in the use of the retarded motion constitutive equation. For example in the calculation of the drag on a solid sphere translating in an unbounded fluid (Giesekus (1963)) the retarded motion expansion gives a correction to the Stokes drag which is quadratic in the Deborah numbers and therefore of limited use at intermediate or high Deborah numbers. Also calculations of interaction effects or the effects of container boundaries becomes somewhat tedious and the predictions are at odds with currently available experiments, Brunn (1980). For this reason we turn in this section to applications of more realistic constitutive equations. One empirical constitutive equation that is receiving considerable attention is the factorized K-BKZ equation:

$$\underline{\underline{\tau}}(t) = \int_{-\infty}^{t} M(t-t')\left[\frac{\partial W}{\partial I_1}\,\underline{\underline{B}}(t,t') + \frac{\partial W}{\partial I_2}\,\underline{\underline{B}}^{-1}(t,t')\right]dt' \qquad (4.1)$$

Here $\underline{\underline{B}}(t,t')$ and $\underline{\underline{B}}^{-1}(t,t')$ are finite strain tensors defined in terms of the fluid motion. If a fluid particle has coordinates x_i at time t, and coordinates x_i' at some previous time t' then a displacement gradient tensor $\underline{\underline{E}}$ can be defined with components $E_{ij} = \partial x_i/\partial x'_j$, and then the finite strain tensor $\underline{\underline{B}}$ is given as $\underline{\underline{B}} = \underline{\underline{E}}\cdot\underline{\underline{E}}^{\dagger}$, and $\underline{\underline{B}}^{-1}$ as the inverse to $\underline{\underline{B}}$. Also in this equation $I_1 = \mathrm{tr}\,\underline{\underline{B}}$ and $I_2 = \mathrm{tr}\,\underline{\underline{B}}^{-1}$. Finally $M(t-t')$ is the linear viscoelastic memory function and $W = W(I_1, I_2)$ is a dimensionless potential function. The functions M and W may be adjusted to approximate experimental measurements as shown for example by Bach and Hassager (1984).

Flow calculations with integral constitutive equations such as the factorized K-BKZ equation above are more complicated than calculations based on the retarded motion expansion mainly because of the need to evaluate the memory integral. Here we will briefly outline the Lagrangian finite element technique for the simulation of flow situations as developed by Hassager and Bisgaard (1983). In this method a given amount of fluid is discretized by a mesh of finite elements. Each node corresponds to a fluid particle and all fluid particles inside the elements are denoted by a unique set of element coordinates. Thus if in a reference state the coordinates of node q are X^o_{iq}, where i = 1,2,3, then the coordinates of an arbitrary particle may be approximated by

$$\tilde{x}^o_i = \sum_{q=1}^{N} X^o_{iq} \phi^q \tag{4.2}$$

where N is the total number of nodes and the ϕ^q are the global interpolation functions. During flow the element mesh deforms but the fluid particles retain their element coordinates. Thus at times t' and t the coordinates of the given arbitrary particle are approximated by

$$\tilde{x}_i(t') = \sum_{q=1}^{N} X_{iq}(t')\phi^q \quad ; \quad \tilde{x}_i(t) = \sum_{q=1}^{N} X_{iq}(t)\phi^q \tag{4.3}$$

A Lagrangian calculation is inherently an initial value problem. In the present formulation this means that the motion of each node is known until some initial time. A flow simulation consists then of formulating and solving a set of integral equations for the node coordinates X_{iq} as functions of time.

The technique has been used to simulate the flow around a sphere moving along the axis of a cylindrical tube. Rather than the complete factorized K-BKZ equation, Hassager and Bisgaard used a simple two constant version of it, namely the Lodge rubberlike liquid with a single time constant. This model is obtained from Eq. 4.1 with the functions:

$$W = I_1 \quad ; \quad M(s) = \frac{\eta_0}{\lambda^2} e^{-s/\lambda} \tag{4.4}$$

The two parameters of the model are the zero-shear-rate viscosity η_0 and the time constant λ. The motion of the sphere is caused by a force F_s acting in the axial direction, and the velocity of the sphere is obtained as a function of time. The steady state velocity is denoted V_s. Hassager and Bisgaard considered the situation in which a sphere of radius R_s is initially at rest on the axis of a cylinder of radius R_c

filled with the fluid. At some initial time t_0 a force F_s pointing in the axial direction is exerted on the sphere and the transient motion of the sphere and fluid is simulated.

One purpose of the calculations is to find the eventual steady velocity V_s of the sphere obtained when the transient behavior is negligible. This velocity in nondimensional form is the Deborah number:

$$De = \frac{V_s \lambda}{R_s} \tag{4.5}$$

Also when the force F_s is made nondimensional one obtains the "wall correction factor"

$$K = \frac{F_s}{6\pi\eta_0 R_s V_s} \tag{4.6}$$

Computed values for K as function of (R_s/R_c) and De are given in Table 1. We see that the K-factor for spheres with R_s/R_c = 0.01 (that is, practically an unbounded fluid) changes very little with Deborah number. On the other hand for larger values of R_s/R_c the K-factor depends strongly on De.

The simulated K-values were compared with experiments for (R_s/R_c) = 0.3, 0.4 and 0.5 and a 1% solution of polyacrylamide in glycerine at 293 K as shown in Figure 1. The measurements have been reduced to non-dimensional form with the parameters η_0 = 21.0 Pa·s and λ = 12.0 sec. We see that the simulation results have made it possible to characterize at least this particular fluid by a zero shear-rate viscosity and a time constant in an inhomogeneous flow experiment.

TABLE 1

Simulation Results for $K(De, R_s/R_c)$ for the Single Time Constant Lodge Rubberlike Liquid

Crochet (1982):						
$R_s/R_c = 0.01$						
De	0	0.5	1.0			
K	1.010	1.006	1.001			
Bisgaard and Hassager (1983):						
$R_s/R_c = 0.01$						
De	0	0.95	2.18	3.17		
K	1.022[a]	1.054	1.022	0.998		
ε[b]	-	0.003	0.002	0.47		
$R_s/R_c = 0.10$						
De	0	0.21	1.04	1.59	2.39	
K	1.263[a]	1.33	1.28	1.26	1.16	
ε[b]	-	0.0001	0.0003	0.007	0.17	
$R_s/R_c = 0.20$						
De	0	0.032	0.39	0.59	0.92	1.58
K	1.680[a]	1.73	1.71	1.69	1.62	1.58
ε[b]	-	0.001	0.002	0.001	0.02	0.17
$R_s/R_c = 0.30$						
De	0	0.093	0.47	0.96		
K	2.371[a]	2.38	2.36	2.31		
ε[b]	-	0.001	0.001	0.26		
$R_s/R_c = 0.40$						
De	0	0.02	0.33	0.81		
K	3.596[a]	3.60	3.36	3.08		
ε[b]	-	0.009	0.008	0.035		
$R_s/R_c = 0.50$						
De	0	0.011	0.12	0.21	0.57	
K	5.970[a]	5.99	5.67	5.34	4.35	
ε[b]	-	0.003	0.001	0.02	0.11	

[a] The values of K for De=0 are those for a Newtonian liquid, Happel and Brenner (1973).

[b] The values of $\varepsilon = -\lambda(dV_s/dt)/V_s$ are computed at the termination of each simulation. These values <u>may</u> give an indication of the extent to which a steady-state motion has been obtained.

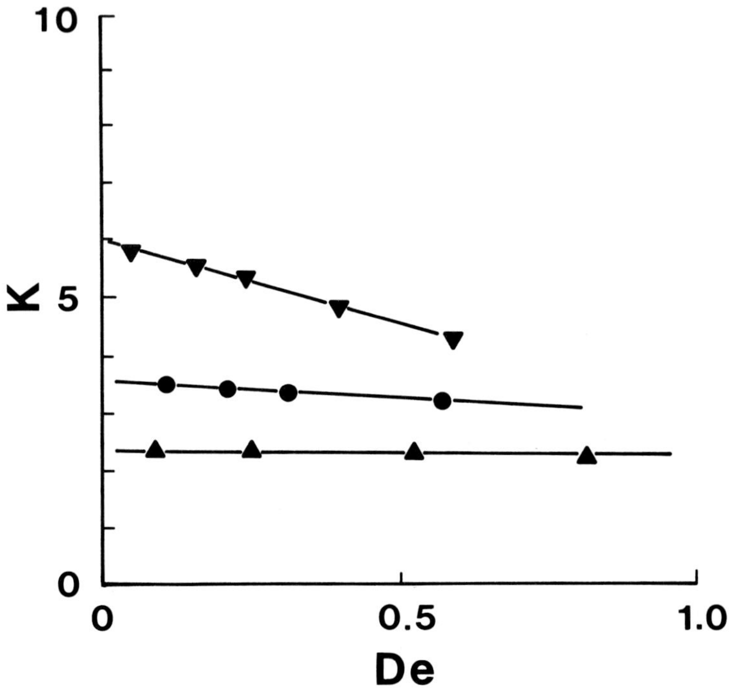

FIGURE 1 Comparison of simulations and experimental values for the wall correction factor K defined in Eq. (4.6) (Bisgaard and Hassager 1983).

——— Simulations

■, ▲, ● Experiments with a 1% solution of polyacrylamide in glycerol at 293 K, plotted with parameters η_0 = 21.0 Pa·s and λ = 12.0 sec.

REFERENCES

1. Astarita, G. and Apuzzo, G. (1965), AIChE Journal, 11, 815-820.
2. Bach, P. and Hassager, O. (1984), MRC Technical Summary Report No. 2755, University of Wisconsin, Madison.
3. Bird (1984) Symposium on Viscoelasticity, Proceedings, MRC, University of Wisconsin, Madison.
4. Bisgaard, C. (1983) J. Non-Newtonian Fluid Mech., 12, 283-302.
5. Brunn, P. (1980) J. Non-Newtonian Fluid Mech., 7, 271-288.
6. Calderbank, P. H., Johnson, D. S. L., and Loudon, J. (1970) Chem. Eng. Sci., 25, 235-256.
7. Coutanceau, M. and Hajjam, M. (1982) Appl. Sci. Res., 38, 199-207.
8. Crochet, M. J. (1982), "The Flow of a Maxwell Fluid Around a Sphere" in Finite Elements in Fluids, Vol. 4 (Ed., Gallagher, R.H.) Wiley, New York.
9. Giesekus, H. (1963) Rheol. Acta, 3, 59-71.
10. Happel, J. and Brenner, H. (1973), "Low Reynolds Number Hydrodynamics", Noordhoff, Leyden.
11. Hassager, O. (1977) in "Chemical Engineering with Per Soltoft" (Eds. Østergaard, K. and Fredenslund, Aa.) Teknisk Forlag, Copenhagen.
12. Hassager, O. (1979) Nature, 279, 402-403.
13. Hassager, O. and Bisgaard, C. (1983) J. Non-Newtonian Fluid Mech., 12, 153-164.
14. Leal, L. G., Skoog, J. and Acrivos, A. (1971) Can. J. Chem. Eng., 49, 569-575.
15. Tiefenbruck, A.F. and Leal, L.G. (1980) J. Non-Newtonian Fluid Mech., 7, 257-264.

The manuscript was prepared while the author held a position as visiting professor at the Mathematics Research Center and the Department of Chemical Engineering, The University of Wisconsin.

WAVE PROPAGATION IN VISCOELASTIC SOLIDS

E. K. Walsh

Department of Engineering Sciences
University of Florida
Gainesville, Florida

1. INTRODUCTION

An important aspect in the prediction of the dynamic behavior of solid viscoelastic materials concerns the ability to characterize their response to large dynamic loads. An essential part of this characterization is a constitutive theory which accounts for the effects associated with finite deformation and long-range memory. Along with this is the need to develop experimental techniques to generate and observe well-defined large amplitude waves that closely reproduce the same kinematical configuration and dynamical situation.

In this report we concentrate on a series of interrelated studies that include theoretical work involving waves in a very general class of viscoelastic materials, some specific constitutive models contained within this class, and comparison of the predicted results with those of experimental studies.

The analytical studies were concerned with the propagation of both steady and unsteady shock and acceleration waves in a general class of simple materials with fading memory and included determination of wave speeds and conditions on the amplitude behavior. More specific models were used in studying conditions on the propagation of steady waves and for use in modeling the results from experimental studies of steady and unsteady shock and acceleration waves in a particular non-linear viscoelastic solid. Experimental developments in the field of shock wave physics provide the generation of large-amplitude plane waves using a plate - impact configuration. In addition, instrumentation techniques have been developed which allow the fine structure of the wave profile to be observed and recorded with high resolution and accuracy.

The emphasis throughout will be on those aspects of the subject which relate to motions of one-dimensional strains. Earlier work involving waves in viscoelastic rods, i.e., in a one-dimensional stress configuration, has been reviewed by Kolsky [24]. Also, while the basic formulation will be limited to mechanical considerations, we will indicate at the end several results which include thermodynamic influences.

2. KINEMATICS OF WAVES

Consistent with the experimental configuration, we assume the motion to consist of a one-dimensional strain. In this case, we identify the body in its fixed reference configuration with an interval R of the real line and each material point with its position X in R. The motion is described by the function $\chi(\cdot\,,\cdot)$ which gives the spatial position

$$x = \chi(X,t)$$

at time t of the material point X. The displacement is

$$u = \chi(X,t) - X$$

and

$$F = \frac{\partial}{\partial X}\chi(X,t)$$

and

$$v = \dot{x} = \frac{\partial}{\partial t}\chi(X,t)$$

define the <u>deformation gradient</u> and <u>particle velocity</u> of X at time t, respectively. The <u>strain</u> ε of X at time t is given by

$$\varepsilon = -\frac{\partial}{\partial X}u(X,t) = 1 - F(X,t)$$

taken <u>positive in compression</u>. Since the response of a visco-elastic material depends on the load or deformation history, we define the <u>history of the strain</u> ε^t up to time t of the material point X as

$$\varepsilon^t(s) = \varepsilon(X,t-s) = 1 - F(X,t-s), \qquad s\,\varepsilon[0,\infty)$$

Note that a knowledge of ε^t is equivalent to knowledge of the present value $\varepsilon = \varepsilon^t(0)$ and the past history ε^t_r, where ε^t_r is the history $\varepsilon^t(s)$ on $s\varepsilon(0,\infty)$.

By a wave we mean a motion which contains a propagating disturbance. We are particularly interested in waves that involve jump discontinuities in certain kinematical and thermal fields. These waves can be closely approximated experimentally and will be shown to be useful in the evaluation of the constitutive behavior of viscoelastic materials.

The intrinsic velocity U of the discontinuity (wave front) is defined by

$$U = \frac{dY(t)}{dt}$$

where Y(t) is the material point in R at which the front is located at time t. The discontinuity in a function f(X,t) across the wave front at time t is shown in the usual jump notation: i.e.,

$$[\underset{\sim}{f}](\underset{\sim}{t}) = f^-(t) - f^+(t)$$

where

$$f^{\pm}(t) = \lim_{X \to Y^{\pm}(t)} f(X,t) .$$

With $U > 0$, f^+ and f^- are the limiting values of f(X,t) immediately ahead of and behind the wave front. If $\dot{f}(X,t)$ and $\partial f(X,t)/\partial X$ are also discontinuous at the wave but continuous everywhere else, then the following kinematical condition of compatibility holds,

$$\frac{d[\underset{\sim}{f}]}{dt} = [\underset{\sim}{\dot{f}}] + U\underset{\sim}{[} \frac{\partial f}{\partial X} \underset{\sim}{]} .$$

We assume in what follows that the motion $\chi(\cdot\ ,\ \cdot)$ is a continuous function of X and t for all X,t.

shock waves: A shock wave is a motion containing a wave front across which the strain ε, the particle velocity v, and their derivatives are discontinuous. From the compatibility condition it follows that

$$[\underset{\sim}{v}] = U[\underset{\sim}{\varepsilon}]$$

acceleration waves: An acceleration wave is a motion containing a wave front across which the strain ε, and the particle velocity v, are continuous, but their derivatives are not. It follows from the compatibility condition that, at $X = Y(t)$

$$[\dot{v}] = U[\dot{\varepsilon}] = -U^2\left[\frac{\partial \varepsilon}{\partial X}\right]$$

steady waves: A motion is said to be steady if the displacement u can be represented by

$$u(X,t) = \hat{u}(\xi)$$

where $\xi = t - X/V$ and $V > 0$ is a steady wave velocity. The corresponding particle velocity and strain become

$$v(\xi) = \frac{d\hat{u}(\xi)}{d\xi}$$

and

$$\varepsilon(\xi) = \frac{1}{V}\, v(\xi) \ .$$

3. SIMPLE MATERIALS WITH FADING MEMORY

Here we characterize the dynamic response of nonlinear viscoelastic solids by the constitutive equation of a simple material with fading memory as used by Coleman and Gurtin [14]. This equation expresses the statement that the stress σ at time t is determined by the entire history of the strain $\varepsilon^t = \varepsilon(t-s)$, $o \leqslant s < \infty$:

$$\sigma(t) = \zeta(\varepsilon^t) = \zeta(\varepsilon, \varepsilon_r^t) \ . \tag{1}$$

The statement that the material exhibits fading memory implies that the stress at time t is influenced more by changes in strain in the recent past (small s) than by those occurring in the distant past (large s), [18].

It will be useful in what follows to evaluate the stress for two particular strain histories. The first,

$$\varepsilon^t(s) = \varepsilon_I(s) = \begin{cases} \varepsilon, & s = 0 \\ 0, & s > 0 \end{cases} \tag{2}$$

corresponds to a strain jump ε suddenly applied to a material point which has been unstrained for all past time. In this case, the stress depends only on the jump in strain ε and the response

$$\sigma = \zeta(\varepsilon_I) = \sigma_I(\varepsilon) \tag{3}$$

is called the instantaneous stress-strain curve for the material. We assume that the function σ_I is continuously differentiable and define the instantaneous tangent modulus E_I and second-order modulus $\tilde{E}_I$ by

$$E_I = \frac{d}{d\varepsilon} \sigma_I(\varepsilon), \qquad \tilde{E}_I = \frac{d^2}{d\varepsilon^2} \sigma_I(\varepsilon) \quad . \tag{4}$$

The second special history is related to the equilibrium response associated with the constant strain

$$\varepsilon^t(s) = \varepsilon_E(s) \;, \quad 0 \leq s < \infty \quad . \tag{5}$$

The stress again depends only on the value of the strain ε and

$$\sigma = \zeta(\varepsilon_E) = \sigma_E(\varepsilon) \tag{6}$$

is called the <u>equilibrium stress-strain curve</u>. Similarly the equilibrium tangent modulus and equilibrium second-order modulus are given by

$$E_E = \frac{d}{d\varepsilon} \sigma_E(\varepsilon) \quad , \quad \tilde{E}_E = \frac{d^2}{d\varepsilon^2} \sigma_E(\varepsilon) \quad . \tag{7}$$

It will be appropriate to impose certain curvature conditions on the stress functional ζ which are expected to be consistent with the observed response behavior of most viscoelastic solids. In particular,

$$\begin{aligned} &\sigma_I(\varepsilon) > \sigma_E(\varepsilon) > 0 \quad , \\ &E_I(\varepsilon) > E_E(\varepsilon) > 0 \quad , \\ &\tilde{E}_I(\varepsilon) > 0, \ \tilde{E}_E(\varepsilon) > 0 \quad . \end{aligned} \tag{8}$$

Another important property of nonlinear materials with memory is that for small relative strain histories γ^t the response is approximated by the constitutive equation of linear viscoelastic, i.e., as $\|\gamma^t\| \to 0$

$$\zeta(\varepsilon^t + \gamma^t) = \zeta(\varepsilon^t) + G(0)\gamma^t(0) + \int_0^\infty G'(s)\gamma^t(s)ds + o\ \|\gamma^t\| \tag{9}$$

where $\|\gamma^t\|$ is the norm of the history γ^t [19]. Here $G(s)$ is the stress relaxation function corresponding to the underlying strain history ε^t and $G'(s) = dG(s)/ds$. We assume

$$G(s) > 0 \quad , \quad G'(s) \leq 0 \ .$$

Bowen and Chen [5] have shown that this latter inequality, evaluated for $\varepsilon^t(s) = \varepsilon_E(s)$ and $s = 0$, is a consequence of the second law of thermodynamics. Note that $E_I(\varepsilon) = G(0)$.

4. STEADY WAVES

Existence of steady waves.

The existence of steady shock and acceleration waves in nonlinear viscoelastic materials was studied by Pipkin [32] motivated in part by results of Coleman, Gurtin, and Herrera [17] and Varley [36] on propagation conditions of waves in these materials. Using the specific constitutive assumption of finite linear viscoelasticity formulated by Coleman and Noll [19], Pipkin obtained exact solutions to the one-dimensional steady field equation which exhibited smooth structured waves, shock waves, and acceleration waves. Somewhat later, Greenberg [22] obtained similar results for shock wave solutions in a large class of nonlinear materials with fading memory. (Shock waves in linear viscoelastic solids were studied by Lee et al [21, 25] and others [23] earlier). We will see that the propagation of steady waves in dissipative materials is important in subsequent considerations of the modeling of nonlinear viscoelastic materials.

Using the steady wave relations and assuming that natural equilibrium conditions ($\sigma = 0$, $\varepsilon = 0$) exist far ahead of the wave, the momentum equation takes the form

$$\sigma(\varepsilon) = \rho_R V^2 \varepsilon(\xi) \tag{10}$$

which holds for all ξ. It follows from this that each point in a steady wave lies on a secant (Rayleigh line) in the stress-strain plane connecting the initial and final states. The slope of the Rayleigh line is $\rho_R V^2$. Combining (10) with the constitutive equation (1) results in a functional equation of the form

$$\rho_R V^2 \varepsilon(\xi) = \zeta\big(\varepsilon(\xi-s)\big) \tag{11}$$

for the steady strain field $\varepsilon(\xi)$.

We wish to consider the existence of solutions $\varepsilon(\xi)$, $0 < \varepsilon < 1$, of this equation which are monotone increasing functions of ξ. Greenberg [22] gave sufficient conditions for the existence of such solutions which are essentially the curvature conditions given in the previous section along with the requirement that the instantaneous longitudinal sound speed C_I, given by

$$\rho_R C_I^2 = \frac{\partial}{\partial \varepsilon} \zeta\left(\varepsilon(\xi-s)\right)$$

must be monotone increasing with ξ. This is consistent with the intuitive notion that a steady wave in a nonlinear material with memory results from a balance between the tendency of the wave to decay due to internal dissipation and to grow due to nonlinear response. Thus, each part of the wave must tend to overtake the part ahead of it.

With these conditions, it can be shown that steady waves do exist and that there are three types of solutions which depend on the steady wave velocity V and the instantaneous and equilibrium sound speeds corresponding to the natural equilibrium state; i.e.

$$\rho_R (C_I)_o^2 = \frac{d}{d\varepsilon} \sigma_I(\varepsilon)\Big|_{\varepsilon=o} \quad , \quad \rho_R (C_E)_o^2 = \frac{d}{d\varepsilon} \sigma_E(\varepsilon)\Big|_{\varepsilon=o} .$$

The solutions are

(i) $(C_I)_o > V > (C_E)_o$,

(ii) $(C_I)_o = V$,

(iii) $(C_I)_o < V$.

Thus, the type of solution depends upon the position of the Rayleigh line with respect to the instantaneous and equilibrium stress-strain curves, $\sigma_I(\varepsilon)$ and $\sigma_E(\varepsilon)$. In case (i), shown in Fig. 1a, the Rayleigh line lies below the instantaneous curve, the solution is a smooth structured wave with the final equilibrium particle velocity $v(\infty)$ and strain $\varepsilon(\infty)$ given by

$$\rho_R V^2 \varepsilon(\infty) = \sigma_E\left(\varepsilon(\infty)\right) \quad , \quad v(\infty) = V\varepsilon(\infty) . \tag{12}$$

In the second case (ii) the Rayleigh line is tangent to the instantaneous curve (Fig. 1b). The steady wave velocity is the acceleration wave velocity and the solution is an acceleration wave with $\varepsilon(\infty)$, $v(\infty)$ satisfying (12). The third solution (iii) is of particular interest and corresponds to a Rayleigh line with a slope greater than the initial slope of the instantaneous curve (Fig. 1c). Here the equilibrium conditions ahead of the wave are satisfied if $v(\xi) = 0$, $\varepsilon(\xi)=0$ for $\xi < 0$ and there is a jump discontinuity in $v(\xi)$ at $\xi = 0$, with $v(0)$ and $\varepsilon(0)$ satisfying

$$\rho_R V^2 \varepsilon(0) = \sigma_I\left(\varepsilon(0)\right) \quad , \quad v(0) = V\varepsilon(0).$$

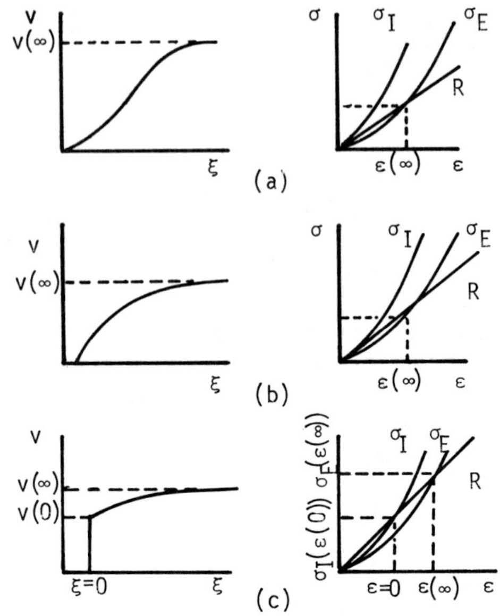

Fig.1. Steady wave solutions

The solution is a steady shock wave with the particle velocity and strain undergoing a smooth transition from v(0) to v(∞).

Experimental generation of steady shock waves

The evolutionary behavior of shock waves in viscoelastic materials has been studied experimentally using a gas gun and plate impact configuration [3, 33]. A typical target sample configuration is shown in Fig. 2. The experiments are designed such that the observation of the resulting wave profiles along the centerline of the target plate is completed before the arrival of unloading waves from the lateral surfaces of the target. For an isotropic material, symmetry requires the motion to be purely longitudinal. Experimental results will be shown for tests on polymethyl methacrylate (PMMA) samples. The targets were fabricated from two discs cemented together with a thin aluminum mirror vapor-deposited on the surface in the center of one of the discs. For most of the tests a velocity interferometer system [3] was used which provided a direct measure of the time history of the particle velocity at a fixed Lagrangian position.

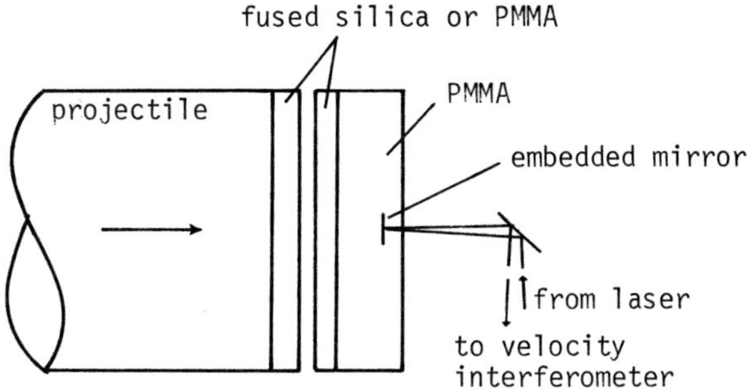

Fig.2. Experimental plate impact configuration for shock and acceleration wave studies

A typical particle velocity history in PMMA is shown in Fig. 3. The profile consists of a shock, in which the particle velocity jumps to a value v_o, followed by a smooth transition to a particle velocity v_∞. At time t_b an unloading wave produced by the reflection of the compressive wave from the free surface of the projectile noseplate arrives at the mirror and reduces the particle velocity. The wave speed V is determined from an accurate measurement of the time span between impact and arrival of the wave front at the mirror.

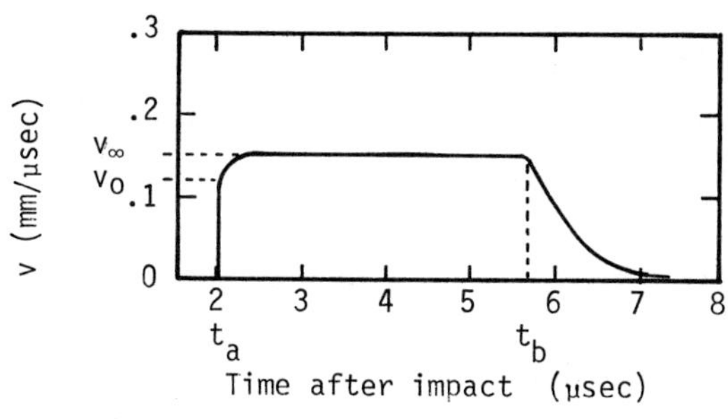

Fig.3. Typical particle velocity profile

Using these techniques, Barker and Hollenbach [4] and Schuler [33] carried out a series of tests to investigate the evolutionary behavior of shock waves in PMMA. The series involved varying the initial shock strength and the distance of propagation of the wave. Figure 4 shows particle velocity histories corresponding to impact conditions of 0.06, 0.15, and 0.30 mm/μsec, with measurements made at varying distances of travel. The records of the 0.06mm/μsec impact level show a noticeable decay in the strength of the front and an accompanying smoothing of the wave shape with increasing distance of propagation. For the 0.15mm/μsec and the 0.30mm/μsec impact levels the particle velocity histories between 6 and 37mm from the impact surface are nearly identical. Thus, it

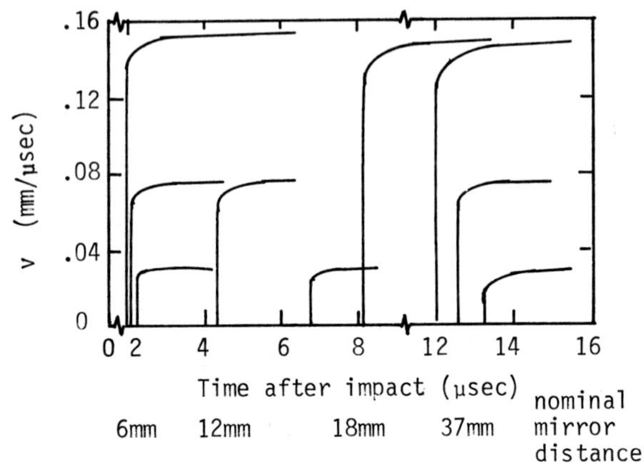

Fig.4. Measured particle velocity histories for PMMA

appears that within 6mm from the impact surface, the waves generated at these levels became steady. From the analysis of steady waves, it is evident that measurement of the steady wave velocity V and the particle velocity jump v_o for various impact levels serves to determine the instantaneous response of the material, while measurement of V and the maximum equilibrium particle velocity v_∞ for various impact levels serves to determine the equilibrium response. Thus, from a series of steady wave experiments two curves corresponding to the wave velocities $V_I(v)$ and $V_E(v)$ are obtained and are represented by least squares polynomial fits. For PMMA

$$V_I(v) = (C_I)_o + b_I v + a_I v^2 \tag{13}$$

and

$$V_E(v) = (C_E)_o + b_E v + a_E v^2 \; . \tag{14}$$

These can be written in terms of the strain and then, using the steady wave equations, as instantaneous and equilibrium stress-strain functions, of the form

$$\sigma_I(\varepsilon) = \rho_R V_I^2(\varepsilon)\varepsilon \quad , \quad \sigma_E(\varepsilon) = \rho_R V_E^{\,2}(\varepsilon)\varepsilon \; . \tag{15}$$

These curves for PMMA are shown in Fig. 5

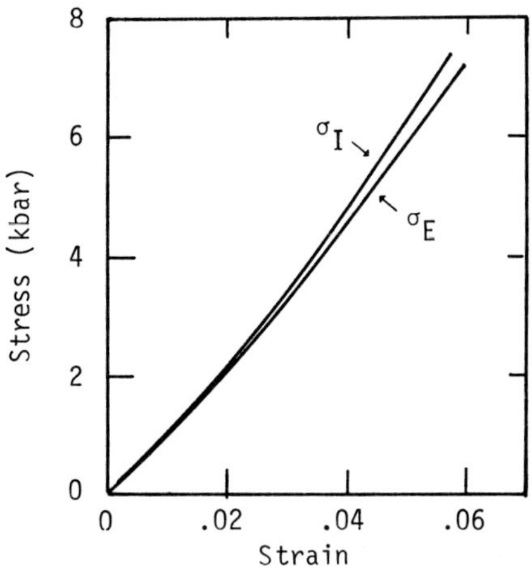

Fig.5. Instantaneous and equilibrium stress-strain curves for PMMA

Analysis of steady shock waves

To analyze the steady wave profiles in PMMA, it is necessary to assume a specific constitutive model. Following Pipkin [32], Schuler [33] used the one-dimensional form of the constitutive equation for finite linear viscoelasticity described by Coleman and Noll [19]:

$$\sigma(t) = \sigma_I\left(\varepsilon(t)\right) + \int_0^\infty G'(\varepsilon,s)\left[1 - (1 - \varepsilon^t(s))^2\right]ds \qquad (16)$$

for all compressive strain histories $\varepsilon^t(s)$. $G(\varepsilon,s)$ is the stress relaxation function and has the property $\lim_{s \to \infty} G'(\cdot,s) = 0$. Evaluating (1) for the equilibrium history yields a relation between $\sigma_I(\varepsilon)$ and $\sigma_E(\varepsilon)$.

At this point it is necessary to assume a form for $G(\varepsilon,s)$. Consistent with the conditions necessary to insure the existence of steady waves, Schuler [33] assumed the form

$$G(\varepsilon,s) = \left[G(\varepsilon,0) - G(\varepsilon,\infty)\right]\exp\left(-\frac{s}{\tau}\right) + G(\varepsilon,\infty), \qquad (17)$$

in terms of a single relaxation time τ. Using (16) and (17) steady wave solutions of (11) were obtained, which contained a shock discontinuity shown above as solution (iii). Figure 6 shows a comparison between the calculated and experimental wave profile for the 0.15mm/μsec impact condition.

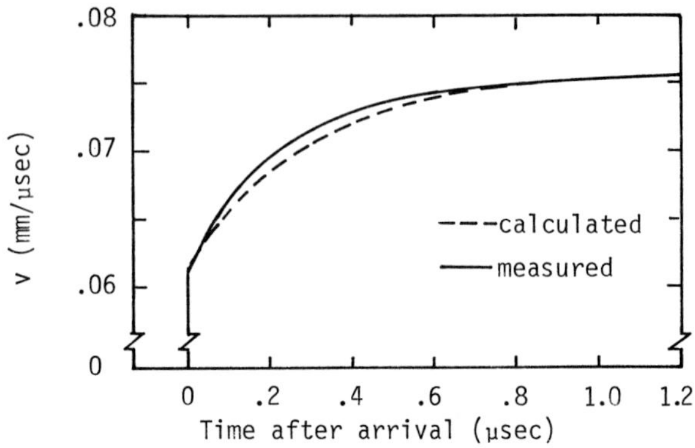

Fig.6. Comparison of measured and calculated wave profiles for PMMA; impact velocity 0.15mm/μsec

Determination of the stress relaxation function from acoustic wave experiments

It is known tnat the stress relaxation function G(.) for a viscoelastic solid can be evaluated from experimental measurements of the frequency dependence of the phase velocity and attenuation of acoustic waves. Of particular interest here is the form of G(.) which is appropriate for describing the one-dimensional shock wave response of viscoelastic materials. If the relaxation function is known for a polymer for a wide range of times, then by specifying the time scale of a given experiment one can determine the function appropriate for that experiment. Nunziato and Sutherland [29] did this for the time scale characteristic of plate impact experiments to deduce a relaxation function for polymethyl methacrylate.

Using velocity and attenuation dispersion data for PMMA over an extended range of frequencies they specified the time scale appropriate for impact experiments, namely 10^{-2} to 1μsec. That is, approximately the lower limit of experimental time resolution to the time at which apparent equilibrium is achieved for wave profiles of infinite duration. Their results showed that, over this time range, the relaxation function could be well represented by the single exponential relation, i.e. (17).

5. SHOCK WAVES

Shock wave amplitude equation

The general theory of the propagation of one-dimensional shock waves in nonlinear materials with fading memory characterized by (1) has been developed by Coleman, Gurtin and Herrera [17] and by Chen and Gurtin [7]. Using the shock jump conditions along with the momentum equation evaluated at the wave, it was shown that the intrinsic velocity U of a compressive shock propagating into a region at rest and unstrained for all past times is given by

$$\rho_R U^2 = \frac{\sigma_I(\varepsilon^-)}{\varepsilon^-} \quad . \tag{18}$$

From the convexity condition (8) it follows that $\rho_R U^2/E_I(\varepsilon^-) = \mu < 1$ and thus the shock velocity is always subsonic with respect to the material behind the shock front. Clearly the shock velocity depends on the strain amplitude ε^- and in fact [7]

$$\frac{dU}{dt} = \left[\frac{(1-\mu)\, E_I(\varepsilon^-)}{2\rho_R U \varepsilon^-}\right] \frac{d\varepsilon^-}{dt} \quad .$$

Thus, the shock velocity increases, decreases, or remains the same according to whether the shock amplitude ε^- increases, decreases, or remains steady.

Using the properties of shock waves and the assumed smoothness of the stress function, Chen and Gurtin [7] derived a <u>shock amplitude equation</u>

$$\frac{dv^-}{dt} = \left[\frac{(1-\mu^2)}{(1+3\mu)}\right] \left[(\dot{v})^- - |\lambda|\right] \quad , \tag{19}$$

where

$$\lambda = \frac{G'(\varepsilon^-, 0)v^-}{E_I^-\,(1-\mu)} \quad . \tag{20}$$

Here $(\dot{v})^-$ is the particle acceleration immediately behind the shock front. Thus the growth or decay behavior of the wave front depends on the acceleration behind the front; i.e.

(i) if $(\dot{v}) > |\lambda|$, $\frac{dv^-}{dt} > 0$ ··· front grows

(ii) if $(\dot{v}) = |\lambda|$, $\frac{dv^-}{dt} = 0$ ··· front remains steady,

(iii) if $(\dot{v}) < |\lambda|$, $\frac{dv^-}{dt} < 0$ ··· front decays.

The parameter λ is called the <u>critical acceleration</u>. Note that in general it is not possible to determine the particle velocity amplitude v^- of a shock from (19) since the particle acceleration $(\dot{v})^-$ is not known in advance. Nevertheless, there are certain problems for which approximate solutions can be obtained. One of these will be discussed below.

<u>Determination of the critical acceleration</u>

The critical acceleration (19) depends upon the strain (or particle velocity) both explicitly and through the strain dependence of the shock velocity U, the instantaneous tangent modulus E_I, and the initial slope of the relaxation function $G'(\varepsilon^-,0)$. Hence to obtain $\lambda(\varepsilon^-)$ for a particular material it is necessary to determine the strain dependence of these functions from experimental measurements. In a study by Schuler and Walsh [35] the critical acceleration was determined for a particular constitutive model and the results compared to experimental shock propagation studies using the solid polymer PMMA.

In the discussion of steady shock waves it was shown how the strain dependence of the shock velocity $V_I(\varepsilon^-) = U(\varepsilon^-)$ was obtained directly from the experimental measurements which yield the shock velocity - particle velocity dependence. Also, from (4), (13) (written in terms of strain), and (15) the strain dependence of the instantaneous tangent modulus can be obtained. This leaves the function $G'(\varepsilon^-,0)$ to be determined.

Toward this end, consider again the constitutive relation for finite linear viscoelasticity (16) and the assumed form for $G(\varepsilon,s)$ (17) discussed previously. The response functions in the definition of the critical acceleration were a result of an analysis which started with the constitutive assumption (1) for materials with memory and which, for small relative strains, took the form (9). To relate (16) with (17) to this general model it is necessary to compute the derivative of this latter equation with respect to an arbitrary strain history and evaluate the result for the appropriate reference history, here the jump history (2). Carrying this through it results that

$$G'(\varepsilon^-,0) = -\frac{2}{\varepsilon^-(2-\varepsilon^-)\tau}\left[\sigma_I(\varepsilon^-) - \sigma_E(\varepsilon^-)\right]. \quad (21)$$

This allows an evaluation of the critical acceleration $\lambda(\varepsilon^-)$ and comparison with the experimental results. Figure 7 shows the particle velocity histories for three steady shock waves resulting from impact velocities of 0.06, 0.15, and 0.30 mm/μsec (curves a,b,c respectively). Since they represent steady waves, $dv^-/dt = 0$ and the value of $(\dot{v})^-$ immediately following the shock front corresponds to the critical acceleration. Figure 8 shows a comparison of these values with those calculated as described above. In view of the sensitivity of λ to variations in the fits for σ_I and σ_E and the relaxation time τ, the correlation is good.

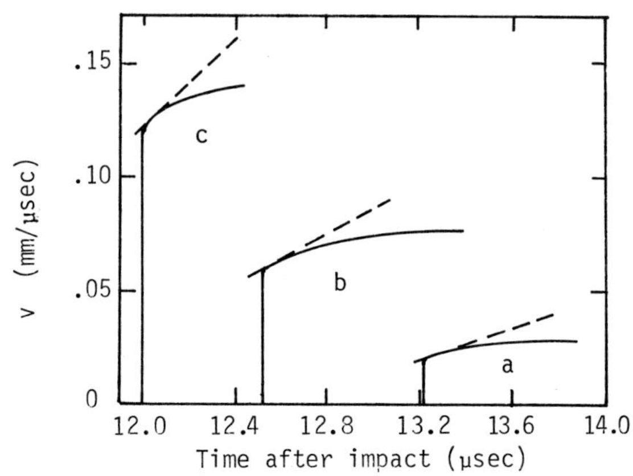

Fig.7. Particle velocity as a function of time

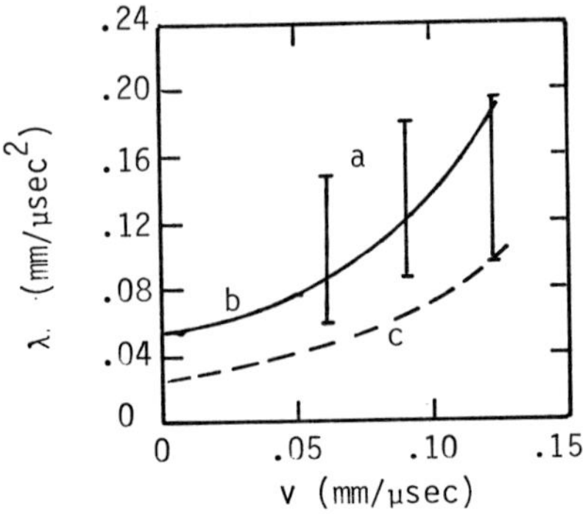

Fig.8. Comparison of observed and calculated critical acceleration: a-experiment: b-thermomechanical theory: c-mechanical theory

Evolution of steady shock waves

In the discussion of experimentally generated steady shock waves it was indicated that the waves became steady after propagating a certain distance. It was also mentioned that the development of steady waves is important since, through a corresponding steady-wave analysis, they provide a means of evaluating a specific viscoelastic constitutive model for the material. Nunziato and Schuler [26] considered the problem of determining a lower bound on the propagation distance required for the development of a steady wave and the dependence of this bound on impact velocity to, e.q., provide an estimate of the minimum sample thickness required for the observation of steady shock waves in plate-impact experiments.

They began with the shock amplitude equation (19) and noted that the strength of the shock front always decays with propagation distance ($dv^-/dt < 0$) while the particle acceleration immediately behind the front ($\dot{v}^-$), after impact, is always positive. Then, assuming, as before, that the material can be characterized as a finite linear viscoelastic material, they obtained bounds on $|\lambda|$, the critical acceleration, and using these, bounds on $|dv/d\iota|$, the magnitude of the rate of decay of the shock front. This was integrated along with the equation for the trajectory of the wave, $dY/dt = U$, to obtain a lower bound for the attenuation curve $v(Y)$. Comparing this with measured values of v at various Y for the same initial value v_o determines the minimum distance δ at which a steady wave can form. Repeating the procedure for several values of v_o yields a v_o - δ graph as shown in Fig. 9. The details of the procedure are given in [26]. The results are in general qualitative agreement with experimental observations. That is, they indicate that, the greater the initial amplitude, the faster the wave evolves toward the steady-wave shape. Further, they show that below a certain initial amplitude a steady wave is not possible, which is consistent with the prediction of the linearized form of the shock amplitude equation that all small-amplitude shocks decay.

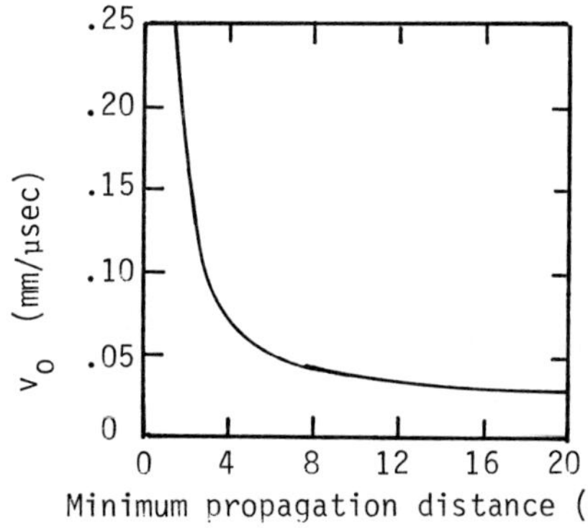

Fig.9. Minimum propagation distance for the evolution of a steady shock wave as a function of initial particle velocity behind the wave for PMMA

Shock pulse attenuation

It was mentioned above that the shock amplitude equation cannot in general be used to determine $v^-(t)$ since the particle acceleration $(\dot{v})^-(t)$ is not known in advance. However, Nunziato and Schuler [27] have shown that (19) can be used to evaluate $v^-(t)$ in the case of an attenuating thin pulse by approximating $(\dot{v})^-(t)$. The problem is to calculate the attenuation of a pulse which consists of a shock front propagating into an unstrained region at rest followed immediately by an unloading wave of arbitrary shape which unloads the material to its unstressed state (Fig. 10).

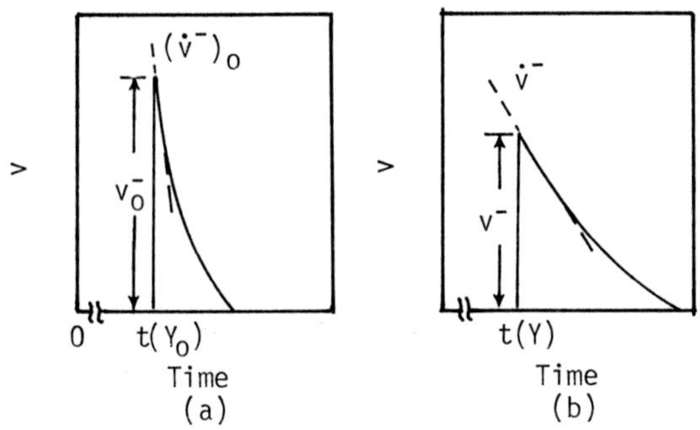

Fig.10. Thin pulse profiles

For viscoelastic materials, the shock loads the material along a Rayleigh line to a point on the instantaneous stress-strain curve $\sigma_I(\varepsilon^-)$, and the unloading wave unloads it along some unknown stress-strain path, but always between the instantaneous and equilibrium stress-strain curves. However, for very thin pulses the strain rates, particularly in the initial portion of the unloading wave, are very high. At such high strain rates the behavior of the material is governed primarily by the instantaneous elasticity of the material. Thus, it might be assumed that the unloading stress-strain path is the instantaneous stress-strain curve $\sigma_I(\varepsilon)$.

This assumption means that the effects of rate-dependence on the unloading wave are neglected and the unloading wave is treated as a wave propagating in a nonlinear elastic material whose stress-strain curve is $\sigma_I(\varepsilon)$. Nunziato and Schuler further assumed that (i) the unloading wave was a simple wave, and (ii) the unloading path in the stress-particle velocity plane is the instantaneous curve σ_{IH} related to v and u by the momentum equation evaluated at the shock wave with U given in terms of the instantaneous shock velocity-particle velocity fit. From assumption (i) each point on the unloading wave propagates at a constant level of particle velocity with the corresponding <u>instrinsic instantaneous sound speed</u> $C_I^2 = (E_I/\rho_R)$.

Let the smooth function $t_0(v)$ represent the initial unloading wave profile at the material point Y_0 (Fig. 10a). Then the unloading portion of a later profile (Fig. 10b) can be represented by the smooth function t(v) which is related to

$t_o(v)$ through the simple wave relation

$$t(v) = t_o(v) + \frac{X - Y_o}{C_I(v)}$$

for each $v \leqslant v^-$ and for all material points $X \geqslant Y_o$. Differentiating this expression with respect to v, inverting and evaluating the result at the shock front yields a relation for the particle acceleration $(\dot{v})^-$ immediately behind the front in terms of $C_I(v)$, Y_o, $Y(t)$ (which is the position of the shock front at time t), and $(\dot{v})_o^- < 0$, the slope of the initial unloading wave profile at the particle velocity v^-. Substituting this into the viscoelastic shock amplitude equation yields an ordinary differential equation for dv^-/dt involving quantities all of which can be formulated in terms of the particle velocity [27]. It follows that the shock wave amplitude is monotone decreasing and will approach a limiting amplitude (zero) in an infinite time.

These results were compared to experiments in which the pulse was produced by the impact of the viscoelastic specimen (PMMA) against a thin (0.0216 mm) gold foil target. For several tests at repeated impact levels, the pulse profile was observed at increasing distances from the impact surface (0.48, 1.26, 3.20, 5.97 mm). The particle velocity history measured at the position X = 0.48mm from the impact surface was taken as the initial pulse shape and was fitted as

$$t_o(v) = - k \ln \left(\frac{v}{v_o^-}\right) , \quad v \leqslant v_o^- .$$

The subsequent attenuation profiles were then obtained by integration of the differential equation for the amplitude. The solution is compared to the experimental results in Fig. 11 where it is seen that the comparison is quite favorable. By assuming unloading along the instantaneous curve, it might be expected that the predicted attenuation would be less than that observed. However, the good comparison seems to indicate that in these experiments the unloading stress-strain path does lie very close to the instantaneous curve. Finally, with the simple wave description and the initial pulse shape, the wave shape at any other point can be calculated.

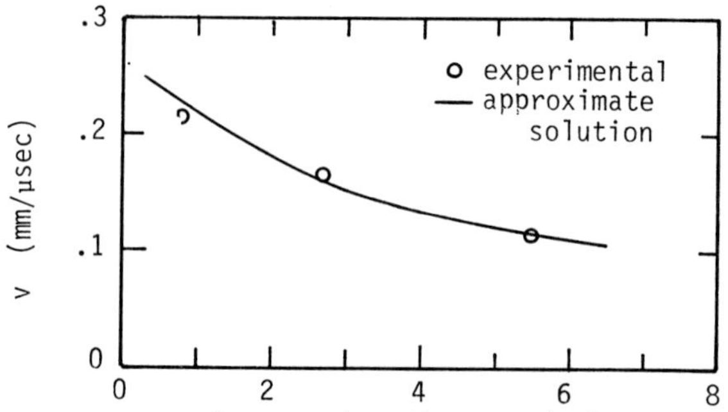

Fig.11. Comparison of measured and calculated shock pulse amplitudes in PMMA

6. ACCELERATION WAVES

Acceleration wave amplitude equation

The velocity of propagation of acceleration waves in materials with memory defined by (1) was studied by Coleman, Gurtin and Herrera [17]. They showed that the intrinsic velocity U of an acceleration wave propagating into a region with an arbitrary strain history ε^t satisfies

$$U^2(t) = \frac{1}{\rho_R} \frac{\partial \zeta(\varepsilon^t)}{\partial \varepsilon} \quad . \tag{22}$$

Coleman and Gurtin [11, 14] also showed that if the acceleration wave amplitude a(t) is given at t = 0, then it is possible to derive an expression for a(t) at later times, the form of the expression depending upon the strain history of the region traversed by the wave front up to time t. Here we give the results only for waves propagating into unstrained regions in equilibrium, i.e. for $X > Y(t)$, $\varepsilon^t(s) = 0$, $0 \leqslant s < \infty$. . In this case, let

$$(E_I)_o = \frac{d}{d\varepsilon}\sigma_I(\varepsilon)\Big|_{\varepsilon=o}, \quad (\tilde{E}_I)_o = \frac{d^2}{d\varepsilon^2}\sigma_I(\varepsilon)\Big|_{\varepsilon=o}, \quad G'_o(0) = G'(0,0), \tag{23}$$

and note that the intrinsic velocity $U = U_o$ is a constant given by

$$\rho_R U_o^2 = (E_I)_o \quad . \tag{24}$$

Under these conditions it can be shown that the acceleration wave amplitude a(t) satisfies the equation

$$\frac{da}{dt} = - \beta a + \frac{\beta}{a_c} a^2 \tag{25}$$

where β and a_c are constants given by

$$\beta = - \frac{G_o'(0)}{2\,(E_I)_o}, \quad a_c = - \frac{G_o'(0)U_o}{(\tilde{E}_I)_o} . \tag{26}$$

This differential equation is a Bernoulli equation with the solution

$$a(t) = \frac{a_c}{\left(\frac{a_c}{a(0)} - 1\right) \exp\,(\beta t) + 1} . \tag{27}$$

Thus, in the case of an acceleration wave, an exact closed-form expression for the time-dependent behavior of the jump in particle acceleration can be obtained which depends only on the properties of the material (cf. [2] for a general discussion of acceleration waves).

From the curvature assumptions given above, it follows that $\beta > 0$ and $a_c > 0$. Thus, for a compressive wave with $a(0) > 0$ it follows that

(i) if $a(0) < a_c$, $a(t) \to 0$ as $t \to \infty$,

(ii) if $a(0) = a_c$, $a(t) \equiv a(0)$,

(iii) if $a(0) > a_c$, $a(t) \to \infty$ monotonically in a finite time t_∞,

where t_∞ is given by

$$t_\infty = - \frac{1}{\beta} \ln\left(1 - \frac{a_c}{a(0)}\right) . \tag{28}$$

The parameter a_c is called a <u>critical amplitude of acceleration</u>. If the initial amplitude is less than a_c, then the internal dissipation of the material ($G_o'(0) < 0$) is the dominant effect and the wave front decays. However, if the initial amplitude is greater than a_c, then the nonlinearity of the instantaneous stress-strain curve is the dominant effect and the acceleration wave amplitude increases without bound in a finite time (which suggests the formation of a shock discontinuity).

Acceleration wave velocities

It would appear from the wave profile shown in Fig. 3 that an unloading wave may be considered as an expansive wave propagating into a deformed region. Schuler [34] experimentally measured the speeds of these release waves in PMMA and compared the results with calculations of acceleration wave speeds based on the constitutive equation for a finite linear viscoelastic material (16).

The velocity of an acceleration wave propagating in a finite linear viscoelastic material may be obtained by substituting the constitutive equation (16) into (22) with the result

$$U^2 = \frac{1}{\rho_R}\left\{\frac{d}{d\varepsilon}\sigma_I(\varepsilon) + \int_0^\infty \frac{d}{d\varepsilon} G'(\varepsilon,s)\left[1 - \left(1 - \varepsilon^t(s)\right)^2\right] ds\right\}. \quad (29)$$

Note that in general the intrinsic velocity U is a function of the current strain state and the past history of the strain. Thus, to compare the calculated wave speeds with the experimental observations of unloading waves, it seems necessary to determine the history of the deformation at each particle through which the wave passes and then use (29) to determine U as a function of time at each particle. Clearly this is not a viable procedure. For example, the deformation history experienced by a particle at the free surface of the flyer plate consists of a sudden jump in strain caused by the arrival of the shock and hence the acceleration wave velocity at this particle can be found from (29) using the jump history (2). However, a particle located at the impact surface has a relatively long time to reach equilibrium prior to the arrival of the unloading wave. If this particle does reach equilibrium then the appropriate history is the equilibrium history (5). At other particles in the path it is not clear which history to assume. Nevertheless, the jump and equilibrium histories represent extremes and so consideration of these two histories should provide bounds on the variation of the acceleration wave velocity.

For the jump history, U is given by

$$U^2 = \frac{1}{\rho_R} E_I(\varepsilon) \quad (30)$$

while for the equilibrium history

$$U^2 = \frac{1}{\rho_R}\left\{\frac{d\sigma_E(\varepsilon)}{d\varepsilon} + \frac{2(1-\varepsilon)}{(2-\varepsilon)\varepsilon}\left[\sigma_I(\varepsilon) - \sigma_E(\varepsilon)\right]\right\}. \quad (31)$$

Formulating σ_I and σ_E in terms of strain one may calculate the acceleration wave velocities predicted by (30) and (31). Figure 12 shows the results, labeled I and E. The velocity corresponding to the jump history, labelled I, is always greater than that corresponding to the equilibrium history, curve E.

Of course, the measured unloading wave velocities are average velocities calculated from $U = (X_m + X_p) [t_a - X_p/U_s]^{-1}$ where X_m is the mirror location, X_p is the thickness of the flyer plate, t_a is the time after impact at which the unloading wave arrives, and U_s is the measured shock velocity. It should be noted though that, for a given impact condition, no significant variation of unloading wave velocity was observed over a range of propagation distances from 6 to 42 mm. The measured unloading wave velocity and corresponding equilibrium strain values are graphed in Fig. 12. For strains less than about 6 percent there is good agreement between the calculated velocity corresponding to the jump history, curve I, and the measured velocities. The calculated wave velocity corresponding to the equilibrium history tends to fall below the experimental points. The lack of agreement above 6 percent strain (7.5 kbar) is a result of an inflection in the equilibrium behavior of PMMA.

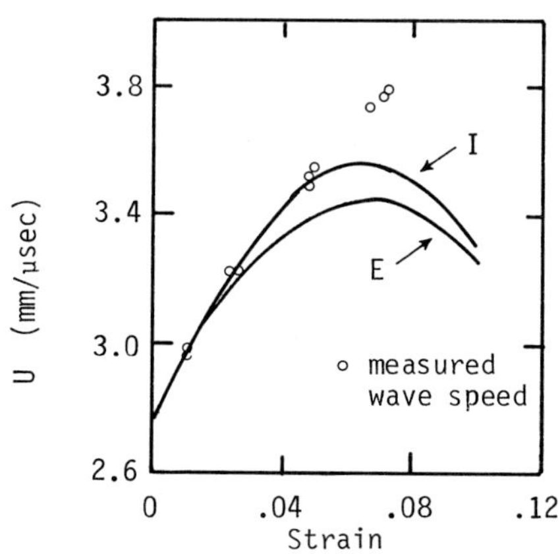

Fig.12. Comparison of observed and calculated acceleration wave speeds

The study of unloading waves suggests an alternative method of formulating a constitutive model. The apparent weak dependence of acceleration wave velocity on strain history suggests assuming that the wave speed is a function of the current value of the strain only as given by (30). Thus, measurements of unloading wave speed determine the instantaneous stress-strain curve. Over the range 0-7.5kbar the instantaneous curve obtained in this manner differs little from that determined from the steady wave analysis.

The growth of acceleration waves

It was outlined above that the growth or decay of an acceleration wave propagating into an unstrained region at rest is governed by a critical amplitude a_c and a constant β. Both of these parameters depend on the acceleration wave velocity U_o, the instantaneous tangent modulus $(E_I)_o$, the instantaneous second order modulus $(\tilde{E}_I)_o$, and the initial slope of the stress relaxation function $G_o'(0)$. Here we describe methods for determining these values from the results of steady shock wave experiments as suggested by Walsh and Schuler [37] and by Chen and Gurtin [8].

As already indicated, the quantities $(E_I)_o$ and $(\tilde{E}_I)_o$ are determined directly from the instantaneous stress-strain curve $\sigma_I(\varepsilon)$ using (23). The velocity U_o can be calculated by $\rho_R U_o^2 = (E_I)_o$ and, as mentioned, represents the lead coefficient of the steady shock velocity-particle velocity fit (13). In terms of these coefficients, $(\tilde{E}_I)_o = 4\rho_R (C_I)_o^2 b_I$.

The final parameter to be evaluated is $G_o'(0)$, the initial slope of the stress relaxation function corresponding to the equilibrium strain history $\varepsilon^t(s) \equiv 0$. This value is obtained by evaluating the relaxation function $G'(\varepsilon^-,0)$ [eqn. (21)], which is based on the constitutive equation for finite linear viscoelasticity, at $\varepsilon = 0$. Thus

$$G_o'(0) = -\frac{\rho_R}{\tau}\left[(C_I)_o^2 - (C_E)_o^2\right].$$

where $(C_E)_o^2 = (E_E(0)/\rho_R)$ is the equilibrium sound speed. With these results β and a_c can be calculated using (26) and the critical time t_∞ is determined as a function of the initial amplitude $a(0)$ from (28). The result for PMMA is

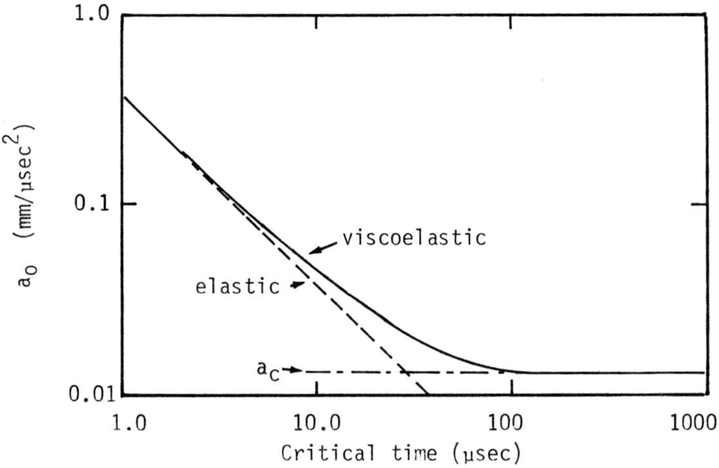

Fig.13. Critical time vs initial acceleration wave amplitude

shown in Fig. 13. Also shown is $t_\infty(a(0))$ for an elastic material with the same values of $(E_I)_o$ and $(\tilde{E}_I)_o$.

In their study, Walsh and Schuler [37] compared these predicted results with experimental observations of the growth of acceleration waves in PMMA. The experimental configuration is shown in Fig. 2 where the projectile faceplate is fused silica and a fused silica buffer is mounted before the PMMA sample. The generation of the acceleration wave is a result of the material properties of fused silica [4]. Unlike most materials the one-dimensional stress-strain curve of this elastic material has a response curvature such that compressive discontinuities tend to spread, i.e., $\tilde{E} < 0$. Thus, as the shock discontinuity produced at the impact surface propagates into the fused silica, it spreads continuously, forming an acceleration wave which then moves into the PMMA target. The amplitude of the acceleration discontinuity at the target varies inversely with the thickness of the fused silica. Of course it is of paramount importance to maintain the one-dimensional nature of the motion for the duration of the experiment and as a result the thickness of the fused silica plate is limited to about 4 cm. This results in a lower limit on the acceleration discontinuity which can be generated with this technique to approximately 0.27mm/μsec^2. It happens that this value of a(0) is greater than the value of a_c for PMMA and thus, as expected, only acceleration waves which increased in amplitude were observed.

Figure 14 shows the results for a series of six experiments in PMMA. The six tests are grouped into two series depending on the thickness of the fused silica plate used. The first series (input A) used a 25.5mm thick disc of fused silica which resulted in an initial acceleration wave amplitude of 0.41mm/μsec^2. From Fig. 13 this input corresponds to a critical time t_∞ = 0.94 μsec. The transit time from the interface to the mirror, determined from $t = X_m / U_o$, was 1.06, 1.63 and 2.25 μsec for the three tests labeled A1, A2, A3, respectively. Since each of these times is greater than t_∞ a shock discontinuity is expected to have formed. The wave profiles shown in Fig. 14 indicate the formation of such a discontinuity in each case.

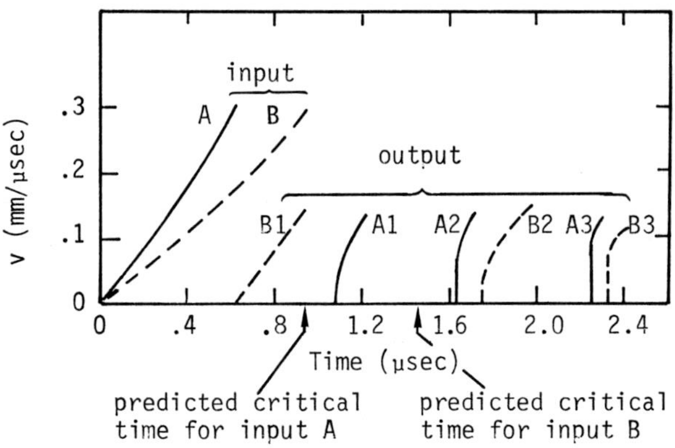

Fig.14. Observed acceleration wave profiles

The second series (input B) used a 38.1mm thick disc of fused silica which gave an initial acceleration level of a(0) = 0.27mm/μsec^2. From Fig. 13 the corresponding critical time is 1.45μsec. Two of the tests in this series (B2,B3) had propagation times greater than t_∞ (1.74, 2.32 μsec, respectively). Thus, for B2 and B3 a shock wave is expected to form and this is evident in Fig. 14. In test B1 the wave was measured after a propagation time of 0.63 μsec. Since this time is less than t_∞ the formation of a shock discontinuity is not predicted. The result is shown in Fig. 14 where it is clear that no shock is evident. Further, since $a(0)>a_c$, the acceleration amplitude is increasing and can be predicted from the amplitude equation (27). With $a(0) = 0.27 \text{mm}/\mu\text{sec}^2$ we calculate $a(0.63) = 0.47 \text{mm}/\mu\text{sec}^2$. The measured level of $0.46 \text{mm}/\mu\text{sec}^2$ is in good agreement with this value.

It is worth noting from Fig. 13 that, at these levels of initial acceleration, the predicted t_∞ for PMMA is close to that for a nonlinear solid with the same tangent and second-order moduli, E and $\tilde{E}$. Of course, for elastic materials there is no critical acceleration, nevertheless, an acceleration wave whose initial amplitude is an order of magnitude greater than the critical acceleration will propagate in a nonlinear polymeric solid (such as PMMA) as if the material were elastic. In this case the simpler elastic theory could be used to interpret the results of acceleration wave experiments on nonlinear viscoelastic materials.

In the procedure described above, the parameter $G_o'(0)$ was determined from an evaluation based on the constitutive model of finite linear viscoelasticity. Chen and Gurtin [8] have proposed an alternative approach in which $G_o'(0)$ can be determined directly from experimental steady wave data as follows. In a steady shock wave, $(\dot{v})^- = |\lambda|$, the <u>critical acceleration</u>, and it was shown that in the limit as the amplitude of the shock discontinuity tends to zero, the critical acceleration has as its limit twice the critical amplitude a_c of an acceleration wave propagating into an unstrained equilibrium region. Since the value of $(\dot{v})^-$ can be measured graphically from steady wave profiles for various amplitudes v^-, an extrapolation of this data to $v^- = 0$ should yield the critical amplitude a_c. Then $G_o'(0)$ can be calculated from (26). Although this approach is satisfactory in principle, the limited number of experimentally determined particle accelerations $(\dot{v})^-$ and their range of uncertainty make the value obtained from an extrapolation to zero particle velocity somewhat questionable.

Acceleration waves in precompressed materials

In this section we look at acceleration wave propagation in a nonlinear viscoelastic material which initially is in an equilibrium state of homogeneous compressive strain, and consider the influence of this precompression on the wave velocity and on the growth and decay behavior of the wave amplitude. As in the above discussion of release wave velocities, the precompression is the result of the passage of a steady shock wave in a region which has had a relatively long time to reach equilibrium. Thus, the appropriate history to employ in equation (29) for the wave speed is the equilibrium history (5). Using the representations (15) for the instantaneous and equilibrium response functions, σ_I and σ_E, the velocity of an acceleration wave predicted by (31) can be calculated as a function of initial strain. By (30) this gives the tangent modulus as a function of initial strain, and, differentation yields the curvature modulus as a function of strain. The initial slope of the stress relaxation function in terms of the initial strain takes the form

$$G'(0) = -\frac{2[\sigma_I(\varepsilon) - \sigma_E(\varepsilon)]}{\varepsilon(2 - \varepsilon)\tau} .$$

Thus, using (26) and the above we can determine U, β, and a_c as functions of the initial strain. The results [28] indicate that, as expected, the velocity U of an acceleration wave increases with initial compressive strain. Further a_c, the critical amplitude, also increases with initial compression, i.e., the greater the initial compression the greater the initial amplitude of a compressive wave must be in order to have "shock formation" occur at the same t_∞. In the case of expansive waves, increasing the initial compression of the state ahead of the wave results in a more rapid decay of the wave amplitude.

The experimental observation of the decay of an expansive acceleration wave requires that a series of experiments be conducted where the wave is observed at increasing distances from the impact surface. Also, the impact condition must be carefully controlled to ensure that the equilibrium state into which the wave propagates is nominally the same. The experimental results shown in Fig. 15 represent three experiments on the polymer PMMA with an impact velocity of 0.15mm/μsec. The acceleration wave amplitude into the sample was nominally the same and the measurements were made at approximately 6, 12, and 37 mm within the target. Note the decrease in the amplitude of the wave with propagation distance. Considering the wave at 6 mm as the initial condition a(0), the two later profiles are calculated using the amplitude equation (27).

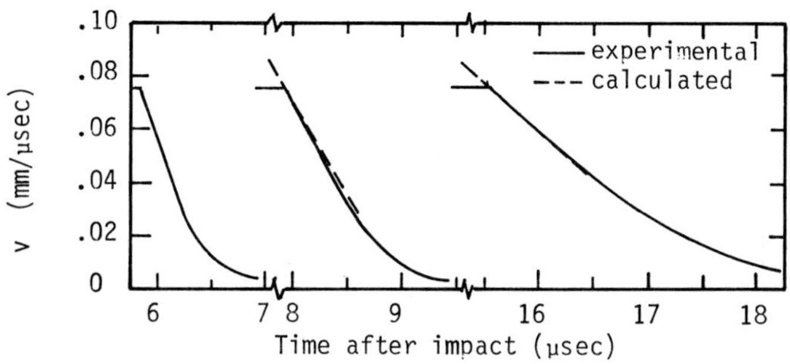

Fig.15. Measured unloading wave profiles as a function of time for PMMA

For this value the calculated wave speed U = 3.19mm/μsec compares well with the observed velocity of 3.22mm/μsec. Also, in this case we have a critical amplitude a_c = 0.028mm/μsec^2 and β = 0.040/μsec. Using these values we calculate the predicted wave amplitudes as a(2.805μsec) = -0.074mm/μsec^2 and a(9.690 μsec) = -0.032mm/μsec^2. The slopes corresponding to these amplitudes are shown in Fig. 15 and are in good agreement with the measured values of -0.070mm/μsec^2 and -0.035mm/μsec^2, respectively. The details of this work are given in [28].

7. THERMODYNAMIC INFLUENCES

To this point we have outlined some aspects of the propagation of shock and acceleration waves in nonlinear viscoelastic materials neglecting thermal influences. Here we give a very brief discussion of the results of several studies regarding viscoelasic waves that include thermodynamic influences. More extensive reviews of much of this work can be found in [6] and [31].

Steady waves

Nunziato and Walsh [30] considered the propagation of steady shock waves in nonlinear thermoviscoelasic materials by specifying a particular form for the free energy $\psi(t)$ in terms of a mechanical and a thermal relaxation function. It follows from results of the thermodynamics of materials with memory (Coleman [10]) that specific constitutive equations for the stress σ and internal energy e are derivable from the free energy. Assuming specific constitutive forms involving experimental mechanical and thermal relaxation functions and applying a steady wave analysis, they calculated the steady wave profiles for comparison with the experimental data of Schuler [33]. The results matched the experimental profiles more closely than the purely mechanical analysis. The details of the study are given in [30].

Shock wave growth and decay

Thermodynamic effects on the propagation of one-dimensional shock waves in materials with memory have been considered by Coleman and Gurtin [12], Chen and Gurtin [9], Achenbach, Vogel, and Herrmann [1], Dunwoody [20], Nunziato and Walsh [30], and others. In [12], Coleman and Gurtin show that most of the classical theory of Hugoniot curves related to shock wave propagation can be generalized to non heat conducting materials with long range, nonlinear, viscoelastic memory. For example, the result that the jump in entropy across a shock wave of small intensity is third-order in the jump in strain. In contrast to this, a result in [1] shows that for viscoelastic materials which do conduct heat the jump in entropy is first-order in the jump in strain.

In [9] Chen and Gurtin derived a differential equation analogous to (19) which relates the strain ε^- and the strain gradient $(\partial_x \varepsilon)^-$ behind the wave when the region ahead of the wave is in an unstrained equilibrium state. This shock amplitude equation serves to establish the existence of a critical strain gradient which governs the growth and decay of the wave. In [30], Nunziato and Walsh applied these results to a particular thermoviscoelastic material and determined the critical strain gradient and shock amplitude equation in terms of material response functions and thermophysical properties. The results were exhibited for the polymeric solid PMMA and compared with experimental results obtained in steady wave studies.

Acceleration wave growth and decay

In their study of thermodynamic influences on the growth and decay of acceleration waves in materials with memory, Coleman and Gurtin [13, 15, 16] showed that every acceleration wave with intrinisic velocity $U(t) \neq 0$ propagating in a non-conductor is homentropic, i.e., the jump in the entropy gradient, $[\partial_X \eta]$, is zero at the wave. Further, if the wave is entering a region at rest in a state of constant and uniform entropy, then (24) and (27) which give the velocity U_0 and acceleration $a(t)$ of an acceleration wave neglecting thermal effects, apply in the more general case if the relaxation function and second-order modulus are taken at fixed entropy. Thus, it would seem possible to use the results of experimental shock wave studies to evaluate the material constants necessary to predict the critical time t_∞ at which an acceleration wave would grow without bound. To test this, the calculation was carried out by Nunziato and Walsh [30] for the solid polymer PMMA. The critical times were calculated and shown to be commensurate with experimental results.

REFERENCES

1. Achenbach, J. D., S. M. Vogel, and G. Herrmann, On stress waves in viscoelastic media conducting heat, Proc. IUTAM Sympos. Vienna (1966), Ed. Parkus, H. and L. I. Sedov.
2. Bailey, P. B. and P. J. Chen, On the local and global behavior of acceleration waves, Arch. Rational Mech. Anal. 41 (1971), 121-131.
3. Barker, L. M., Laser interferometry in shock-wave research, Exp. Mech., May 1972, 1-7.
4. Barker, L. M., and R. E. Hollenbach, Shock wave studies of PMMA, fused silica and sapphire, J. Appl. Phys. 41 (1970), 4208-4226.

5. Bowen, R. M. and P. J. Chen, Thermodynamic restriction on the initial slope of the stress relaxation function, Arch. Rational Mech. Anal. 31 (1973) 278-284.
6. Chen, P. J., Growth and Decay of Waves in Solids, Handbuch der Physik Vol. VIa/3, Ed. C. Truesdell, Springer-Verlag Berlin (1973), 303-402.
7. Chen, P. J., and M. E. Gurtin, On the growth of one-dimensional shock waves in materials with memory, Arch. Rational Mech. Anal. 36 (1970), 33-46.
8. Chen, P. J., and M. E. Gurtin, On the use of experimental results concerning steady shock waves to predict the acceleration wave response of nonlinear viscoelastic materials, J. Appl. Mech. 39 (1972), 295-296.
9. Chen, P. J., and M. E. Gurtin, Thermodynamic influences on the growth of one-dimensional shock waves in materials with memory. ZAMP 23 (1972), 69-79.
10. Coleman, B. D., Thermodynamics of materials with memory, Arch. Rational Mech. Anal. 17 (1964) 1-46.
11. Coleman, B. D., J. M. Greenberg, and M. E. Gurtin, Waves in materials with memory V. On the amplitude of acceleration waves and mild discontinuities, Arch. Rational Mech. Anal. 22 (1966), 333-354.
12. Coleman, B. D. , and M. E. Gurtin, Thermodynamics and one-dimensional shock waves in materials with memory, Proc. R. Soc. A, 292 (1966), 562-574.
13. Coleman, B. D., and M. E. Gurtin, Thermodynamics and wave propagation in non-linear materials with memory, Proc. IUTAM Sympos. Vienna (1966), 54-76.
14. Coleman, B. D., and M. E. Gurtin, Waves in materials with memory II. On the growth and decay of one-dimensional acceleration waves, Arch. Rational Mech. Anal. 19 (1965), 239-265.
15. Coleman, B. D., and M. E. Gurtin, Waves in materials with memory III. Thermodynamic influences on the growth and decay of acceleration waves, Arch. Rational Mech. Anal. 19 (1965), 266-298.

16. Coleman, B. D. and M. E. Gurtin, Waves in materials with memory IV. Thermodynamics and the velocity of general acceleration waves, Arch. Rational Mech. Anal. 19 (1965), 317-338.

17. Coleman, B. D., M. E. Gurtin, and I. Herrera R., Waves in materials with memory I. The velocity of one-dimensional shock and acceleration waves, Arch. Rational Mech., Anal. 19 (1965), 1-19.

18. Coleman, B. D., and W. Noll, An approximation theorem for functionals with applications in continuum mechanics, Arch. Rational Mech. Anal. 6 (1960), 355-370.

19. Coleman, B. D., and W. Noll, Foundations of linear viscoelasticity, Rev. Mod. Phys. 33 (1961), 239-249.

20. Dunwoody, J., One-dimensional shock waves in heat conducting materials with memory, Arch. Rational Mech. Anal. 47 (1972), 117-148.

21. Glauz, R. D. and E. H Lee, Transient wave analysis in a linear time-dependent material, J. Appl. Phys. 25 (1954), 947-953.

22. Greenberg, J. M., The existence of steady shock waves in nonlinear materials with memory, Arch. Rational Mech. Anal. 24 (1967), 1-21.

23. Hunter, S. C., Viscoelastic waves. In: Progress in Solid Mechanics, Vol. I, Amsterdam: North Holland Publ.Co. (1960).

24. Kolsky, H., The propagation of stress waves in viscoelastic solids, Appl. Mech. Rev. 11 (1958), 465-468.

25. Lee, E. H., and I. Kanter, Wave propagation in finite rods of viscoelastic material, J. Appl. Phys. 24 (1953), 1115-1122.

26. Nunziato, J. W., and K. W. Schuler, Evolution of steady shock waves in polymethyl methacrylate, J. Appl. Phys. 44 (1973), 4774-4775.

27. Nunziato, J. W., and K. W. Schuler, Shock pulse attenuation in a nonlinear viscoelastic solid, J. Mech. Phys. Solids 21 (1973), 447-457.

28. Nunziato, J. W., K. W. Schuler, and E. K. Walsh, The influence of precompression on acceleration wave propagation in a nonlinear viscoelastic material, J. Appl. Mech. 42 (1975), 731-732.
29. Nunziato, J. W., and H. J. Sutherland, Acoustical determination of stress relaxation functions for polymers, J. appl. Phys. 44 (1973), 184-187.
30. Nunziato, J. W., and E. K. Walsh, Amplitude behavior of shock waves in a thermoviscoelastic solid, Int. J. Solids Structures 9 (1973), 1373-1383.
31. Nunziato, J. W., E. K. Walsh, K. W. Schuler, and L. M. Barker, Wave Propagation in Nonlinear Viscoelastic Solids, Handbuch der Physik Vol VIa/4, Ed. C. Truesdell, Springer-Verlag Berlin (1974), 1-108.
32. Pipkin, A. C., Shock structure in a viscoelastic fluid, Quart. Appl. Math. 23 (1966), 297-303.
33. Schuler, K. W., Propagation of steady shock waves in polymethyl methacrylate, J. Mech. Phys. Solids 18 (1970), 277-293.
34. Schuler, K. W., The speed of release waves in polymethyl methacrylate. Proc Fifth Symposium on Detonation. Washington, D.C. U.S. Gov't Printing Off. (1970).
35. Schuler, K. W., and E. K. Walsh, Critical induced acceleration for shock propagation in polymethyl methacrylate, J. Appl. Mech. 38 (1971), 641-645.
36. Varley, E., Acceleration fronts in viscoelastic materials, Arch. Rational Mech. Anal. 19 (1965), 215-225.
37. Walsh, E. K., and K. W. Schuler, Acceleration wave propagation in a nonlinear viscoelastic solid, J. Appl. Mech. 40 (1973), 705-710.

This work was supported in part by the National Science Foundation.

OVERVIEW OF MACROSCOPIC VISCOELASTIC FLOW

K. Walters

Department of Applied Mathematics
The University College of Wales
Wales, United Kingdom

1. INTRODUCTION

A possible research programme designed to investigate the behaviour of non-Newtonian elastic liquids in complex flows might involve the following components:

(i) A study of suitable non-Newtonian test fluids in *simple* flows like steady simple shearing flow, small-amplitude oscillatory shear flow and the like. This component may be termed basic Rheometry and is seen as providing a foundation set of data and material functions for the test fluids. The determination of the viscometric functions and dynamic data must be regarded as standard in such an exercise, but an estimate of extensional-viscosity levels and the dependence of these levels on extensional strain rate is also potentially of great interest and importance.

(iia) An experimental study of the behaviour of the non-Newtonian test fluids in the *complex* geometries of interest. The precise experimental technique may vary, involving, for example, force measurements, point velocity measurements (provided by LDA and other techniques 1, 2, 3) or more simple flow visualization devices. In the present work, a laser flow visualization technique is employed (see, for example 4, 5).

(iib) The use of the rheometrical data obtained in (i) to construct suitable constitutive models for the test fluids. These models must of course satisfy the well known constraints imposed by the formulation principles of continuum mechanics (as enunciated, for example, by Oldroyd 6). The rheometrical data may be seen as providing meaningful constraints on constitutive models, where, for reasons of tractability, these models fall outside conditions where *general* equations can be specified. Where flow conditions are more favourable and general equations are applicable, as in slow flow for example (7), the rheometrical data can of course be used to specify the relevant material constants. However, in the present exercise, our main concern is with *complex* flows which certainly fall outside the 'slow flow' regime and the rheometrical data must be seen as simply providing useful constraints on the approximate constitutive equations employed. The objective is to solve these equations, in conjunction with the familiar equations of motion and continuity, for the complex flows of interest in the experimental programme. Analytic solutions are, on the whole, out of the question for reasons of tractability, and numerical solutions of the relevant equations are indispensable (see, for example, 8-10).

(iii) The final component in the research programme is an important one - the comparison of the experimental results of (iia) with the numerical simulations of (iib).

We do not suggest that the above research programme is unique or flawless and we would have sympathy, for example, with an argument in favour of some input from microrheological considerations. However, the research programme as outlined above can be defended as at least providing one possible coherent attempt to predict and understand the often unusual and bizarre behaviour of elastic liquids when these are made to flow in complex geometries. To give some justification for this confidence, we illustrate the research programme by appeal to one particular type of non-Newtonian test fluid - the so called Boger fluids (11), which in our case were dilute (~0.1%) solutions of polyacrylamide in a maltose syrup/water base. (Details of these test fluids are provided in references 4, 5, 11, 12).

2. BEHAVIOUR IN SIMPLE FLOWS (RHEOMETRY)

The original Boger fluid, a dilute solution of polyacrylamide in a highly viscous Newtonian solvent, is now seen as being only one of a class of elastic liquids with the same general type of rheometrical behaviour (see, for example, 11-59). However, of the various possibilities, the original solution seems to be the most elastic, judged on the basis of the relative levels of the first normal stress difference ν_1 and the shear stress τ in a steady simple shear flow. Accordingly, we have confined attention to the maltose syrup/water based solutions in our research programme.

The dilute polyacrylamide solutions are usually found to have a reasonably constant apparent viscosity, when examined in a steady simple-shear flow. This must not be seen as too surprising, since the solvent viscosity may account for 90-95% of the final polymer solution viscosity level and in normal circumstances a maximum of the order of 10% can therefore be anticipated in the viscosity variation with shear rate q. Indeed, it is often found that over a substantial range of shear rates the apparent viscosity has a reasonably *constant* value and any (slight) shear thinning only occurs at relatively high shear rates. This means that a meaningfully constant viscosity η_0 can be defined, which can be used with some confidence in defining an appropriate Reynolds number for a flow.

Associated with the constant viscosity region is usually a quadratic dependence of the first normal stress difference ν_1 on shear rate q, giving rise to what is generally referred to as second-order behaviour. At the higher shear rates, the dependence of ν_1 on q is weaker, in the sense that the so-called first normal stress coefficient (ν_1/q^2) decreases with increasing shear rate.

Judging from the open-trough experiments of Keentok et al (16), the second normal stress difference ν_2 is very small indeed for Boger fluids and we may take $\nu_2 = 0$ as a reasonable estimate for these fluids.

For completeness, we note that at some critical set of conditions, depending on shear rate and possibly other factors, a build-up of structure occurs in Boger fluids, resulting in apparent shear thickening with the viscosity taking values higher than the zero shear value (12, 14). This antithixotropy and shear thickening can be broken down by continuous shearing.

Notwithstanding these anomalies, we still find it convenient to define two simple non-dimensional numbers, R and W, to characterize the Boger-fluid flows. The Reynolds number R is given by

$$R = \frac{\rho U L}{\eta_0} , \tag{2.1}$$

where η_0 is the (constant) viscosity, ρ is the density, U is a characteristic velocity (the mean velocity in most of our experiments) and L is a characteristic length; the elasticity number W, sometimes called the Weissenberg number or Deborah number, is given by

$$W = \frac{\lambda U}{L} , \tag{2.2}$$

where λ is a characteristic time defined from the (second-order) viscometric data through

$$\nu_1 = 2 \eta_0 \lambda q^2 . \tag{2.3}$$

A further important feature of Boger fluids is their substantial resistance to stretching flows as manifested through the very high values of the extensional viscosity η_E. It is usual to talk in terms of the so-called Trouton ratio, which is defined as the ratio of η_E to the shear viscosity η_0. We have recently employed a lubricated die rheometer to estimate the values of the Trouton ratio for Boger fluids (see, for example, 17, 18). The device which is shown schematically in Fig 1, yields the *planar* extensional viscosity in our experiments. Representative experimental data for both a Newtonian and a Boger fluid are given in Fig 2.

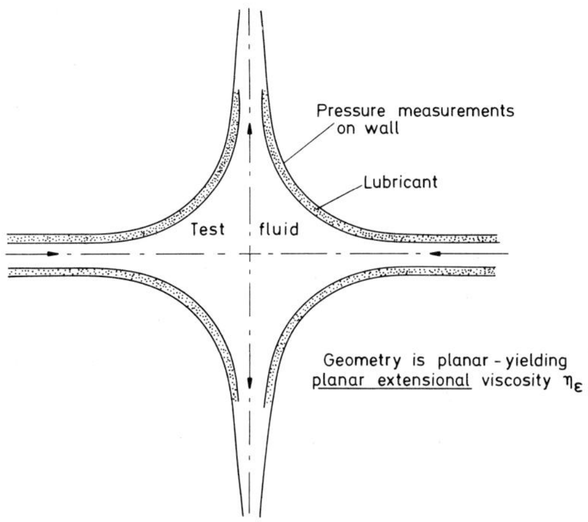

Fig 1. Schematic diagram of the lubricated-die rheometer.

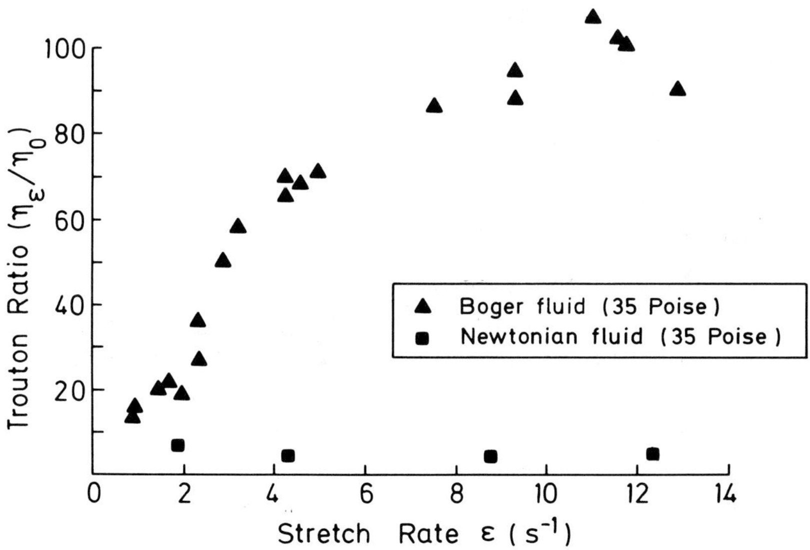

Fig 2. Planar extensional viscosity data for a Newtonian fluid and a Boger fluid of the same shear viscosity.

The Trouton ratio for the Newtonian fluid is seen to be reassuringly near the expected value of 4 and the observed trend for the Boger fluid as the strain rate tends to zero is also in accord with the expectations of continuum mechanics.

Of equal interest are the high values of the Trouton ratio for the Boger fluid as the strain rate increases, with the indication of a maximum in this ratio at the highest attainable strain rates.

Fig 2 is in many ways an important addition to the Rheometry literature, given the extreme difficulty of obtaining meaningful steady-state extensional-viscosity data for mobile liquid systems. At the same time, it must be stressed that the Trouton ratios for the Boger fluid shown in Fig 2 are modest by Boger fluid standards and values in excess of 10^3 are often found in (admittedly) less controllable extensional flow experiments (12, 14, 19).

Other rheometrical information on Boger fluids is provided in references 12-15.

If the time-dependent behaviour of the Boger fluids at the higher shear rates is ignored, we may conveniently characterize the fluids (in a preliminary study at least) by the so-called Oldroyd fluid B model, with equations of state in a rectangular Cartesian frame given by

$$p_{ik} = -p\delta_{ik} + T_{ik} , \tag{2.4}$$

$$T_{ik} + \lambda_1 \overset{\nabla}{T}_{ik} = 2\eta_0 \left[d_{ik} + \lambda_2 \overset{\nabla}{d}_{ik}\right] , \tag{2.5}$$

where p_{ik} is the stress tensor, T_{ik} the extra-stress tensor, d_{ik} is the rate of strain tensor, δ_{ik} is the Kronecker delta and p is an arbitrary isotropic pressure (in the case of incompressible fluids). η_0 is a constant viscosity coefficient λ_1 and λ_2 are a relaxation time and retardation time, respectively, and ∇ denotes the upper convected time derivative introduced by Oldroyd (6). We require (see, for example, 20)

$$9\lambda_2 > \lambda_1 \geqslant \lambda_2 > 0 , \tag{2.6}$$

and in view of the high solvent viscosity in the case of the Boger fluids, we further require the difference $\lambda_1 - \lambda_2$ to be significantly smaller than either λ_1 or λ_2 (12). The viscometric functions for the Oldroyd B model are given by

$$\begin{aligned} \tau &= \eta_0 q \, , \\ \nu_1 &= 2\,\eta_0(\lambda_1 - \lambda_2)q^2 \, , \\ \nu_2 &= 0; \end{aligned} \qquad (2.7)$$

and the *uniaxial* extensional viscosity η_E is given by

$$\eta_E = \frac{3\eta_0[1-\lambda_2 k - 2\lambda_1\lambda_2 k^2]}{[1+\lambda_1 k][1-2\lambda_1 k]} ,$$

where k is the rate of strain.

Disadvantages of such a simple model as (2.5) are well known. Only one relaxation time λ_1 is envisaged, which is a limitation, and, although very high values of η_E are attainable for the model, the prediction of an infinite value at a finite rate of strain ($k = 1/(2\lambda_1)$) is thought to be another significant limitation of the model. It is clear that simulating even the relatively simple rheometrical behaviour of Boger fluids is not a straightforward task, but at least the Oldroyd fluid B can be regarded as a useful first attempt, especially in view of the difficulties encountered in using even models of this apparent simplicity in simulating complex flow behaviour.

3. BEHAVIOUR IN COMPLEX FLOWS

The experimental results summarized in this paper were obtained from a laser flow-visualization technique, which is now well documented in the literature (4, 5). The laser is used to provide a parallel beam of light which illuminates the test section. The elastic liquids are seeded with mutually-buoyant particles (~0.1mm in diameter) and it is expected that these particles follow the streamline paths in all but the strongest flows. A time-exposure photograph of the test section provides a permanent record of the paths of the particles and therefore of the streamlines if the flow is steady.

The geometries used in the present study are shown schematically in Fig 3. All the dimensions are in mm's and in the 'two dimensional' geometries 3 (a)-(c), (f), (g), the third dimension in all cases is 20mm.

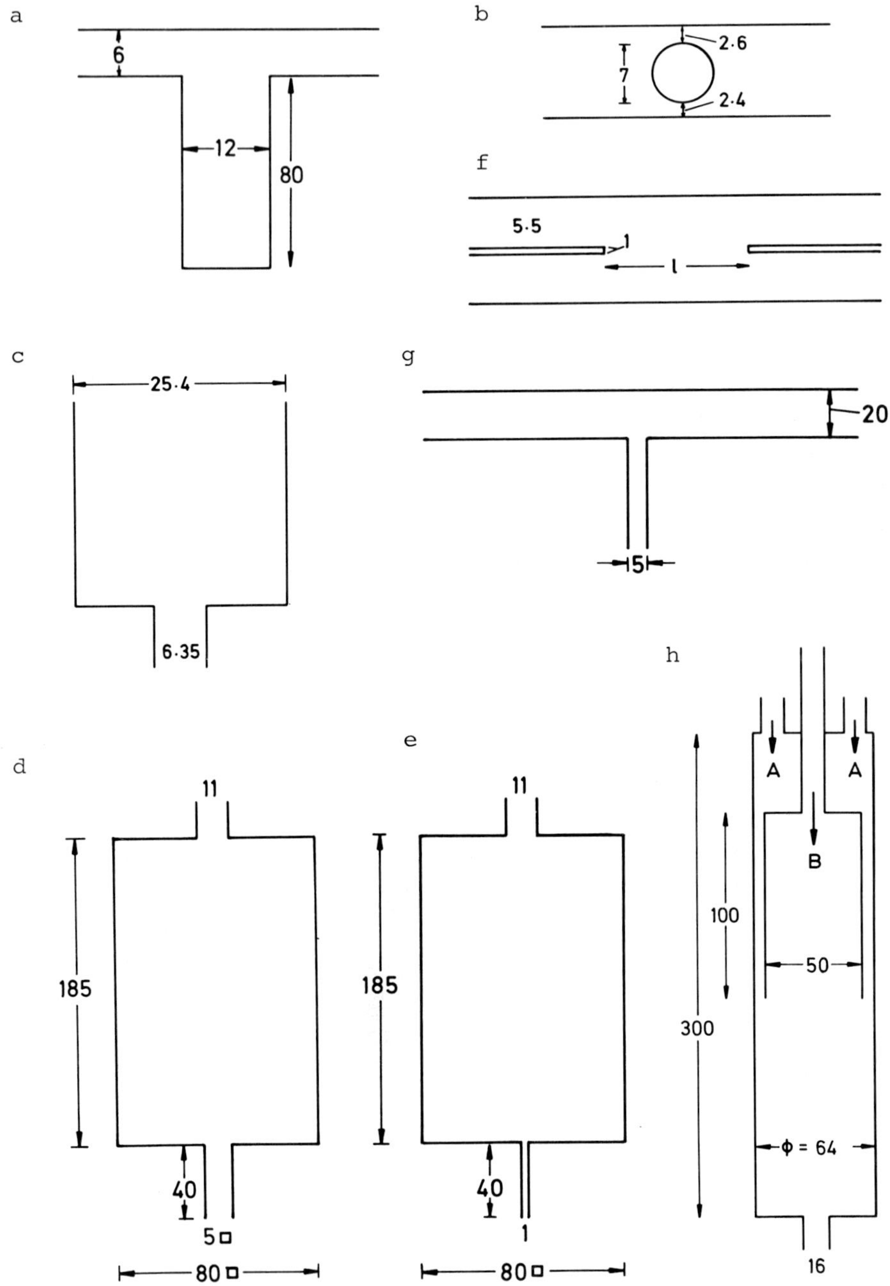

Fig 3. Schematic diagram of the flow geometries used in the current work.

All the flows are pressure driven with long enough entry and exit lengths to generate 'fully-developed' flow conditions at entry and exit.

(i) *Viscoelasticity versus fluid inertia*

Typical streamlines for flow over a deep slot (Geometry 3(a))are shown in Fig 4. The asymmetric shift in the streamlines in the Newtonian case (W = 0) which is caused by fluid inertia and that found in the elastic liquid are seen to be in opposite directions and it is possible to manufacture conditions where the opposing influences of fluid inertia and fluid memory balance and the flow is essentially symmetric. These competing influences are seen in many of our experiments. In contraction flows, for example, fluid elasticity can lead to so-called vortex enhancement with strong recirculating regions in the salient corners. Vortex enhancement can be reduced in intensity by increasing the Reynolds number (see, for example, 5).

Similar conclusions concerning the competing influences of fluid inertia and fluid elasticity are to be found in the literature (see, for example, 21, 22).

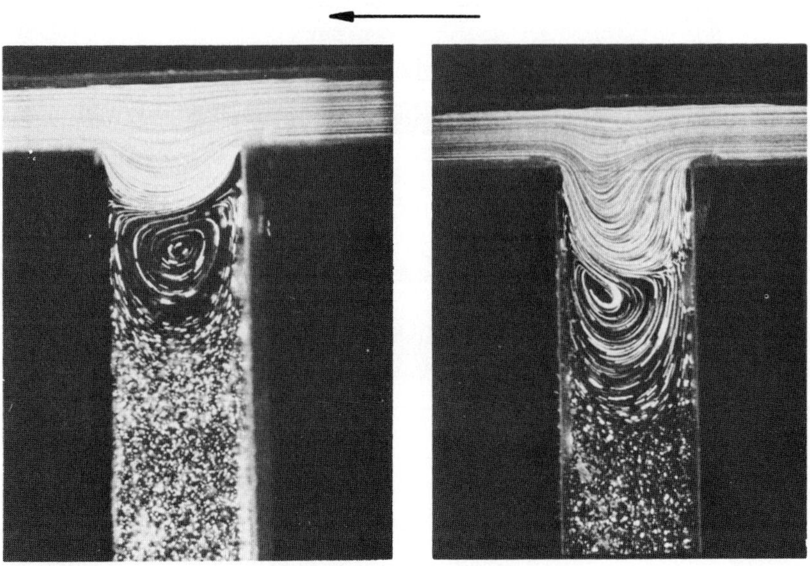

R = 6, W = 0 R = 6, W = 0.75

Fig 4. Streamlines for flow over a deep slot.

Also worthy of note in Fig 4 is the position of the recirculating vortices in the deep slot. Fluid elasticity drives the vortex deeper into the slot.

Not shown in the figures, but certainly present, are lower and much weaker secondary vortices (see, for example, 4).

(ii) *Slight asymmetries in the geometry are exaggerated by fluid elasticity.*

For flow past a cylinder which is in an asymmetrical position with respect to the channel walls (Geometry 3(b)), the elastic liquid is seen to exaggerate the geometrical asymmetry (Fig 5). This is in marked contrast to the Newtonian situation where, without detailed study, it would be very difficult to discern any asymmetry in the flow (4).

R = 5
W = 0

R = 5
W = 0.1

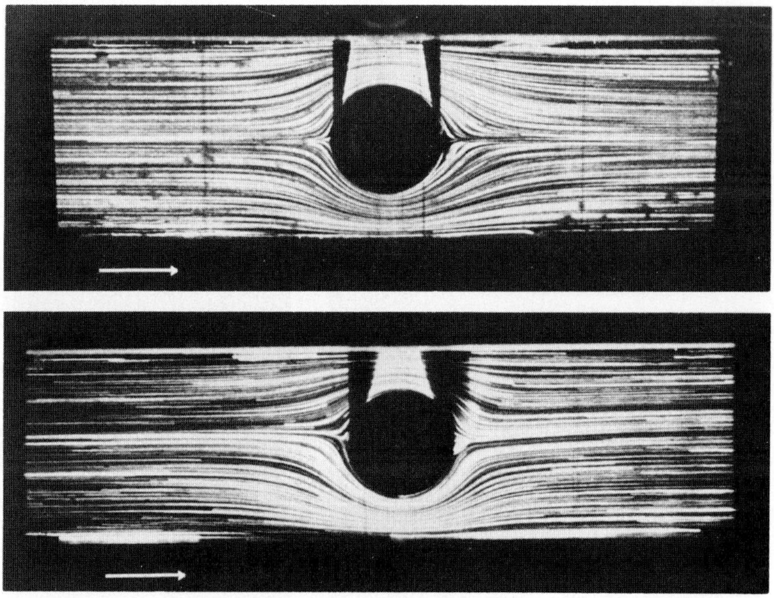

Fig 5. Streamlines for flow past a cylinder (Geometry 3(b)).

In previous work on axisymmetric contraction flows (5), we rounded (slightly) half of the contraction exit lip. This produced major changes throughout the flow field, which we attributed to the importance of the precise sharpness of re-entrant corners in non-Newtonian flows. We now remark that these observations can also be seen as further examples of the point being made in this section, namely, that slight asymmetries in the geometry can be greatly exaggerated by fluid elasticity so far as observable flow characteristics are concerned.

(iii) *Contraction flows*

Contraction flows of Boger fluids provide many provocative observations and those presented here supplement the extensive studies on contraction flows by Boger himself (see, for example, 23).

We note that in our definitions of R and W for contraction flows, we have taken U as the mean flow rate in the upstream section and L is the upstream channel width (or the radius in the axisymmetric case).

Fig 6, for a 4:1 planar contraction, shows that, if anything, the small stagnant regions (or, more correctly, weakly recirculating regions) found in the salient corners in the case of Newtonian liquids (cf. 5) are diminished in size for the Boger fluid.

To illustrate that this is not generally the case for non-Newtonian elastic liquids, we also show the streamlines for an aqueous solution of polyacrylamide, which is dramatically shear thinning. There is clear evidence of vortex enhancement in a planar contraction flow in this case.

Axisymmetric contraction flows of Boger fluids, in contrast, are known to give rise to dramatic vortex enhancement (5,23) and we show in Fig 7 typical streamlines for the associated square/square contraction geometry (3(d)). The behaviour here is in marked contrast to that found in the corresponding planar contraction case (cf. Fig 6).

In an attempt to generate vortices in the planar contraction geometry, we constructed a geometry with a very high contraction ratio (80:1)(Fig 3(e)). Then, working at conditions of maximum flow rate, we were able to generate vortices near the contraction zone (Fig 8). However, the vortex structure was in general asymmetric. The larger vortex could appear on either side of the contraction zone, from which we concluded that this particular asymmetry was not directly attributable to a geometrical asymmetry.

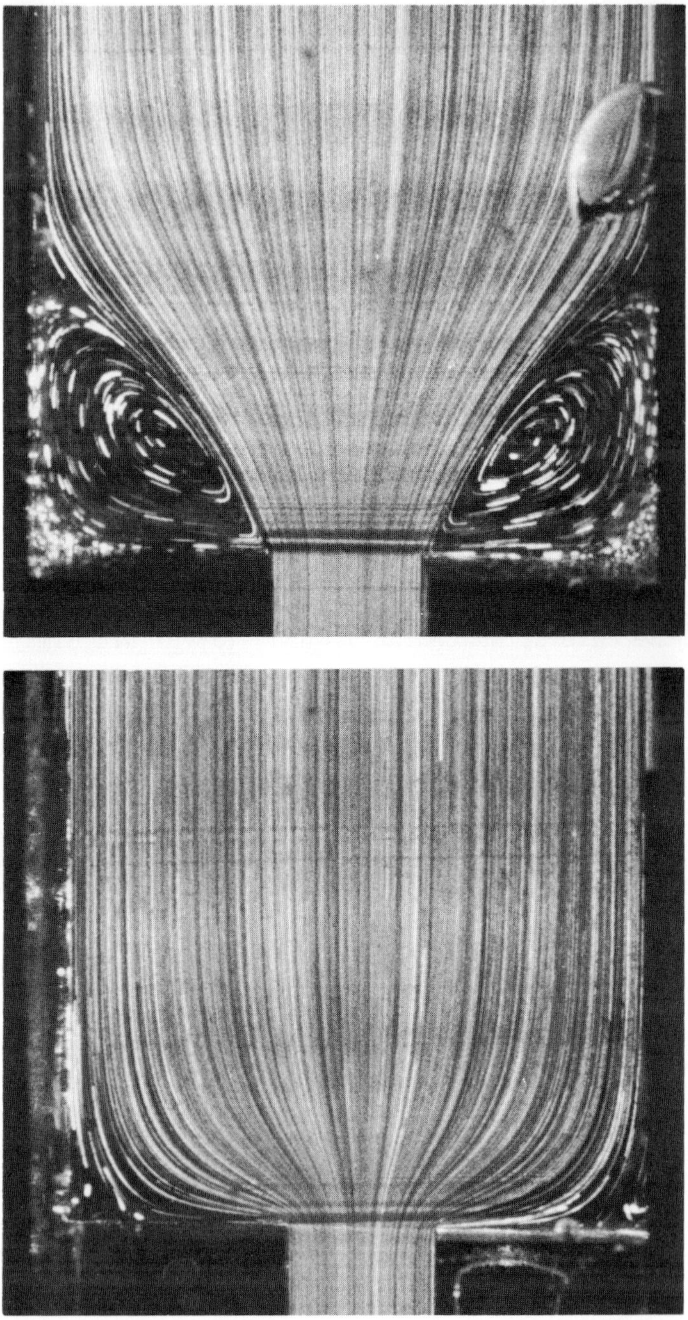

(a)
R=1.33
W=0.38

(b)

Fig 6. Streamlines for a 4:1 planar contraction (a) Boger fluid (b) Aqueous solution of polyacrylamide. Flow rate in (b) is 18.5cc/s.

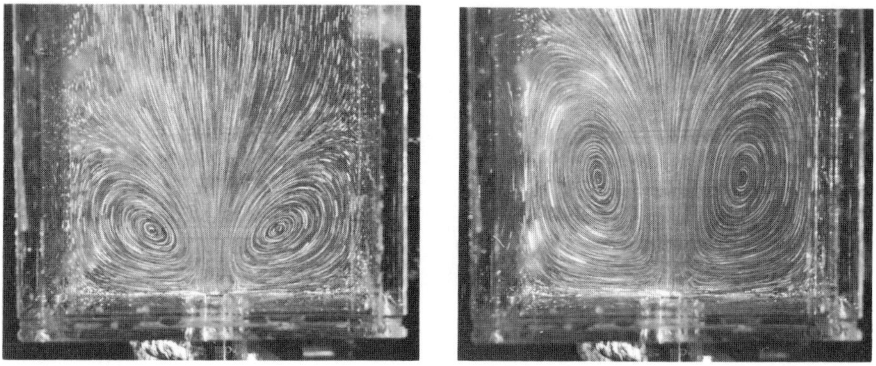

R = 0.094, W = 0.0078 R = 0.188, W = 0.0156

Fig 7. Streamlines for a Boger fluid in contraction geometry 3(d).

R = 0.18, W = 0.0014

Fig 8. Streamlines for a Boger fluid in an 80.1 planar contraction. Geometry 3(e).

R = 0.000412, W = 0.0000424

R = 0.000156, W = 0.00000951

Fig 9. Streamlines for two Boger fluids at very low Reynolds numbers in contraction geometry 3(d).

In many ways, we have found it as difficult to generate vortices for a Boger fluid in a planar contraction as it is to avoid them in the square/square contractions! In the latter connection, we remark that vortices were still present in this geometry at a Reynolds number of 0.0004 and that it was necessary to reach very low Reynolds numbers indeed before the vortices disappeared and the flow approached that expected for a Newtonian fluid (see Fig 9).

In an attempt to study re-entrant corner conditions, we removed the corner from one (or two) of the 4 sides of the contraction exit square as shown in Fig 10. Typically, we obtained the streamlines shown in Fig 11, the symmetrical streamlines corresponding to symmetrical conditions (10(a)) and the asymmetrical streamlines arising from the cut away of the corner in 10(b). Initially we took this as further evidence that the precise shape of the re-entrant corner was

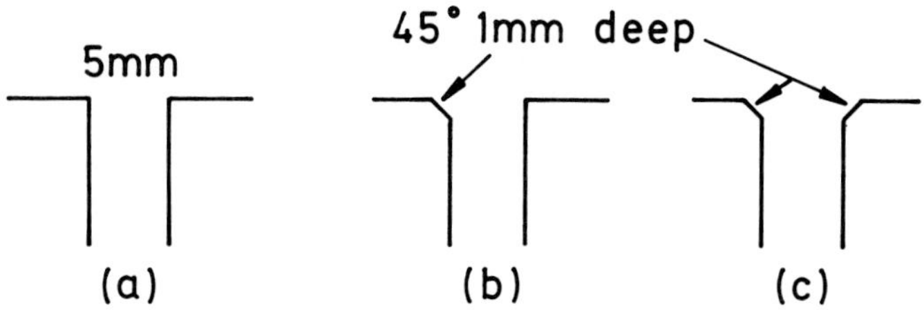

Fig 10. Various re-entrant corner conditions for contraction geometry 3(d).

of crucial importance in dictating the flow, not only in the neighbourhood of the corner, but also over a major part of the flow field. However, when *two* opposite sides were cut away, and the exit geometry consisted of 10(a) and (c), we obtained basically the same symmetrical streamlines for both the sharp cut away combinations (Fig 12). In *this* geometry, therefore, the cutting away of the corners has little effect on the vortex structure, *provided the flow is symmetric*. However, it appears from some of our most recent experiments that the same conclusion cannot be drawn from cutting away the exit lip in an *axisymmetric* contraction (cf.5) and the whole question of re-entrant corner singularities and their effect on flow characteristics is not straightforward and is provocative, to say the least.

(iv) *Unidirectional flow versus Reversed Flow*

The so-called mixing/separating flow, shown schematically in Fig 3(f), provides a dramatic example of the qualitative changes in flow characteristics which can be brought about by the introduction of fluid elasticity. When the flow rates in the various arms are equal, we obtain the type of behaviour shown in Fig 13 for a Newtonian fluid and a Boger fluid. Some of the flow is of the 'unidirectional' type and some of the 'reversed' kind (5). An impressive symmetry is possible in the Newtonian case, but this is somewhat elusive for the Boger fluid, something we might have anticipated from previous comments. Further, there is evidence that the elastic liquid has a stronger preference for unidirectional flow than its Newtonian counterpart.

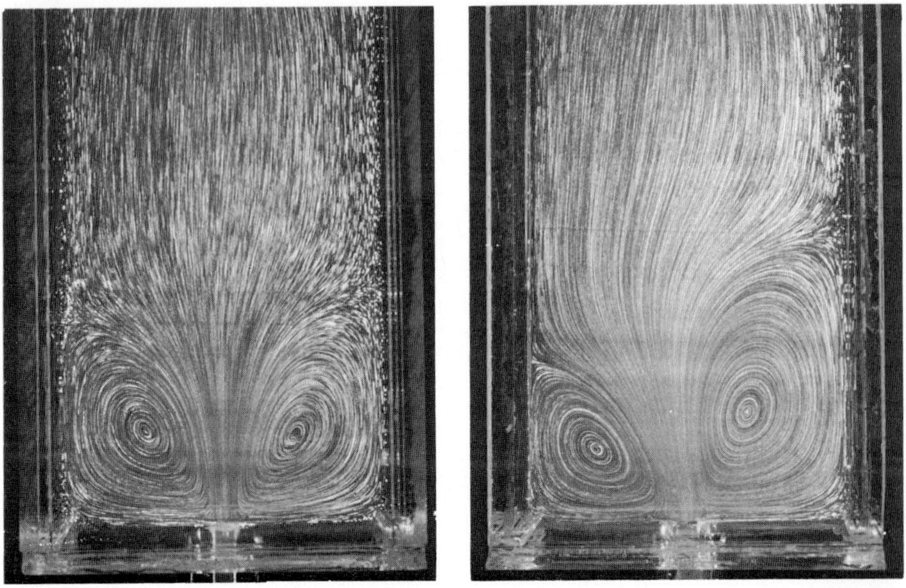

R = 0.0131, W = 0.0008

Fig 11. Streamlines for a Boger fluid in contraction geometry 3(d). (a) corresponds to corner conditions 10(a). (b) corresponds to corner conditions 10(b).

R = 0.0049, W = 0.0005

Fig 12. Streamlines for a Boger fluid in contraction geometry 3(d). (a) corresponds to corner conditions 10(a). (b) corresponds to corner conditions 10(c).

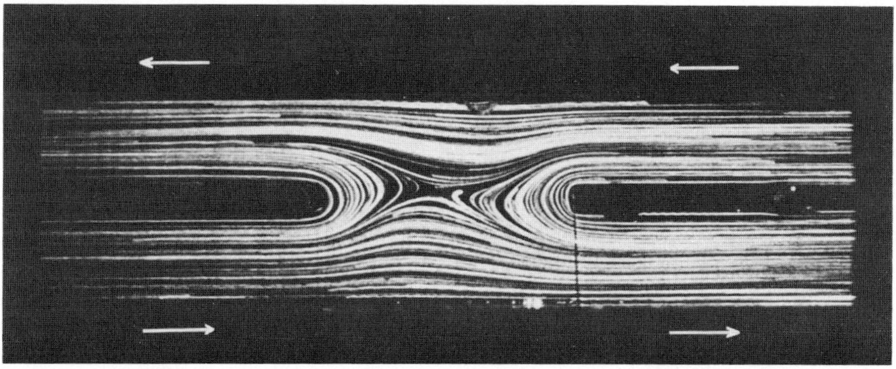

R = 5, W = 0

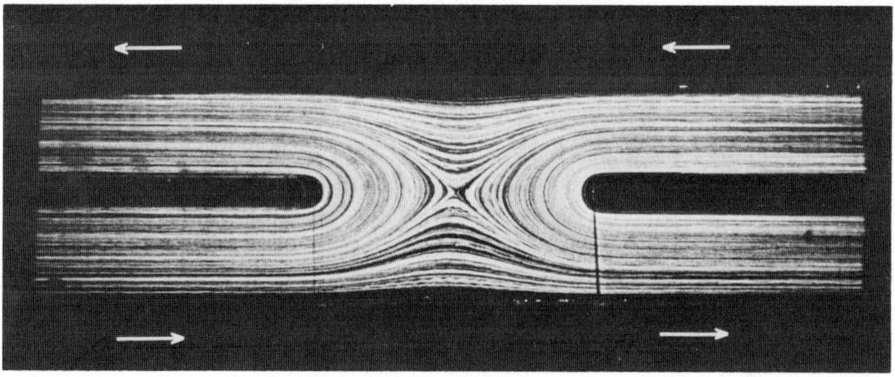

R = 5, W = 0.75

Fig 13. Streamlines for the mixing/separating flow geometry 3(f). l = 15mm.

By varying the geometrical conditions (i.e. the gap between the protruding arms) and the flow conditions (i.e. R and W) it is possible to obtain conditions for which the flow for the Newtonian and Boger fluids is qualitatively different (Fig 14). In the Newtonian case, the flow at relatively high Reynolds numbers (>20) becomes completely of the reversed kind with two recirculating vortices, as shown schematically in Fig 14. On the other hand, the elastic liquid, at comparable Reynolds numbers, strongly favours unidirectional flow with any reversed flow being confined to a boundary-layer type region near the barriers.

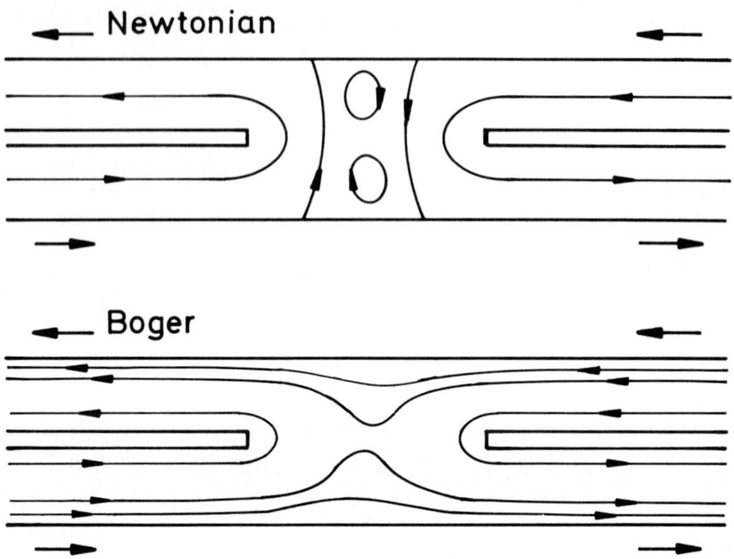

Fig 14. Schematic representation of streamlines in the mixing/separating flow geometry 3(f).

(v) *Interface flows*

To demonstrate interfacial effects in flowing non-Newtonian liquids, we shall find it convenient in the first instance to refer to flow in a T geometry (Fig 3(g)). Facilities were available to conduct experiments with *different* liquids entering the various arms; the flow rates in the entry arms could also be varied.

Representative experimental results for the *one*-liquid situations are given in Fig 15. In the Newtonian case, no significant vortex behaviour was observed in any of the experiments, in contrast to the Boger fluid situation, where a strong vortex was present in the entry arm with the weaker flow rate. Interestingly, in the symmetrical situation (with equal flow rates in the two entry arms) the strong vortex structure essentially disappeared.

(a) N4 R = 0.06, W = 0 N4 R = 0.06, W = 0

(b) B44 R = 0.016, W = 0.055 B44 R = 0.009, W = 0.013

(c) B44 R = 0.014, W = 0.05 B44 R = 0.014, W = 0.05

Fig 15. Streamlines for the T geometry 3(g). N4 is a Newtonian fluid. B44 is a Boger fluid.

B44 R = 0.02, W = 0.06 N4 R = 0.006, W = 0

Fig 16. Streamlines for the T geometry with different fluids entering the arms. Boger fluid on the left. Newtonian fluid on the right.

In Fig 16 we show the streamlines corresponding to a *two*-liquid situation with Newtonian liquid entering one arm and Boger fluid the other. (The fluids could be regarded as immiscible over the flow field of interest). We note with interest that there is now a strong recirculating vortex in the *Newtonian* liquid, which is reminiscent of what we would expect in a Boger fluid. In other words, it would appear that the mechanical properties of the relevant (Newtonian) fluid are not as important in this flow as the boundary conditions along the interface between the two liquids.

To further demonstrate this dramatic interfacial effect, we consider the lubricated contraction flow shown in Fig 3(h). When the liquids A and B are the same (either Newtonian or elastic) we obtain the expected flow fields already discussed, with strong vortex behaviour in the case of the Boger fluid and only a small, almost stagnant,region near the salient corner in the Newtonian case (Fig 17). However, when the main liquid (B) is a Boger fluid and the lubricant (A) is Newtonian, we observe strong vortex behaviour in the *Newtonian* liquid, which again must be due to the influence of the Boger fluid through the interface conditions. In Fig 17(c) the Boger fluid has been dyed to show clearly the recirculating vortex in the Newtonian liquid. Figs 16 and 17 present significant challenges to those working in Numerical Simulation, but these are under active consideration at the present time (24).

N6 16.3ml/s B46 18.5ml/s N6 16.3ml/s B46 dyed

Fig. 17. Streamlines for the lubricated contraction geometry 3(h). (a) Newtonian. Flow rate 16.3ml/s. (b) Boger fluid 18.5ml/s. (c) Liquid (A) is Newtonian. Liquid (B) is a Boger fluid. Combined flow rate 16.8ml/s.

4. THEORETICAL SIMULATION OF OBSERVED BEHAVIOUR IN COMPLEX FLOWS

(i) *Basic equations*

Ideally, we would wish to simulate complex flows by means of a constitutive model of complete generality; for example, one with equations of state given by

$$T_{ik}(t) = \underset{-\infty}{\overset{t}{F}}\left[C_{j\ell}(t')\right] , \tag{4.1}$$

where $C_{j\ell}(t')$ is a suitable strain or strain-rate measure and F is a tensor-valued functional#. Even if sufficient information were available on the functional F, the resulting equations for any complex flow would be too complicated to admit any prospect of a solution and it is difficult to see future developments in non-Newtonian fluid mechanics remedying the situation. To make headway, therefore, approximations have to be made and, basically, these can arise in two ways:

(I) There are approximations arising from simplifications in the flow, so that $C_{j\ell}$ has a relatively simple form, leading to *generally valid equations for restricted classes of flow*. The 'hierarchy equations' of Coleman and Noll (7), valid for sufficiently slow flow (and fading memory), are excellent examples of this type of approximation.

(II) Approximations arising from simplifications to the response functional F leading to *approximate equations which are valid under all conditions of motion and stress*. In this type of approximation, it is the response that is restricted not the class of flows which can be studied. The Oldroyd B model already defined in (2.5) is a popular example of approximate models of this type.

The interrelation between approximations of type (I) and (II) has also been considered in a recent paper by Saut and Joseph (25), which we regard as a very useful recent addition to the 'Formulation' literature.

For the complex flows of interest in the present study, approximations of type (I) are not appropriate and we are forced for pragmatic reasons to apply approximations of type

#In the following, we exclude any attempt to accommodate the *time-dependent* effects associated with the rheometrical behaviour of Boger fluids mentioned in §2.

(II) to make headway. The advantages and disadvantages of such a procedure are now well documented and are discussed in detail by Crochet, Davies and Walters (9).

(ii) *Method of Solution*

To illustrate modern developments in the numerical simulation of non-Newtonian flow, we shall concentrate attention on the Oldroyd B model (2.5), although recent work by various research groups has been able to accommodate models that are significantly more complicated than (2.5) (see, for example, 26).

For a steady two-dimensional flow with velocity components (u,v) in the (x,y) directions, the relevant equations are the equation of continuity

$$\frac{\partial u}{\partial x} + \frac{\partial v}{\partial y} = 0 \ , \tag{4.2}$$

the stress equations of motion

$$\rho\left[u\frac{\partial u}{\partial x} + v\frac{\partial u}{\partial y}\right] = F_x - \frac{\partial p}{\partial x} + \frac{\partial T_{xx}}{\partial x} + \frac{\partial T_{xy}}{\partial y} \ , \tag{4.3}$$

$$\rho\left[u\frac{\partial v}{\partial x} + v\frac{\partial v}{\partial y}\right] = F_y - \frac{\partial p}{\partial y} + \frac{\partial T_{xy}}{\partial x} + \frac{\partial T_{yy}}{\partial y} \ , \tag{4.4}$$

and the rheological equations of state

$$T_{xx}\left[1 - 2\lambda_1\frac{\partial u}{\partial x}\right] + \lambda_1\left[u\frac{\partial T_{xx}}{\partial x} + v\frac{\partial T_{xx}}{\partial y}\right] - 2\lambda_1\frac{\partial u}{\partial y}T_{xy}$$

$$= 2\eta_0\frac{\partial u}{\partial x} + 2\eta_0\lambda_2\left[u\frac{\partial^2 u}{\partial x^2} + v\frac{\partial^2 u}{\partial x\partial y} - 2\left(\frac{\partial u}{\partial x}\right)^2 - \frac{\partial u}{\partial y}\left(\frac{\partial u}{\partial y} + \frac{\partial v}{\partial x}\right)\right], \tag{4.5}$$

$$T_{yy}\left[1 - 2\lambda_1\frac{\partial v}{\partial y}\right] + \lambda_1\left[u\frac{\partial T_{yy}}{\partial x} + v\frac{\partial T_{yy}}{\partial y}\right] - 2\lambda_1\frac{\partial v}{\partial x}T_{xy}$$

$$= 2\eta_0\frac{\partial v}{\partial y} + 2\eta_0\lambda_2\left[u\frac{\partial^2 v}{\partial x\partial y} + v\frac{\partial^2 v}{\partial y^2} - 2\left(\frac{\partial v}{\partial y}\right)^2 - \frac{\partial v}{\partial x}\left(\frac{\partial v}{\partial x} + \frac{\partial u}{\partial y}\right)\right], \tag{4.6}$$

$$-\lambda_1\frac{\partial v}{\partial x}T_{xx} - \lambda_1\frac{\partial u}{\partial y}T_{yy} + \lambda_1\left[u\frac{\partial T_{xy}}{\partial x} + v\frac{\partial T_{xy}}{\partial y}\right] + T_{xy}$$

$$= \eta_0\left[\frac{\partial u}{\partial y} + \frac{\partial v}{\partial x}\right] + \eta_0\lambda_2\left[\left(u\frac{\partial}{\partial x} + v\frac{\partial}{\partial y}\right)\left(\frac{\partial u}{\partial y} + \frac{\partial v}{\partial x}\right)\right.$$

$$- \frac{\partial v}{\partial y}\left(3\frac{\partial u}{\partial y} + \frac{\partial v}{\partial x}\right) - \frac{\partial u}{\partial x}\left(\frac{\partial u}{\partial y} + 3\frac{\partial v}{\partial x}\right)\Bigg] \ , \qquad (4.7)$$

where F_x and F_y are the components of the body force per unit mass in the x and y directions, respectively.

It is immediately obvious that the stress components are given in implicit form by (4.5)-(4.7) and must be treated as dependent variables along with the pressure and the velocity components (u,v). This presented the first obstacle to the numerical simulation of complex flows, but the problem has now been successfully overcome (8,9). In the corresponding integral constitutive model formulation, the stress components may be given as explicit functions of the kinematical variable, but in this case, the displacement functions x'_i need to be determined# and these are now new dependent variables which do not figure explicitly in any *Newtonian* fluid situation. The 'tracking of particles', which is involved in the determination of x'_i, is the subject of current research in various research groups involved with numerical simulation employing integral models. Suffice is to say that differential and integral equations of state provide their own distinctive problems which are under active consideration at the present time. Within reason, the precise difficulty *of detail* of constitutive models is no longer thought to be a significant problem. The major difficulties lie elsewhere, as we shall indicate.

To illustrate one possible method of solution of the basic equations, we follow the authors own finite-difference techniques. Other methods are adequately covered in review articles and text books (8-10).

In line with finite-difference practice in Newtonian fluid mechanics, we eliminate the pressure and body force components between (4.3) and (4.4) by simple differentiation and introduce the stream function ϕ from (4.2) and the vorticity ω through the equations

$$u = \frac{\partial \phi}{\partial y} \ , \qquad v = -\frac{\partial \phi}{\partial x} \ , \qquad (4.8)$$

---

x'_i is the position at time t' of the element that is instantaneously at x_i at time t.

$$\omega = -\left[\frac{\partial^2 \phi}{\partial x^2} + \frac{\partial^2 \phi}{\partial y^2}\right] . \tag{4.9}$$

Further, the decomposition

$$T_{ik} = S_{ik} + 2\eta_0 d_{ik} \tag{4.10}$$

retains the elliptic character of the stress equations of motion:

$$\eta_0 \nabla^2 \omega + \rho\left[\frac{\partial \phi}{\partial x}\frac{\partial \omega}{\partial y} - \frac{\partial \phi}{\partial y}\frac{\partial \omega}{\partial x}\right] = \frac{\partial^2 S_{xx}}{\partial x \partial y} + \frac{\partial^2 S_{xy}}{\partial y^2} - \frac{\partial^2 S_{xy}}{\partial x^2} - \frac{\partial^2 S_{yy}}{\partial x \partial y} . \tag{4.11}$$

The governing equations are now (4.5)-(4.7), (4.9)-(4.11), suitably non-dimensionalized. Various finite-difference methods (e.g. S.O.R. and Conjugate Gradient) have been successfully applied to these equations and for modest values of the elastic parameters numerical simulations have been obtained for many of the flow geometries discussed in §3. In keeping with all work in this field, the numerical solutions break down at frustratingly low values of the elastic parameters; for the case of the Maxwell model ($\lambda_2 = 0$) there is of course only one non-dimensional elastic parameter, W, given by (2.2) with $\lambda \equiv \lambda_1$.

We shall return to the so called 'High Weissenberg number' problem later.

(iii) *Boundary conditions*

In Newtonian fluid mechanics, where the governing equations are the Navier-Stokes equations, it is sufficient to specify the velocity components or the surface-force components over the boundary of the relevant domain plus the pressure at one point if no normal force has been specified anywhere. In the case of materials with memory, extra boundary conditions are required. For example, if the domain contains an entry region, it is necessary to specify the stress field at entry, or alternatively, the deformation history of the fluid entering the domain. Of course the latter completely determines the former (but not vice versa).

To illustrate how the extra boundary conditions enter into non-Newtonian flow problems, we shall concentrate on a

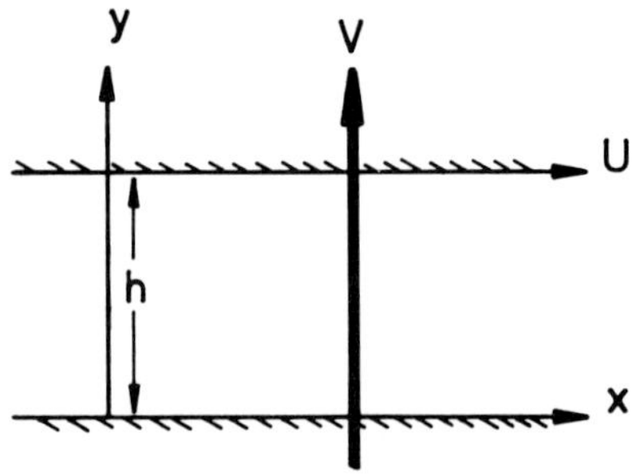

Fig 18. The flow geometry.

flow problem discussed by Giesekus (27). We shall approach the problem from a slightly different standpoint and also restrict attention to the simplified situation of creeping flow.

The two dimensional flow is shown schematically in Fig 18. The bottom plate at $y = 0$ is of infinite extent and is at rest. The top plate at $y = h$, also of infinite extent, moves with a velocity U in the x direction. There is injection with velocity V over $y = 0$ and suction at the same velocity over $y = h$. The boundary conditions are therefore

$$\begin{aligned} u(0) &= 0, \quad u(h) = U\,, \\ v(0) &= V, \quad v(h) = V\,. \end{aligned} \tag{4.12}$$

For illustration purposes, we confine attention to the Oldroyd B model given by (2.5) and remark that the upper-convected Maxwell model is given by $\lambda_2 = 0$.

A simple expansion procedure, taking the operator $\lambda_1 \nabla$ as small, leads to what we might call the second-order equivalent (SOE) model, with equations of state (to first order) given by

$$T_{ik} = 2\eta_0 d_{ik} - 2K\overset{\nabla}{d}_{ik}\,, \tag{4.13}$$

where $K = \eta_0(\lambda_1 - \lambda_2) \geqslant 0$. (4.13) is a simplified case of the second-order model of Coleman and Noll (7).

In view of the boundary conditions (4.12), we are led to take

$$u = u(y), \quad v = V(\text{a constant}), \tag{4.14}$$

in which case the relevant equations for the creeping flow of the Oldroyd B fluid are

$$\left[1 + \lambda_1 V\frac{d}{dy}\right]T_{xy} = \eta_0\left[1 + \lambda_2 V\frac{d}{dy}\right]\frac{du}{dy} , \tag{4.15}$$

$$\frac{dT_{xy}}{dy} = 0. \tag{4.16}$$

If we now introduce the non-dimensional variables

$$\lambda_1^* = \frac{\lambda_1 V}{h} , \quad \lambda_2^* = \frac{\lambda_2 V}{h} , \quad u^* = \frac{u}{U} , \quad y^* = \frac{y}{h} ,$$

$$T = \frac{h}{\eta_0 U} T_{xy} , \tag{4.17}$$

the equations become (on immediately dropping the star notation for convenience)

$$\left[1 + \lambda_1\frac{d}{dy}\right] T = \left[1 + \lambda_2\frac{d}{dy}\right]\frac{du}{dy} , \tag{4.18}$$

$$\frac{dT}{dy} = 0 , \tag{4.19}$$

subject to

$$u(0) = 0, \quad u(1) = 1 . \tag{4.20}$$

We also wish to apply a *stress* boundary condition at $y = 0$:

$$T(0) = T_0 . \tag{4.21}$$

From (4.18) and (4.19) we obtain

$$u' + \lambda_2 u'' = T_0 , \tag{4.22}$$

and for the second-order equivalent model

$$u' - Ku'' = T_0 , \tag{4.23}$$

where the dash refers to differentiation with respect to y and K is the non-dimensional SOE parameter.

In order to recapture the Newtonian solution, which is easily shown to be

$$u = y, \quad T_0 = 1 , \tag{4.24}$$

we require

$$T_0 \to 1 \text{ as } \lambda_2 \to 0 \text{ or } K \to 0 . \tag{4.25}$$

We note with interest that the solution for the Maxwell model ($\lambda_2 = 0$) is also given by

$$u = y, \quad T_0 = 1 , \tag{4.26}$$

and it is *not* possible to apply an arbitrary T_0 (subject of course to (4.25)) in this case. For the Maxwell model, the velocity boundary conditions alone are sufficient in this problem.

For the Oldroyd B model, on the other hand, the solution to (4.22) subject to the boundary conditions has the more general form

$$u = y + (1 - T_0)\left[\frac{1-e^{-\frac{y}{\lambda_2}}}{1-e^{-\frac{1}{\lambda_2}}} - y\right] , \tag{4.27}$$

where T_0 is arbitrary, subject only to $T_0 \to 1$ as $\lambda_2 \to 0$. Here we have a solution which reduces to the Newtonian solution ($u \to y$ when $\lambda_2 \to 0$) and which requires knowledge of the stress at $y = 0$ in its determination. *Without that knowledge the problem is underspecified.*

For the SOE, the solution for u is of the form

$$u = y + (1 - T_0)\left[\frac{1-e^{\frac{y}{K}}}{1-e^{\frac{1}{K}}} - y\right] , \tag{4.28}$$

where we recall that $K \geqslant 0$. The Newtonian solution is again recovered as $K \to 0$, but we note that the solution given in (4.28) does not have a Taylor series expansion in powers of K (and neither does (4.27) in λ_2 for that matter). In fact, a series solution of the differential equation (4.23) in powers of K would yield

$$u = y, \quad T_0 = 1 . \tag{4.29}$$

The above considerations would lead to the conclusion that the consistent use of the hierarchy equations would only be valid for $T_0 = 1$, and that for $T_0 \neq 1$ the hierarchy-type expansion of the general functional constitutive equation could not be used. Here we have a situation where no matter how slow the flow is (as determined by the magnitude of V in this context) or how low the fluid elasticity is, the relevant non-dimensional parameter (which incorporates both V and the material constants) cannot be used as an ordering parameter in the flow under consideration except for $T_0 = 1$. For $T_0 \neq 1$, the first order (Newtonian), second-order, third-order hierarchy of equations is not a permissible sequence of approximations to the general functional equation (4.1).

Finally, we remark that in current work on the numerical simulation of non-Newtonian flow involving differential models, the extra-boundary-condition problem is usually accommodated automatically, since the stress components are treated as dependent variables in such developments.

5. COMPARISON OF THEORY AND EXPERIMENT

Figs 19 and 20, which contain experimental results and numerical simulations for the same complex flows, show that significant progress has been made in marrying theory and experiment, at least so far as *qualitative* features of the flow are concerned. The numerical simulations were obtained using a finite-*difference* method (4), but finite-element and other techniques like the boundary-integral method have also been used by other research groups with at least equal success (9). However, it must be admitted that severe problems remain and numerical simulation has been unable to meet many of the challenges posed in the experimental sections (§3). The major problem is well known - all numerical algorithms break down at some (frustratingly low) value of W. The fundamental problem is now well documented and is under active consideration in many research groups at the present time (see, for example, 8,9). The unsuitability of certain constitutive models, bifurcation phenomena, change of type of the governing equations as W increases, and the interaction between discretization error and non-linear iteration have all been suggested as possible causes of the breakdown, but in many ways the problem is still an open one in urgent need of attention. Until this high-W problem is resolved, we cannot expect to be able to simulate numerically the dramatic changes brought about by fluid elasticity which formed the basis of the discussion in §3.

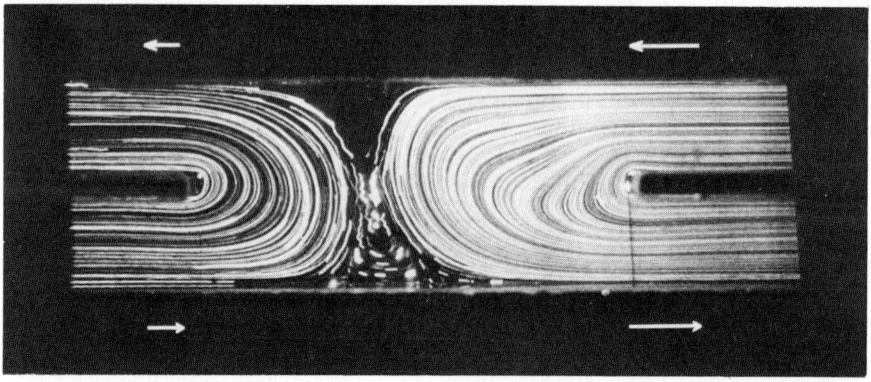

R = 1, W = 0.23

R = 0.75, W = 0.17

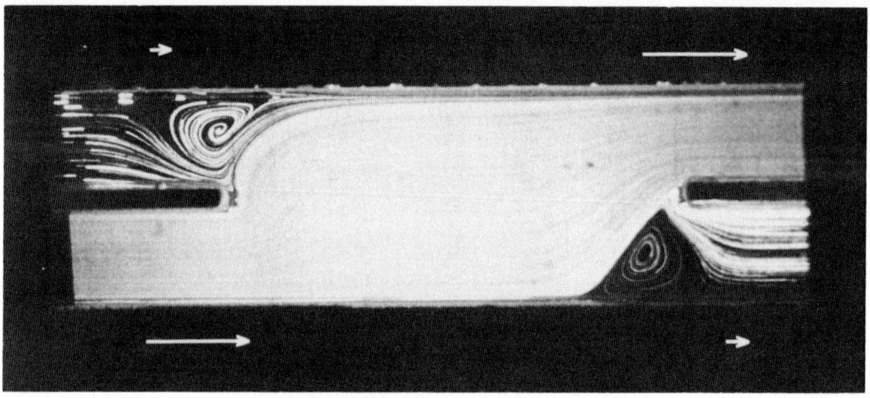

R = 1, W = 0.23

Fig 19. Mixing/separating flow for a Boger fluid; relative flow rates indicated by lengths of arrows; Reynolds number and Weissenberg number based on flow rate measurement in upper right arm (cf.4).

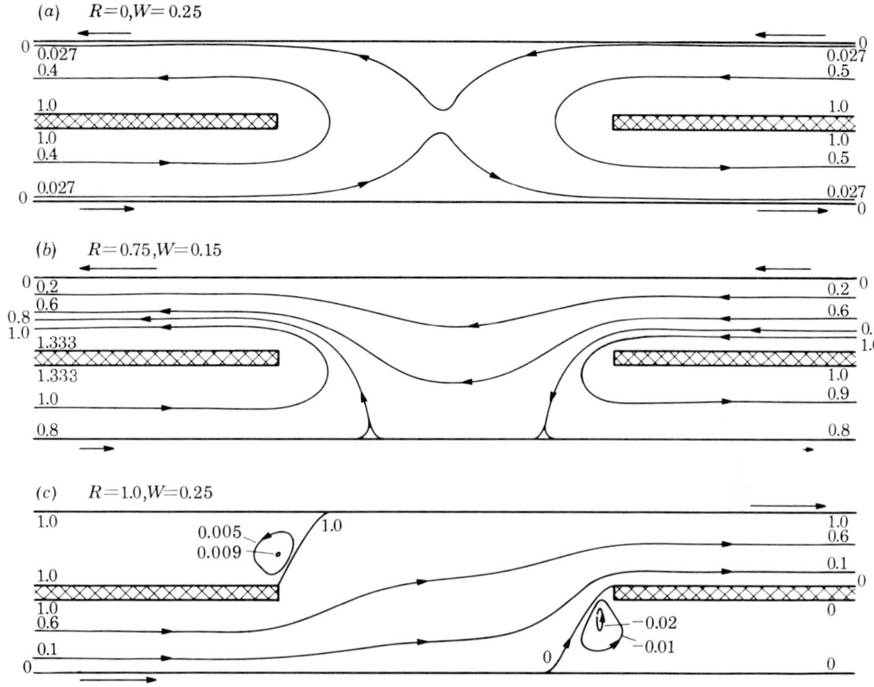

Fig 20. Numerical simulation for mixing/separating flow with relative flow rate in each arm indicated by length of arrows (cf.4).

REFERENCES

1. Lawler, J.V. et al. Paper presented at 9th Int. Congress on Rheology, Acapulco, Mexico, (1984).
2. Binnington, R.J., G.J. Troup and D.V. Boger, A low cost laser-speckle photographic technique for velocity measurement in slow flows. J. non-Newtonian Fluid Mechanics 12 (1983) 255-267.
3. Cochrane, T., K. Walters and M.F. Webster. Newtonian and non-Newtonian flow near a re-entrant corner. J. non-Newtonian Fluid Mechanics 10(1982) 95-114.
4. Cochrane, T., K. Walters and M.F. Webster. On Newtonian and non-Newtonian flow in complex geometries. Phil. Trans. Roy. Soc. A301(1981) 163-181.

5. Walters, K. and M.F. Webster. On dominating elastico-viscous response in some complex flows. Phil. Trans. Roy. Soc. A308(1982) 199-218.
6. Oldroyd, J.G. On the formulation of rheological equations of state. Proc. Roy. Soc. A200(1950) 523-541.
7. Coleman, B.D. and W. Noll. An approximation theorem for functionals, with applications in continuum mechanics. Arch. Rat. Mech. Anal. 6(1960) 355-370.
8. Crochet, M.J. and K. Walters. Numerical methods in non-Newtonian fluid mechanics. Ann. Rev. Fluid Mech. 15(1983) 241-260.
9. Crochet, M.J., A.R. Davies and K. Walters. '*Numerical Simulation of non-Newtonian Flow*' (Elsevier)(1984) xiii + 352.
10. Pearson, J.R.A. and S.M. Richardson (Eds) '*Computational Analysis of Polymer Processing*' (Applied Science Publishers) (1983) xiv + 343.
11. Boger, D.V. A highly elastic constant-viscosity fluid. J. non-Newtonian Fluid Mechanics 3 (1977) 87-91.
12. Jackson, K.P., K. Walters and R.W. Williams. A rheometrical study of Boger fluids. J. non-Newtonian Fluid Mechanics 14(1984) 173-188.
13. Choplin, L., P.J. Carreau and A. AitKadi. Highly elastic-constant viscosity fluids. Polymer Eng. and Science 23 (1983) 459-464.
14. Martischius, F-D. Das rheologische Verhalten von Polymerlösungen : Strömungsverfestigung in Scher-und Dehnströmungen. Rheologica Acta 21(1982) 288-310.
15. Prilutski, G., R.K. Gupta, T. Sridhar and M.E. Ryan. Model Viscoelastic liquids. J. non-Newtonian Fluid Mechanics 12(1983) 233-241.
16. Keentok, M., A.G. Georgescu, A.A. Sherwood and R.I. Tanner. The measurement of the second normal stress difference for some polymer solutions. J. non-Newtonian Fluid Mechanics 6(1980) 303-324.
17. Walters, K. Some modern developments in non-Newtonian Fluid Mechanics. Proc. 9th International Congress on Rheology, Acapulco (1984), Vol 1, 31-38.
18. Williams, P.R. and R.W. Williams. To be published.

19. Rawlinson, D.M. Ph.D. thesis, University of Wales, (1985) (In preparation).

20. Oldroyd, J.G. Non-Newtonian effects in steady motion of some idealized elastico-viscous liquids. Proc. Roy. Soc. A245(1958) 278-297.

21. Thomas, R.H. and K. Walters. The motion of an elastico-viscous liquid due to a sphere rotating about its diameter. Quart. J. Mech. Applied Maths. 17(1964) 39-53.

22. Richards, G.D. and P. Townsend. A finite-element computer model of the hole pressure problem. Rheol. Acta 20 (1981) 261-269.

23. Boger, D.V. Circular entry flows of inelastic and viscoelastic fluids. In '*Advances in Transport Processes*' Ed. A.S. Mujundar and R.A. Mashelkar (Wiley Eastern Ltd) (1982) 43-104.

24. Binding, D.M., M.J. Crochet, J. Dheur and K. Walters (To be published).

25. Saut, J.C. and D.D. Joseph. Fading memory. Arch. Rat. Mech. Anal. 81(1983) 53-95.

26. Malkus, D.S. and B. Bernstein. Flow of a Curtiss-Bird fluid over a transverse slot using the finite-element drift-function method. J. non-Newtonian Fluid Mechanics 16(1984) 77-116.

27. Giesekus, H. Zur formulierung der Randbedingungen in Strömungen viskoelastischer Flüssigkeiten mit Injektion und Absaugung an den Wänden. Rheol. Acta 9(1970) 474-487.

ACKNOWLEDGMENT

I wish to acknowledge the helpful assistance of a number of my colleagues, notably Dr. D. M. Binding, Mr. M. Brennan, Dr. A. R. Davies, Mr. R. E. Evans, Dr. R. S. Jones and Mr. D. M. Rawlinson. Other collaborative ventures are covered by the list of references.

NECKING PHENOMENA AND COLD DRAWING

L. J. Zapas and J. M. Crissman

Polymers Division
National Bureau of Standards
Gaithersburg, Maryland

1. Introduction

Many semicrystalline polymeric materials when subjected to uniaxial extension may at some point of the stretch history exhibit the phenomenon of 'necking'. That is, a tensile bar in uniaxial extension initially undergoes a homogeneous deformation which then changes into a nonhomogeneous motion. The occurrence of a neck is influenced by factors such as the ambient temperature, the rate at which the specimen is deformed, and the specimen preparation history which determines the degree of crystallinity. For a given chemical specie, the molecular weight and molecular weight distribution also influence the occurrence of necking. Until quite recently most of the published work has been concerned with experiments in which the necking phenomenon was accomplished using a 'hard' machine in which the specimen was extended at a constant rate of clamp separation. In early work [1,2][1] it was proposed that the necking resulted from a local temperature rise within the specimen. Later, Vincent [3] observed that necking occurred in polyethylene even at very low rates of

----

[1] Numbers in brackets correspond to references found at the end of this paper.

extension at room temperature. By measuring the temperature rise he concluded that heating effects become important only at strain rates in excess of $10^{-1}s^{-1}$. More recently Crissman and Zapas [4-6] in a series of three papers presented the results of uniaxial creep experiments on several different polyethylenes under varied load and temperature histories. In some instances the experiments lasted for several months before necking occurred. They also reported [7] results for isotactic polypropylene under a variety of loading histories, including constant rate of stress experiments. In most of the experiments the rate of deformation was small enough that the adiabatic heating was negligible. Several of these results will be discussed in this work in connection with the theory of Bernstein and Zapas [8] on the stability and cold drawing of viscoelastic bars which is an extension of the work of Ericksen [9] on the discontinuous deformations of solid elastic bars. From these works it is not possible to predict a priori when the material will exhibit the phenomenon of necking, but they do give an explanaton for the occurrence of the neck which depends upon the stretch history. In the viscoelastic case, neck formaton is predicted for a material which is subjected to long time creep under a constant engineering stess, the stress level being above a certain minimum value.

2. Theoretical Considerations

In this work we shall be concerned only with deformation histories involving uniaxial extension. We begin by considering a bar of a viscoelastic material which has been at rest at all times τ up to time $\tau=0$; its length in the undeformed state is ℓ_o. At time $\tau \geq 0$ it is subjected to a given stretch history such that it has a length $\ell(\tau)$. The stretch at time τ is given by $\mu(\tau)=\ell(\tau)/\ell_o$, where $\tau<t$. As was done in reference [8] we now define a jump stress $\sigma_J(\lambda,t;\mu(\tau))$ which is the value of the instantaneous engineering stress that the material would produce at time $\tau=t$ upon being subjected to an instantaneous superposed stretch of magnitude $\lambda-\mu(t)$ following a uniaxial strain history $\mu(\tau)$. In the remainder of the paper we shall for

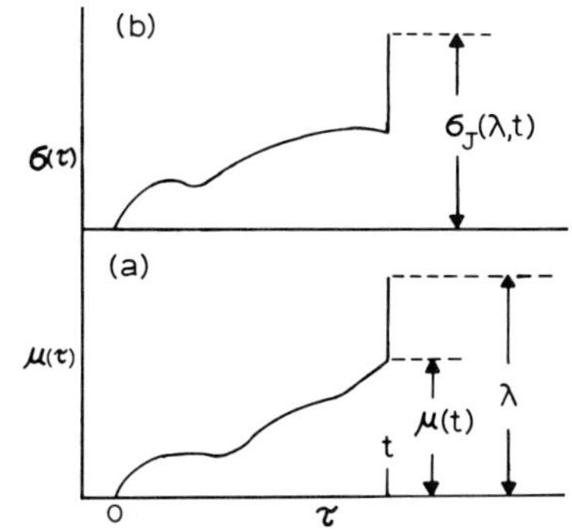

Figure 1. Schematic representation of (a) an instantaneous superposed stretch ($\lambda-\mu(t)$) and (b) the resulting jump stress $\sigma_J(\lambda,t)$.

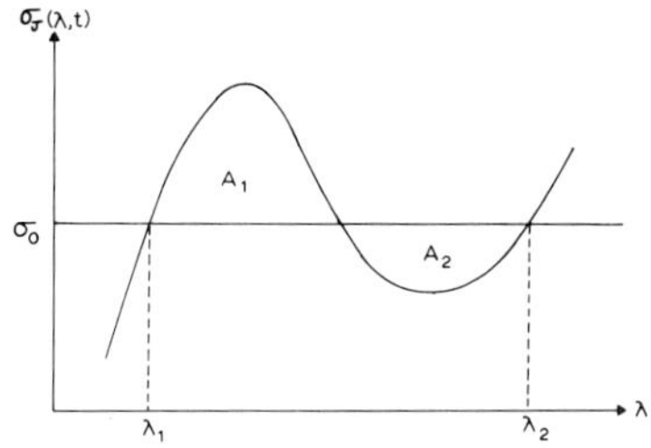

Figure 2. Schematic of an isochronal jump curve.

brevity leave the dependence of σ_J on the stretch history $\mu(\tau)$ understood and write simply $\sigma_J(\lambda,t)$. Such a behavior is shown schematically in Figure 1.

For a particular stretch history $\mu(\tau)$ we can obtain the value of $\sigma_J(\lambda,t)$ for different values of λ. If the isochronal jump curve can be represented schematically as in Figure 2, where σ_o is the response of the stress at $\mu(t)$, then for the case where the area A_1 is larger than the area A_2 the material will be stable. However, when A_1 becomes smaller than A_2 necking will occur. If now the quantity B is defined to be the following relation,

$$B = \int_{\ell_1}^{\lambda_2} \sigma_J\ (\lambda,t)\ d\lambda - (\lambda_2-\lambda_1)\ \sigma_o \tag{2.1}$$

then when B is positive the bar will continue to elongate in a homogeneous fashion. When B becomes zero, the bar reaches an incipient point of instability, at which point small flaws or inhomogeneities in the specimen will cause it to neck. When B becomes negative (in Figure 2 the condition $A_2>A_1$) the material will find itself in multiple stages of elongation. The analogy between cold drawing and a phase transition emerges clearly from this work and the work of Ericksen [9]. In the case of semicrystalline polymers, a material which is highly oriented can have a modulus as much as 10 to 30 times greater than that of the original material [4,7]. The main question is concerned with how one can obtain values for the jump stress, which experimentally is not possible to determine. One could use either a molecular theory, if one existed, or assume a constitutive equation and then design a behavioir which is consistent with the experiments. For materials for which the mechanical behavior can be described with the BKZ theory [10], hereafter referred to as BKZ materials, the jump stress is given by the following relation:

$$\sigma_J(\lambda,t)=w'(\lambda,t) - {}_0\int^t \frac{1}{\mu(\tau)} w_*'\left(\frac{\lambda}{\mu(\tau)},t-\tau\right) d\tau \tag{2.2}$$

In equation (2), $w'(\lambda,t)$[2] is the engineering stress that one obtains from a single step stress-relaxation experiment at time t after the imposition of a step in strain of magnitude $\lambda-\mu(t)$, and $\omega_*'(\cdot,\cdot)$ is the partial derivative of $\omega'(\cdot,\cdot)$ with respect to the second argument. Throughout the discussion we have assumed that all of the motions take place under isothermal conditions, and where applicable equation (2) can be used to describe instabilities in situations where the deformation $\mu(\tau)$ is a very slow one.

3. Experimental Procedures

Five different semicrystalline polymers were used in this study. The first was an experimental sample of isotactic polypropylene provided by the resin manufacturer. It had a viscosity average molecular weight of about 207,000 and contained 0.02 percent of stabilizer. Three of the samples were commercial grade high density linear polyethylenes having different molecular weights. Their weight average molecular weights were 99,000, 160,000, and 192,000, while their number average molecular weights were very nearly the same, being in the range from 15,000 to

----

[2] In this work we shall follow the same symbolic representation of the various quantities used by Bernstein and Zapas [8].

16,000. The fifth sample was a commercial grade linear ultra high molecular weight polyethylene (UHMWPE) which, based on the manufacturer's method of estimating molecular weight from dilute solution viscometry measurements, had a molecular weight of approximately $4X10^6$. As-received the first four samples were in the form of pellets, while the UHMWPE was in the form of a fine powder.

Flat sheets approximately 0.1 cm in thickness were prepared from each type of polymer by compression molding. The molding procedures differed depending upon the type of polymer, and the details of each molding operation can be found in the following references: Polypropylene-reference [7], the three high density linear polyethylenes-reference [6], and the UHMWPE-reference [11]. With the exception of the single step stress-relaxation experiments, the experiments were done at 23±0.5°C using a dumbbell shaped specimen cut with a die from the flat sheets. Because of the large deformations to which most of the specimens were subjected, the strain was determined with the aid of marks placed on the specimen and a cathetometer rather than an extensometer. The creep experiments were done under conditions of dead loading (constant applied engineering stress). Deformation histories involving constant rate of clamp separation and constant rate of loading were carried out on a servo-controlled hydraulic test machine.

For the single step stress-relaxation experiments the specimens were cut with a die which conformed to the geometry of the 'T-50' bar described in ASTM D599-61 [12]. In this geometry the width of the narrow section is constant over the entire portion of the specimen exposed between the grips. In order to avoid the possiblity that the attachment of an extensometer could contribute to the premature necking or failure of the speciment the strain was determined using a cathetometer.

4. Experimental Results

4.1 Single Step Stress Relaxation

According to the BKZ theory, a knowledge of the single step stress-relaxation behavior in uniaxial extension at various strains and times is sufficient to describe the behavior for any other strain history in uniaxial extension, within the range of strains and times for which the experimental data are available. In such an experiment the specimen, at a time t=0, is subjected to a sudden step in strain and the stress response is measured as a function of time. In the present work we have examined the stress relaxation behavior of the quenched polypropylene subjected to a series of steps in strain up to 7.6%. At strains greater than 7.6% the specimens always necked upon application of the step.

In Figure 3, isochrones of log true stress versus log strain are shown. The isochrones cover four decades in time. The behavior is highly nonlinear over nearly the entire range of strains examined. For the quenched polypropylene, the stress-strain behavior becomes linear only in the region of strain below 1%. Observe that at a strain of 7.6% the isochrones at long times are flattening to a zero slope, and the modulus has been reduced by a factor of at least 3 from what it would have been if the behavior were linear.

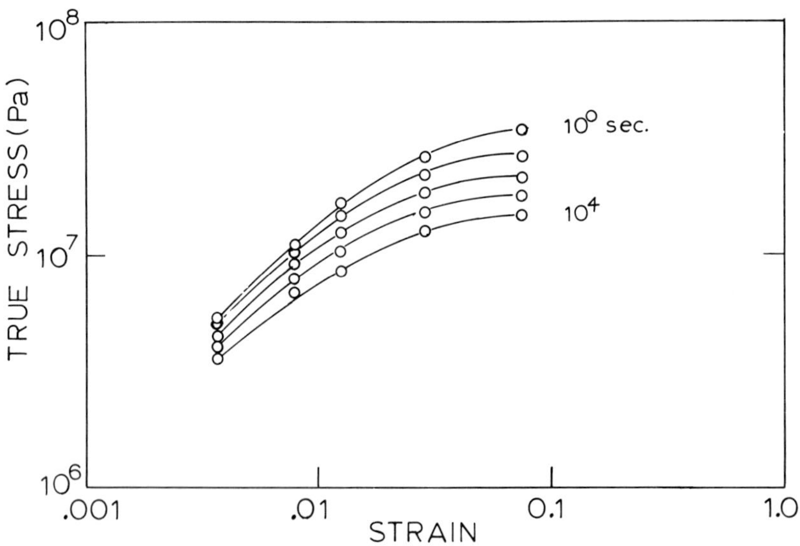

Figure 3. Isochrones of true stress versus strain covering 4 decades in time (quenched isotactic polypropylene).

4.2 Constant Rate of Strain Experiments

In this set of experiments the specimens were extended at a constant rate of clamp separation. In Figure 4, we show for one experiment (at a rate of ~$5.3 \times 10^{-5} s^{-1}$) the percent elongation of the specimen, as measured between fiducial marks placed on the gage portion of the dumbbell, as a function of time. Also shown is the elongation which the specimen would have had in the absence of end effects. Although the dashed line does not pass through the point (0,0), it can be seen that at strains greater than about 3% the actual data points deviate only very little from this line. Therefore, we can assume that the experiments were done at a constant rate of strain. In Figure 5, the measured engineering stress versus strain is shown for four different rates which vary by a factor of 10^3. The first result of interest is that the stress (for a given strain) varies only by a factor of 1.7 even though the rate varied by a factor 10^3. Also, the specimens did not neck at the strain, ε_M (the strain at the maximum stress), but necked at a larger strain, ε_N, as is indicated by the letter N in Figure 5. The maximum in the stress, σ_M, and the strain at the point of necking, ε_N, both occur at successively larger strains as the rate of elongation decreases. After necking occurred the stress showed a plateau before again rising at the very large strains. Values of ε_M, σ_M, ε_N, and σ_p as a function of the rate κ are summarized in Table 1.

Using equation (2), we have calculated the stress for a constant rate of elongation experiment, and the values obtained were somewhat lower than the experimental values, the maximum deviation being about 7%. According to the BKZ theory the existence of the maximum in the engineering stress suggests that $w'(\mu,t)$, at constant t, has a maximum.

4.3 Constant Rate of Loading

This experiment involves increasing the engineering stress at constant rate and the strain is determined as a function of time. In our experiments one characteristic of the constant rate of loading experiments was that as soon as the specimen necked it broke. In Figure 6 log strain versus log time is shown for specimens subjected to rates

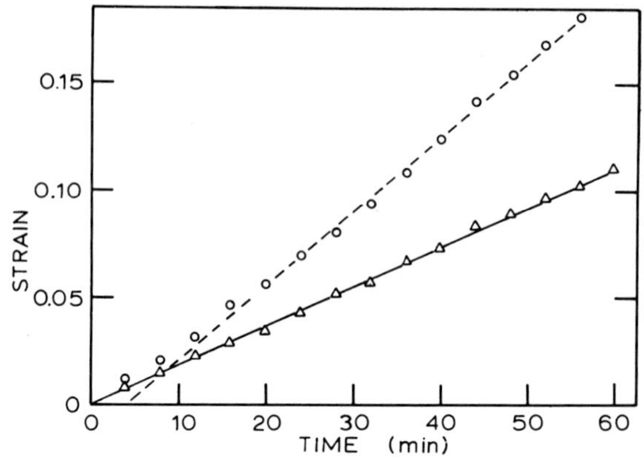

Figure 4. Strain versus time for a constant rate of clamp separation experiment ($\approx 5.3 \times 10^{-5}$). Circles - strain as determined in the specimen gage section only, triangles - strain as determined from the overall clamp separation.

TABLE 1

SUMMARY OF CONSTANT RATE OF STRAIN EXPERIMENTS

$\kappa^{(a)}$	$\sigma_M^{(b)}$	$\varepsilon_M^{(c)}$	$\varepsilon_N^{(d)}$	$\sigma_P^{(e)}$
6.8×10^{-3}	29.6	.08	.16	19.2
6.0×10^{-4}	27.0	.10	.19	17.4
5.1×10^{-5}	22.9	.10	.21	14.8
7.6×10^{-6}	19.2	.12	.24	12.8

(a) rate of strain per second

(b) maximum stress prior to necking (MPa)

(c) strain at the point of maximum stress

(d) strain at the point where necking occurred

(e) plateau value of stress (MPa)

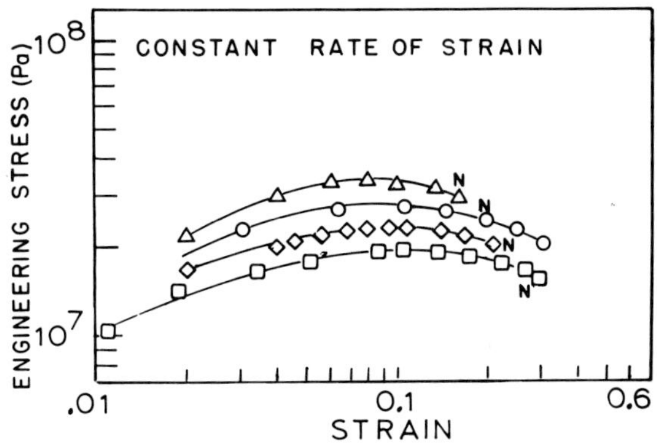

Figure 5. Log engineering stress versus log strain from constant rate of strain experiments. Triangles - $6.8\times10^{-3}s^{-1}$, circles - $6.0\times10^{-4}s^{-1}$, diamonds - $5.1\times10^{-5}s^{-1}$, squares - $7.6\times10^{-6}s^{-1}$. The letter n indicates the point at which a neck became visible.

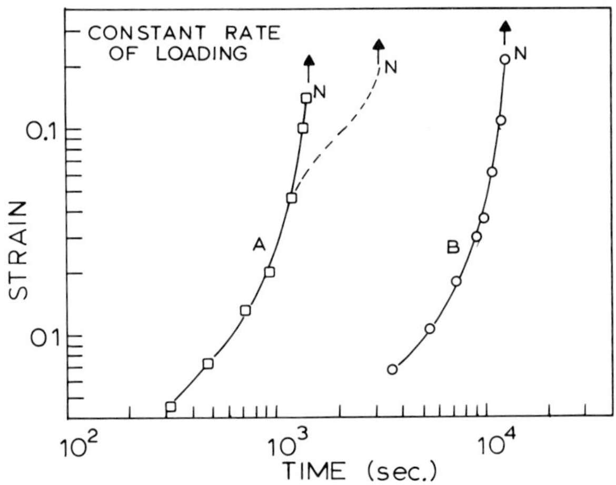

Figure 6. Log strain versus log time from constant rate of loading experiments. Squares - $1.83\times10^{-3}s^{-1}$, circles - $1.78\times10^{-4}s^{-1}$. The dashed line corresponds to an experiment done at a rate of $1.83\times10^{-3}s^{-1}$ until a stress of 22 MPa was reached at which time the rate of loading was reduced to $1.8\times10^{-5}s^{-1}$. The arrows indicate necking and fracture.

of loading which differed by a factor of ten. The curves A and B can be superimposed by a shift along the time axis alone. At a rate of $1.83 \times 10^{-4} s^{-1}$ fracture occurred at a strain of 0.22. As in the constant rate of strain experiment, necking occurred at a higher strain in the specimen subjected to the lower rate of loading. Since the two curves A and B can be superposed, it can be shown from the BKZ theory that σ versus ε can also be superposed by a shift along the stress axis. That this is true can be seen in Figure 7 where we have shown a plot of log stress versus log strain for the same two rates of loading. We note that for the constant rate of loading experiments the strain was determined with the aid of a cathetometer rather than an extensometer in order to avoid the possibility that the attachment of the extensometer could contribute to the premature necking of the specimen.

In a third constant rate of loading experiment, shown by the dashed line in Figure 6, the data were obtained using two different rates of loading. The initial rate was the same as that for the specimen subjected to a rate of $1.83 \times 10^{-3} s^{-1}$. Then when a stress of 2.20×10^{7} Pa was reached, the rate was changed to a much slower one

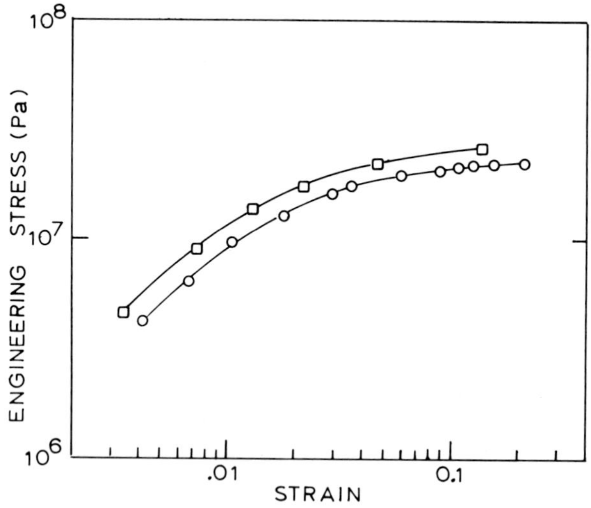

Figure 7. Log engineering stress versus log strain from constant rate of loading experiments. Squares - $1.83 \times 10^{-3} s^{-1}$, circles - $1.28 \times 10^{-4} s^{-1}$.

($\cong 1.8 \times 10^{-5} s^{-1}$). For this specimen necking and fracture occurred at a strain of 0.19, which was intermediate to the strains at which the other two specimens necked and then broke.

4.4 Creep

Shown in Figure 8 are creep curves for eight specimens subjected to engineering stresses in the range from 14.2 to 23.2 MPa. The letter N indicates the time and strain at which the specimens necked and the letter β the elongation at which the neck had propagated through the straight portion of the specimen. Except for the specimen having the smallest load (filled triangles) the creep experiment was continued up until the point at which fracture occurred. The two specimens having the largest loads (open circles and squares) both fractured almost immediately upon necking. The four specimens having intermediate loads all drew substantially before fracture occurred. For these four the maximum stretch (λ) was 14 to 15, which means that the point β (λ=6) represents only about 40% of the total creep attainable in this material at room temperature. The specimen under a stress of 15.2 MPa (hexagons) necked and then fractured during the early stages of drawing. Even though this specimen required much longer to neck than did three of the specimens having higher loads, it is interesting that it fractured sooner. A general observation for this material is that the point at which necking occurs shifts to a larger strain with decreased load.

Somewhat as an aside, we note that the creep behavior of the polypropylene is in many respects similar to that of the linear polyethylenes which have a weight average molecular weight in the range from 0.9×10^5 to 2.0×10^5, and a number average molecular weight of around 1.5×10^4. We have shown in previous work [5] that in uniaxial creep polyethylene exhibits at least two distinct failure modes. In one case, that of relatively large stresses, the material undergoes an instability which leads to necking, and then, depending upon the molecular weight and molecular weight distribution, may cold-draw substantially before breaking. In the second case, that of relatively small stresses, the specimen no longer exhibits necking, but rather fails by cracking. In the latter case the fracture can occur at strains of 5% or less. In the case of polyethylene, the transition from a mechanism of necking and cold-drawing to one of cracking occurs over the range of stresses from about 13.5 to 16.5 MPa, and at times near 10^5 seconds. We have observed that in the transition region a specimen may exhibit both necking and cracking at the same time, and in this region it becomes difficult to predict just which mechanism will ultimately cause the failure.

If the same two failure modes occur in the case of polypropylene, then it is clear from Figure 8 that the mechanism of cracking does not appear until times of nearly

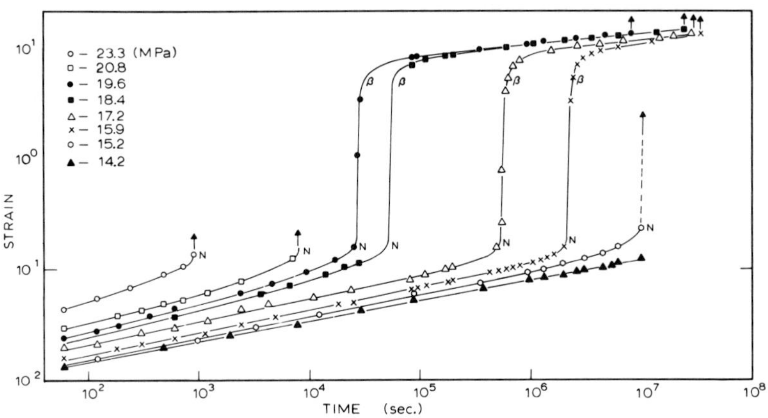

Figure 8. Log creep strain versus log time for specimens of polypropylene under a constant applied tensile load. The letter n indicates the strain at which necking occurred and β the elongation at which the neck had propagated the entire specimen. The arrows indicate fracture.

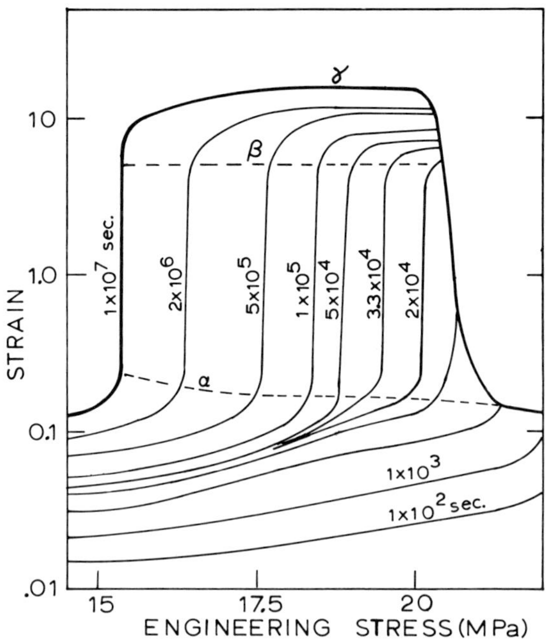

Figure 9. Isochrones of log strain versus engineering stress obtained from the creep data shown in Figure 8. The line α indicates necking and the line β has the same meaning as in Figure 8. γ denotes the failure envelope.

10^7 seconds are reached. For example, in the specimen loaded to a stress of 15.2 MPa numerous small cracks and/or crazes appeared beginning at a strain of 12-14 percent, and these may have contributed to its apparent early fracture. It is also true that the specimen stressed at 14.21 MPa (filled triangles) exhibited small cracks and/or crazes at a strain of about 10 percent. From our experience with the behavior of other semi-crystalline polymers it is possible that this specimen would have failed by cracking had the experiment been continued, but may have required up to two years longer under load. The creep data in Figure 8 have been replotted in Figure 9 in terms of isochrones of strain versus the applied engineering stress. Because of the high values of strain which this material achieves in the fully drawn state, the data are plotted semilogarithmically. We distinguish three regions bounded by the lines α, β, and γ. Below line α is the region in which the material deforms homogeneously, and upon reaching line α necking occurs. Between the lines α and β lies a region in which both necked and unnecked material coexist. Above line β is a third region where now the drawn material deforms more or less homogeneously. Line γ represents the envelope for fracture. One can consider Figure 9 to be a phase diagram for creep, where in the region between α and β two phases are present, one drawn material and the other undrawn material.

5. Discussion

We stated earlier that, within the range of strains and times covered in our experiments, the stresses in a constant rate of strain experiment calculated using the BKZ theory were in good agreement with the experimental data. In all cases the calculated values were lower than the observed values, but the maximum deviation was 7%. This rather good agreement encouraged us to examine more closely the model proposed by Bernstein and Zapas concerned with the stability and cold-drawing of viscoelastic bars. Using all the results presented here we have constructed by trial and error the diagram shown in Figure 10. This diagram is

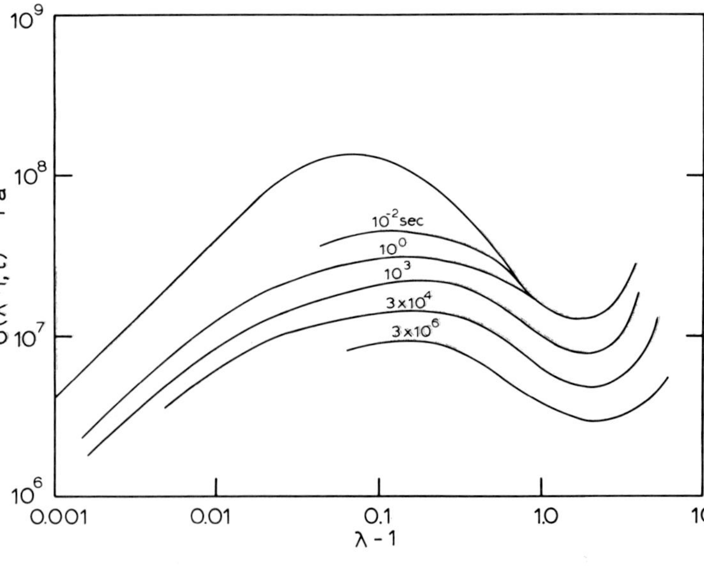

Figure 10. Isochronal stress-strain curves constructed from all of the available experimental results.

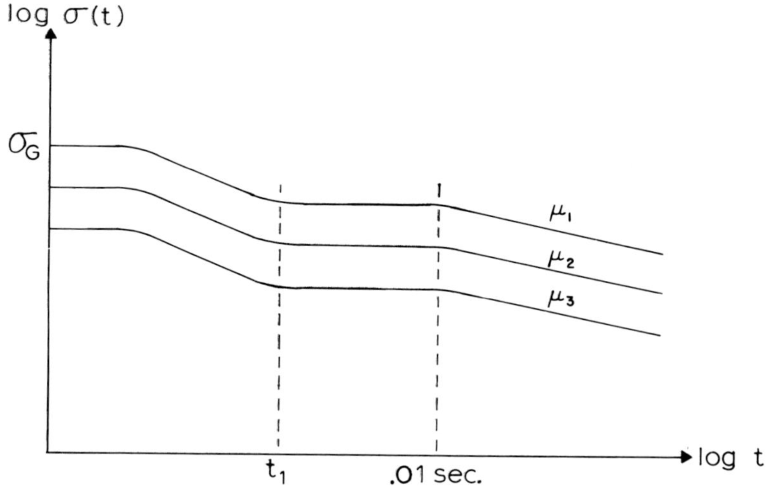

Figure 11. Log $\sigma(t)$ versus log time showing the transition region.

different than the one given in reference [7] in that it has the restriction that the experiment to obtain the single step stress-relaxation behavior remains stable up to strains of 7.6%, and also describes very closely the incipient point of instability for the other experiments.

We may rewrite equation (2) as follows:

$$\sigma_J(\tau,t)=w'(\lambda)\ -\int_0^t \frac{1}{\mu(t-\xi)}\ w_*'\left(\frac{\lambda}{(\mu(t-\xi)},\xi\right)\ d\xi \qquad (5.1)$$

where $\xi=t-\tau$. It can be observed that for values of ξ approaching 0, $\lambda/\mu(t-\xi)$ is different than unity and one needs to know the values of the instantaneous modulus or the modulus of the 'glassy state'. From the work of Passaglia and Martin [13] we estimate the value of the glassy state to be about 4.1 GPa, assuming the Poisson ratio to be 0.4. From dynamic measurements in flexure carried out at room temperature and at frequencies from 150 to 2500 Hz it was found that the flexure modulus was constant to within one percent and equal to about 1.62 GPa. Since this value is significantly lower than the modulus corresponding to the glassy state, and it is constant over the range examined, we assume that there is a transition region as is shown schematically in Figure 11, where the plateau starts at a time smaller than 0.01 second. In the constant rate of strain experiments the material, having been at rest up to time $\tau=0$, is strained such that for times $\tau>0$ $\mu(\tau)=1+\kappa\tau$, where κ is the rate of strain. In our experiments the maximum values of $\kappa\tau$ prior to the occurrence of necking were less than 0.3, so that to a first approximation we can take $\mu(^t/_{30})$ to be equal to $\mu(t)$. Equation (3) can then be written as:

$$\sigma_J(\lambda,t)=w'(\lambda,t) - \int_0^{\mathfrak{t}} \frac{1}{\mu(t)} w_*'(\frac{\lambda}{\mu(t)},\xi)d\xi - \int_{\mathfrak{t}}^{t} \frac{1}{\mu(t-\xi)} w_*'(\frac{\lambda}{\mu(t-\xi)}),\xi)d\xi \qquad (5.2)$$

or

$$\sigma_J(\lambda,t)=w'(\lambda,t) + \frac{1}{\mu(t)} w'(\frac{\lambda}{\mu(t)},0) - \frac{1}{\mu(t)}w'(\frac{\lambda}{\mu(t)},\mathfrak{t}) - \int_{\mathfrak{t}}^{t} \frac{1}{\mu(t-\xi)} w_*'(\frac{\lambda}{\mu(t-\xi)},\xi)d\xi \qquad (5.3)$$

where $\mathfrak{t} = t/30$.

We can make the further assumption that in the range of time between $\mathfrak{t}$ and t, and stretches λ and $\lambda/\mu(\tau)$, the isochrones of $\log w(\lambda,t)$ versus $\log t$ are very nearly parallel straight lines having a slope $-\eta_s$. We can then write for the last integral:

$$-\int_{\mathfrak{t}}^{t} \frac{1}{\mu(t-\xi)} w_*'(\frac{\lambda}{\mu(t-\xi)},\xi)d\xi = \eta_s \int_{\ln\mathfrak{t}}^{\ln t} \frac{1}{\mu(t-\xi)} w'(\frac{\lambda}{\mu(t-\xi)},\xi)d\ln\xi \qquad (5.4)$$

The right hand integral can then be approximated using the theorem of the mean.

For the single step stress-relaxation behavior we again assume that the material, having been at rest up to time $\tau=0$, is deformed at a constant rate of strain, reaching the maximum strain in a time of one second, after which it relaxes at the final strain. Starting with these assumptions we have, by trial and error, constructed a potential function (some isochrones are shown in Figure 10) from which the incipient points of instability were calculated reasonably well for all of our strain histories.

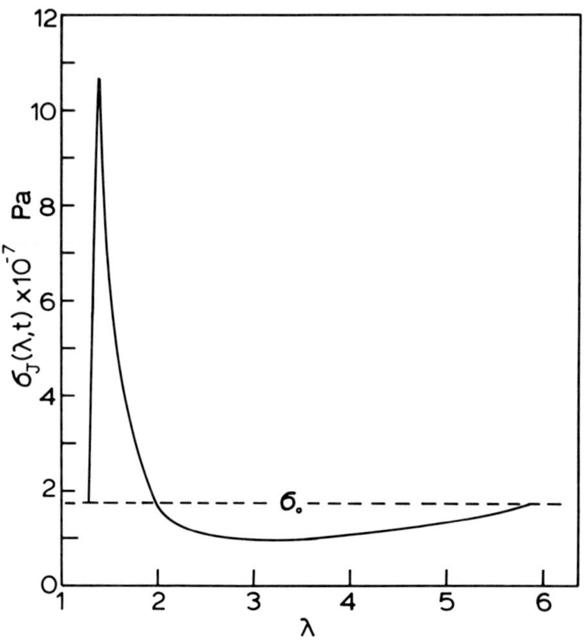

Figure 12. Jump curve showing $\sigma_J(\lambda, t)$ versus λ where $A_1=A_2$.

This result should not be surprising since to a large extent the same data was used to design the potential function. In Figure 12 the jump stress is shown for a constant rate of clamp separation experiment in which $\kappa=7.6\times10^{-6}s^{-1}$ and the instablility occurs at $\mu(t)=1.26$. It can be seen that in all of the experiments just described the strain at which necking occurs becomes larger the longer the time required to form a neck. To show that this behavior is not always the case we have performed two additional experiments on polypropylene. In one, a single step constant load was applied to the specimen until necking occurred. In the second experiment a multistep haversquare wave form of the same amplitude in stress (load) used in the first experiment was applied to the specimen until necking occurred. In each case the strain was monitored as a function of time. As was expected, in the multistep experiment the maximum strain (at the end of each cycle) for the cumulative time under load was always smaller than that for the corresponding time in the single step experiment. (The total time in the second experiment will in all cases be almost double the cumulative time under load.) Using some simplifying assumptions, as was done in reference [7], it was found that in the multistep experiment the strain at which necking occurs should be smaller than that for the single step case while the cumulative time under load in the multistep case should be much greater than the time at which necking occurs for the single step. From the above we conclude that the instability theory of Bernstein and Zapas may provide a reasonable description of the phenomenon of necking, but does not describe the further process of cold-drawing.

Next we present diagrams of the type shown in Figure 9 for several linear polyethylenes of different moecular weights and molecular weight distributions. All of the data shown in this paper were obtained at room temperature (23 ± 0.5°C). Again, the data are presented in terms of isochrones of log strain versus engineering stress. In the first three cases the number average molecular weight is

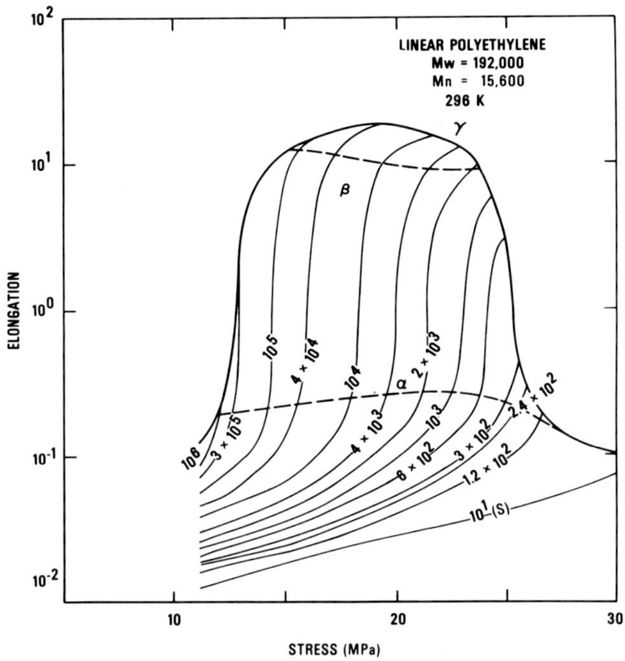

Figure 13. Isochrones of log strain versus engineering stress at 23°C for a linear polyethylene having M_n=15,600 and M_w=192,000 (from creep data). The symbols α, β, and γ having the same meaning as in Figures 7 and 8.

very nearly the same ($M_n \approx 15,500$) whereas the weight average molecular weight (M_w) varied by about a factor of 2. The creep behavior of the linear polyethylene having an M_w=192,000 is presented in Figure 13. As before, the region below the dashed line α represents the region of homogeneous deformations. At the line α necking occurs and between lines α and β the neck propagates through the specimen. Between β and λ the fully necked material again undergoes a 'homogeneous' deformation up to the line γ at which point fracture occurs. In this region a significant amount of creep can occur. For example, at an engineering stress of about 18 MPa the specimen doubled in length (λ changed from 10 to 20) before the specimen broke. At the highest loads the specimen necks and then breaks almost immediately, whereas as the opposite end of the load range where the load is small the specimen breaks before necking can occur.

A similar diagram is shown in Figure 14 for a second linear polyethylene having an M_w=160,000. In this case the range of applied stress for which the neck can propagate the entire length of the specimen is significantly narrower than in the previous example, and the maximum extension possible in the cold-drawn material is about 2.5 times the values it had at the point crossed by the line β (λ≅25). The behavior of a still lower molecular weight linear polyethylene is presented in Figure 15. In this case the specimens break as soon as necking occurs. Whereas in the case of polypropylene the strain at necking (line α in Figure 9) was found to increase monotonically as the time of the experiment increased, the locus of the point of necking for the three polyethylenes goes through a maximum.

Further examination of the properties of the fully extended fibers was done for the polyethylene having an M_w of 160,000. Specimens were cut from the transversely isotropic fibers and were relaxed at 23°C for one month. Single step stress-relaxation experiments were then done at strains of the order of one percent. It was found that the percent relaxation per decade of time was about the same as that of the undrawn material, although the absolute value of the stress-relaxation modulus was 30 to 50 times larger depending upon the amount of cold-drawing. Similar specimens, which were clamped with fixed ends on an Invar bar and placed in an oven for about 45 minutes at a temperature just below the melting point of the undrawn material, showed the following interesting behavior. While in the oven the fiber remained taut. However, upon removal from the oven the fiber actually bowed indicating that elongation had occurred during the cooling. The stress-relaxation modulus after the heat treatment was found to be lower by a factor of about 3 than that of the initial unheat-treated fiber. Over a period of one week

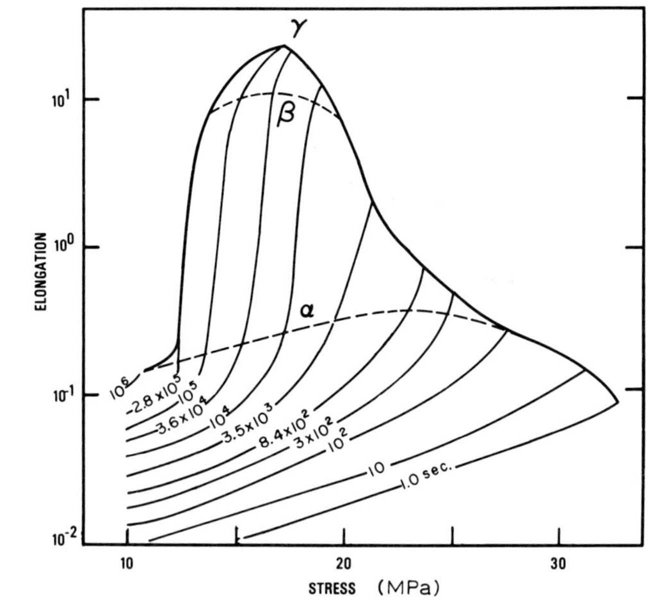

Figure 14. Isochrones of log strain versus engineering stress at 23°C for a linear polyethylene having M_n=15,000 and M_w=160,000 (from creep data). The symbols have the same meaning as in Figure 8.

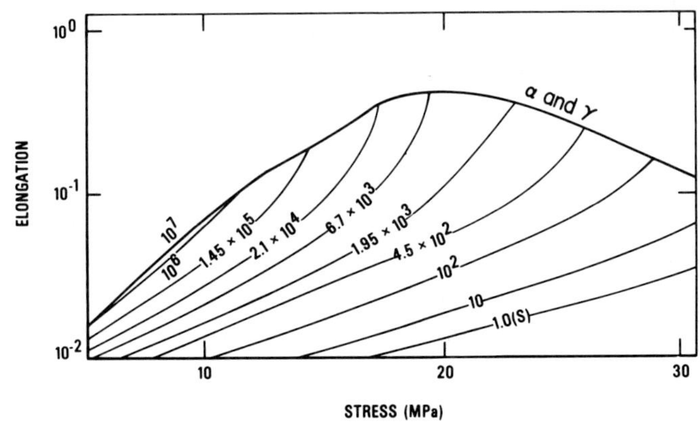

Figure 15. Isochrones of log strain versus engineering stress at 23°C for a linear polyethylene having M_n=15,300 and M_w=99,000 (from creep data). The symbols have the same meaning as in Figure 8.

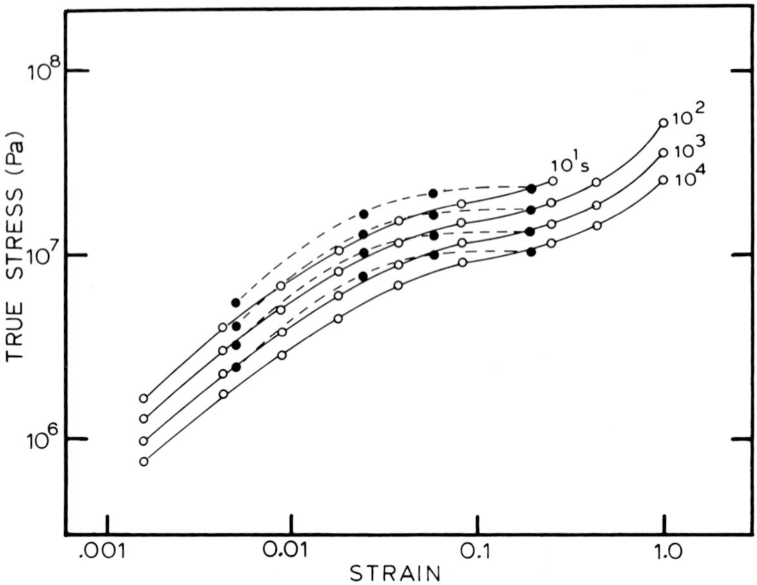

Figure 16. Log true stress versus log strain for ultra high molecular weight (UHMW) polyethylene at 23°C (from stress-relaxation data). Open circles - as molded, filled circles - annealed.

after the heat treatment the modulus showed a gradual increase and after one week had in some cases recoverd to a value nearly one half its original value.

Finally, an ultra high molecular weight linear polyethylene (UHMWPE), having a molecular weight of about 4.5×10^6 and prepared in a manner similar to that used to prepare the lower molecular weight polyethylenes, did not exhibit the phenomenon of necking at 23°C when subjected to the same stress or strain history described earlier. However, upon annealing for several days at a temperature slightly below the melting point, (which resulted in a somewhat higher percent crystallinity than that of the unannealed material) it did exhibit necking under the same conditions. In Figure 16 are shown isochrones of true stress versus strain obtained from single step stress-relaxation experiments. It can be observed that for the unannealed material the stress increases monotonically with increased strain. This result at least shows that the assumption made in previous instability theories [4,8,14], that necking will not occur if such a behavior exists, is consistent with our experiments. In the case of the annealed material, for which the isochrones tend to flatten, it can be concluded that the engineering stress will show a maximum, but little more can be said since at yet higher strains the material necks.

The various experimental results we have just presented suggest strongly that the instability leading to neck formation is associated with the nonlinearity of the mechanical behavior coupled with the time dependence. In experiments involving constant rate of clamp separation, necking occurs well beyond the point where the stress-strain curve goes through a maximum, and the amount beyond depends on the rate of strain. For polypropylene (see Figure 5), at the lowest rate of clamp separation shown, the strain at which the neck became visible was about three times the value at which the maximum in the stress occurred. At constant rate of loading, the stress-strain behavior is monotonically increasing (see Figure 7) up to the point of necking where immediately thereafter the specimens break. To describe the instability behavior in a hard device one might in some way use elastic behavior. However for a soft device it becomes very clear that one needs to consider the viscoelastic behavior of the material.

REFERENCES

1. Müller, F. M., Kolloidzeitschrift, 114, 59 (1949).
2. Marshall, I. and A. B. Thompson, Proc. Roy. Soc. A221, (1954).
3. Vincent, P. I., Polymer, 1, 7 (1960).
4. Zapas, L. J. and J. M. Crissman, Poly. Eng. & Sci., 19, No. 2, 104 (1974).
5. Crissman, J. M. and L. J. Zapas, Poly. Eng. & Sci., 19, No. 2., 99 (1974).
6. Zapas, L. J. and J. M. Crissman, ACS Symposium Series, No. 95, Durability of Macromolecular Materials, R. K. Eby, Editor, 301 (1978).

7. Crissman, J. M. and L. J. Zapas, Polymer, 24, 351, (1983).
8. Bernstein, B. and L. J. Zapas, J. Rheology, 25, 83 (1981).
9. Ericksen, J. L., J. Elasticity, 5, 191 (1975).
10. Bernstein, B., E. A. Kearsley, and L. J. Zapas, Trans. Soc. Rheol., 7, 391 (1963).
11. Zapas, L. J. and J. M. Crissman, Polymer, 25, 57 (1984).
12. American Society for Testing and Materials 1965 Book of ASTM Standards, Part 28.
13. Passaglia, E. and G. M. Martin, J. Res. NBS, 68A, 519 (1964).
14. Coleman, B. D. and L. J. Zapas, J. Poly. Sci., Poly. Phys. Ed, 17, 2215 (1979).

POLYMERIC LIQUIDS: FROM MOLECULAR MODELS TO CONSTITUTIVE EQUATIONS

R. Byron Bird

Chemical Engineering Department and
Rheology Research Center
University of Wisconsin-Madison
Madison, Wisconsin

1. INTRODUCTION

In order to solve flow problems in polymer fluid dynamics, it is necessary to have a constitutive equation that gives the stress tensor in terms of various kinematic tensors. This central problem can be partially solved by three different approaches: (a) Experimental measurements of rheological properties contribute in a major way, since the experimental data can be used to construct empirical constitutive equations that reproduce the measured quantities within reasonable limits; this is a tedious activity, however, because in any one rheometric experiment usually only one or two combinations of stress tensor components can be measured. (b) Continuum mechanics can supply information about constitutive relations in the form of various types of ordered expansions that follow from some general postulates about the stress tensor; however fluid dynamicists find expressions obtained in this way to be of limited value, because their use is restricted to a rather small range of the kinematical variables. (c) Molecular theories can be used to obtain rheological properties, and in some cases, complete constitutive equations, in terms of the parameters that characterize the mechanical model that purports to describe the most important features of the polymer molecule; however, the extent to which the kinetic theory can describe the polymer liquid response depends on how realistic the mechanical model is, but an increase in the realism of the model is accompanied by a formidable increase in the mathematical and computational complexity.

At the present time the best we can do is to combine the three approaches above. Molecular models and kinetic theory can suggest forms for constitutive equations that hold out the promise of being useful over wide ranges of kinematic variables. In obtaining the constitutive equations we make use of continuum mechanics relations involving various strain and rate-of-strain tensors. The constants or functions that occur in the final constitutive equation cannot be computed a priori, and hence they are evaluated by comparing the computed rheological properties with those obtained from experiments.

Here we give a brief introduction to the kinetic theory of polymeric liquids, including the molecular modeling, the setting up of the basic equations, the development of constitutive equations, and the uses of the final results. For more information see Volume 2 of "DPL" [1].

2. MOLECULAR MODELS

Polymers are molecules of very large molecular weight, and there is an enormous variety in their chemical architecture. In this discussion we consider for the most part those polymers which are formed by stringing together certain repeating groups of atoms in such a way that an extremely long chain is formed. Such a chain can be oriented in space, and in addition it has a lot of flexibility resulting from the large number of internal degrees of freedom. The chain can appear in a coiled-up configuration, or it can be stretched out into a long string-like configuration. In Figure 1 we show several mechanical models that have been used in the kinetic theory of flexible, chainlike polymers: (a) the Kirkwood-Riseman freely rotating chain is made up of "beads" joined together by massless "rods", but the angles between successive rods are fixed; (b) the Kramers freely jointed chain is composed of beads joined by rods, with universal joints at the beads; (c) the Rouse-Zimm chain is constructed from beads and "springs", with universal joints; and (d) the elastic dumbbell is made up of just two beads and a spring. These models have been listed in decreasing order of

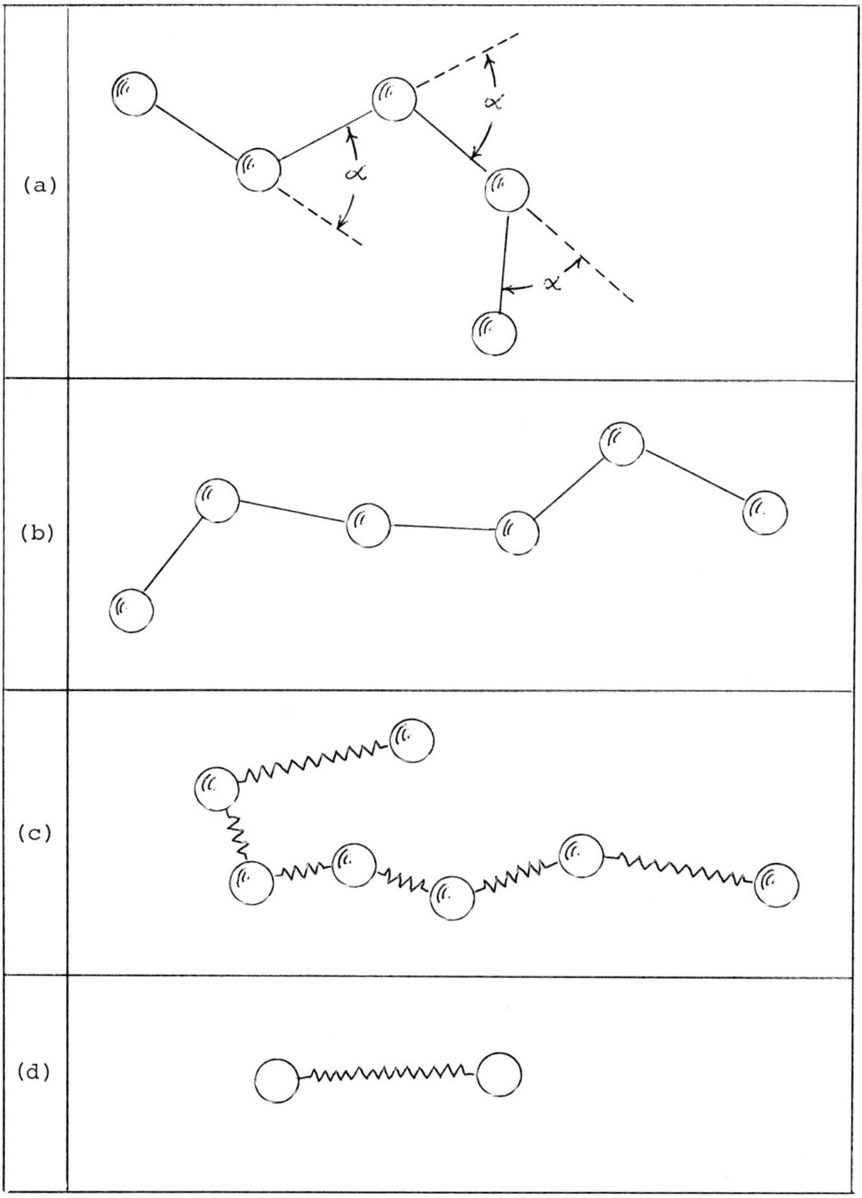

Figure 1. Mechanical models for representing chainlike polymer molecules in kinetic theory: (a) Kirkwood-Riseman chain, (b) Kramers, chain, (c) Rouse-Zimm model, (d) elastic dumbbell model

complexity, but each model is supposed to represent a long, extensible molecule that can undergo rotational and uncoiling motions. The constant angles between the rods in the Kirkwood-Riseman chain describes the inherent stiffness in polymer chains, which the other models do not describe. The elastic dumbbell model clearly is incapable of mimicking the responses associated with the many internal degrees of freedom of a polymer molecule; on the other hand, its very simplicity has made it a favorite for study, and it has in fact been quite useful in the development of an understanding of the relation between molecular motions and rheological phenomena.

To some extent the choice of molecular model depends on the contemplated use of the kinetic theory results. To describe the small-amplitude oscillatory experiment used for studying linear viscoelastic responses, one needs a model with many internal degrees of freedom, particularly if it is desired to describe effects at high frequencies; the Rouse-Zimm chain model has proven to be quite useful in this instance. To describe the steady-state shear flow experiment, where the overall rotation of the molecule is the principal molecular motion involved, it is not particularly necessary to use a model with many beads since the small scale motions are not activated in the flow. To describe an elongational flow experiment, in which the molecules are being stretched out considerably, a model that has a finite extensibility is required, such as the Kramers chain or an elastic dumbbell with a nonlinear spring that can be stretched only to a finite limit. If one wishes to be able to describe all three of these types of experiments quantitatively, then a more comprehensive model such as the Kirkwood-Riseman chain has to be used, since it can describe chain orientation, small scale motions, finite extensibility, and chain stiffness.

3. KINETIC THEORY FOR DILUTE POLYMER SOLUTIONS, ILLUSTRATED WITH THE ELASTIC DUMBBELL MODEL [DPL, Chapter 11]

In this section we show by example how one sets up the basic equations for the kinetic theory of polymers in a dilute solution. This presentation follows a simple, intuitive procedure, which enables one to formulate the problem by using only configuration-space ideas. This procedure is, however, inadequate if it is desired to use molecular models that contain constraints (fixed distances and/or angles); for that purpose it is preferable to use a more general phase-space treatment [2]. In addition the phase-space treatment makes it much clearer what assumptions are inherent in the more traditional configuration-space treatments.

The simplest model we can select is the elastic dumbbell of Fig. 1 in which two beads are joined by a spring. The interbead vector, from bead "1" to bead "2" is called $\underline{Q}$, and the tension in the spring is called $\underline{F}^{(c)}$. Each bead is presumed to experience a Stokes' drag force with a friction coefficient ζ (the latter is the drag force divided by the bead velocity relative to the fluid velocity). The polymer solution has a velocity field that is homogeneous: $\underline{v} = \underline{v}_0 + [\underline{\underline{\kappa}}\cdot\underline{r}]$, where $\underline{\underline{\kappa}}$ is in general a function of time, and $\underline{r}$ is the position vector. We want to know the probability, $\psi(\underline{Q},t)d\underline{Q}$, that a dumbbell in the solution will have a configuration in the range $d\underline{Q}$ about $\underline{Q}$ at time t.

To get the configurational distribution function $\psi(Q,t)$, we first write down the equation of continuity in the configuration space:

$$\frac{\partial\psi}{\partial t} = -\left(\frac{\partial}{\partial \underline{Q}} \cdot [\![\dot{\underline{Q}}]\!]\psi\right) \tag{1}$$

Here $[\![\dot{\underline{Q}}]\!]$ is the velocity-space average of the time rate of change of the configuration vector $\underline{Q}$. This quantity can be obtained by making a force balance on each of the beads and then subtracting the two equations in order to get the balance of the forces for the internal motion:

$$-\zeta\left([\![\dot{\underline{Q}}]\!] - [\underline{\underline{\kappa}}\cdot\underline{Q}]\right) - 2kT\frac{\partial}{\partial \underline{Q}}\ln\psi - 2\underline{F}^{(c)} = \underline{0} \tag{2}$$

The first term accounts for the hydrodynamic drag forces on the beads, the second term describes the Brownian motion forces, and the third term accounts for the forces transmitted through the connecting spring. When the force balance is solved for $[\![\dot{\underline{Q}}]\!]$, and when the latter is substituted into the continuity equation, we get the "diffusion" equation--the second order partial differential equation for the configurational distribution function:

$$\frac{\partial\psi}{\partial t} = -\left(\frac{\partial}{\partial \underline{Q}} \cdot \left\{[\underline{\underline{\kappa}}\cdot\underline{Q}]\psi - \frac{2kT}{\zeta}\frac{\partial}{\partial \underline{Q}}\psi - \frac{2}{\zeta}\underline{F}^{(c)}\psi\right\}\right) \tag{3}$$

Once the flow pattern and the spring force law have been chosen (i.e., $\underline{\underline{\kappa}}$ and $\underline{F}^{(c)}$ are specified), then the "diffusion equation" can in principle be solved to get the distribution of configurations of the dumbbells.

Macroscopic properties of the solution--mechanical, electrical, or thermal--can then be calculated by taking appropriate averages using the distribution function, $\psi(Q,t)$. Here we are concerned with the mechanical (i.e., rheological) properties, and the macroscopic property of interest is then the stress tensor, $\underline{\underline{\tau}}$, which appears in the equation of motion $\rho D\underline{v}/Dt = -\nabla p - [\nabla \cdot \underline{\underline{\tau}}] + \rho \underline{g}$. The stress tensor accounts for the various mechanisms by which forces are transmitted through the fluid: by the motion of the solvent molecules, by the tensions in the springs of the dumbbells, and by the momentum transport of the beads of the dumbbells. The contribution due to the tensions in the springs must be averaged over all the possible configurations of the dumbbells. The final expression is the "Kramers form" of the stress tensor:

$$\underline{\underline{\tau}} = -\eta_s \underline{\underline{\gamma}}_{(1)} - n < \underline{QF}^{(c)} > + nkT\underline{\underline{\delta}} \tag{4}$$

in which η_s is the solvent viscosity, $\underline{\underline{\delta}}$ is the unit tensor, n is the number density of dumbbells (i.e., polymer molecules) in the solution, $\underline{\underline{\gamma}}_{(1)} = \nabla\underline{v} + (\nabla\underline{v})^\dagger$ is the rate-of-strain tensor, and < ... > stands for $\int (\ldots)\psi(\underline{Q},t)d\underline{Q}$. Another expression for the stress tensor can be obtained by eliminating the spring force from the above expression by using the second tensorial moment of the diffusion equation. The expression thus obtained is in the "Giesekus form" of the stress tensor:

$$\underline{\underline{\tau}} = -\eta_s \underline{\underline{\gamma}}_{(1)} + \frac{n\zeta}{4} < \underline{QQ} >_{(1)} \tag{5}$$

The subscript on $< \underline{QQ} >_{(1)}$ indicates the "contravariant convected time derivative"

$$\underline{\underline{\tau}}_{(1)} = D\underline{\underline{\tau}}/Dt - \{(\nabla\underline{v})^\dagger \cdot \underline{\underline{\tau}} + \underline{\underline{\tau}} \cdot \nabla\underline{v}\}$$

The discussion above for the elastic dumbbell model has displayed the two main parts of the kinetic theory: (i) the derivation of a diffusion equation for the configurational distribution function, and (ii) the establishment of an expression for the stress tensor (or other macroscopic quantity).

For more complex models the kinetic theory has the same general structure; however, for models that have built-in constraints (fixed interbead distances, or fixed angles between the rods) generalized coordinates have to be used and the formulation of the theory becomes somewhat more involved [2]. It should be evident that in the modeling the hydrodynamics has been treated only very roughly, with an isotropic, linear hydrodynamic drag law being postulated. Further refinements can be introduced in which one accounts for the velocity field perturbation at one bead owing to the motion of the other beads ("hydrodynamic interaction"); there is a rather extensive literature on this particular modification of the elementary theory [DPL, §10.6]. Other modifications are nonisotropic drag coefficients [3], ζ varying with interbead distance [4], and inclusion of a dashpot in parallel with the spring [5].

4. DISTRIBUTION FUNCTION AND CONSTITUTIVE EQUATION FOR HOOKEAN DUMBBELLS [DPL, §10.4]

The simplest dumbbell model is that with a linear (Hookean) spring, so that the spring connector force law is $\underline{F}^{(c)} = H\underline{Q}$, in which H is the Hookean spring constant. For this particular type of dumbbell the diffusion equation for $\psi(Q,t)$ has been solved to give [6,7,8]:

$$\psi(\underline{Q},t) = \frac{(H/2\pi kT)^{3/2}}{\sqrt{\det \underline{\underline{\alpha}}}} \exp[-(H/2kT)(\underline{\underline{\alpha}}^{-1}:\underline{Q}\underline{Q})] \tag{6a}$$

$$\underline{\underline{\alpha}}(t) = \underline{\underline{\delta}} - \frac{1}{\lambda_H}\int_{-\infty}^{t} e^{-(t-t')/\lambda_H}\, \underline{\underline{\gamma}}_{[0]}(t,t')dt' \tag{6b}$$

in which $\lambda_H = \zeta/4H$ is the characteristic (relaxation) time for the elastic dumbbell solution, and $\underline{\underline{\gamma}}_{[0]}$ is a finite strain tensor. The finite strain tensor is defined in terms of the motion of the fluid. If a fluid particle is at a position $\underline{r}$ at time t , then its position $\underline{r}'$ at some previous time t' is given by the "displacement function": $\underline{r}' = \underline{r}'(\underline{r},t,t')$; a displacement gradient tensor $\underline{\underline{E}}$ can then be defined , with components $\underline{\underline{E}} = (\partial/\partial \underline{r}')\underline{r}$, and finally the finite strain tensor is given as $\underline{\underline{\gamma}}_{[0]} = \underline{\underline{\delta}} - \{\underline{\underline{E}}\cdot\underline{\underline{E}}^{\dagger}\}$. It should be noted that the kinematic tensor that appears in the differential equation is

$\underline{\underline{\kappa}} = (\nabla \underline{v})^{\dagger}$, but that the kinematic tensor appearing in the solution (i.e., the finite strain tensor) is related to $\underline{\underline{E}}$. The two tensors are not simply related as may be seen from:

$$\underline{\underline{E}}(t,t') = \underline{\underline{\delta}} + \int_{t'}^{t} \underline{\underline{\kappa}}''dt'' + \int_{t'}^{t}\int_{t''}^{t} \{\underline{\underline{\kappa}}''' \cdot \underline{\underline{\kappa}}''\}dt'''dt'' + \ldots \quad (7)$$

which is valid for the homogeneous flow field being considered here; the abbreviated notation $\underline{\underline{\kappa}}'' = \underline{\underline{\kappa}}(t'')$ has been used.

Once the solution to the partial differential equation for $\psi(\underline{Q},t)$ has been obtained, the configurational distribution function can then be substituted into the expression for the stress tensor to get the following constitutive equation:

$$\underline{\underline{\tau}} = -\eta_s \underline{\underline{\gamma}}_{(1)} + \int_{-\infty}^{t} \{\frac{nkT}{\lambda_H} e^{-(t-t')/\lambda_H}\} \underline{\underline{\gamma}}_{[0]}(t,t')dt' \quad (8)$$

This gives the stresses in the model fluid in terms of the strains that the fluid has suffered throughout all of the past history of the motion.

Alternatively one can get the constitutive equation by eliminating $< \underline{Q}\underline{Q} >$ between Eqs. (4) (written with $\underline{F}^{(c)} = H\underline{Q}$) and (5), to get:

$$\underline{\underline{\tau}} = -\eta_s \underline{\underline{\gamma}}_{(1)} + \underline{\underline{\tau}}_p \quad (9a)$$

$$\underline{\underline{\tau}}_p + \lambda_H \underline{\underline{\tau}}_{p(1)} = -nkT\lambda_H \underline{\underline{\gamma}}_{(1)} \quad (9b)$$

When this procedure is used, the constitutive equation for $\underline{\underline{\tau}}_p$, the polymer contribution to the stress tensor, is given as a differential equation; this result can be converted into the integral form above by using standard continuum mechanical arguments.

We have now shown, for one very elementary model, how one goes from the mechanical model to the complete constitutive equation. We see exactly how the various model parameters (ζ, H, n, η_s) arise in the final constitutive relation; two of the parameters occur grouped together in the time constant $\lambda_H = \zeta/4H$, and this quantity has to be determined ultimately from experimental data.

Having obtained a constitutive equation we next must test it against rheological measurements. From the Hookean-dumbbell constitutive equation we can obtain, for example, the stresses in a steady-state shear flow $v_x = \dot{\gamma}y$ with shear rate

$\dot{\gamma}$, and from these we can get the three viscometric functions: the viscosity, $\eta = \tau_{yx}/(-\dot{\gamma})$; the first normal stress coefficient, $\Psi_1 = (\tau_{xx} - \tau_{yy})/(-\dot{\gamma}^2)$, and the second normal stress coefficient, $\Psi_2 = (\tau_{yy} - \tau_{zz})/(-\dot{\gamma}^2)$. These functions are known experimentally to be monotone decreasing functions of the shear rate, but the model prediction gives constant values for all three of them. One can also make comparisons with data for unsteady-state shear flows, for elongational flows, and still other flows, and one finds that in most of these other experiments, the model prediction is rather poor. As a consequence one has to conclude that the simple elastic dumbbell model, although mathematically tractable, is not sufficiently good to describe even qualitatively some of the observed phenomena.

The next question is: how can the model be improved upon? At this point one can try several different modifications. One modification is to join N beads by N-1 Hookean springs to form a chain [DPL, Ch. 12]; this model does give much better results in linear viscoelasticity (for example, in small amplitude oscillatory motions), but it still yields viscometric functions that are independent of the shear rate. Another modification is the use of a finitely extensible nonlinear spring in the elastic dumbbell model since real molecules cannot be stretched indefinitely; we see in the next section that this simple modification of the basic model gives much better results. However, replacing the linear spring by a nonlinear one is only one of many modifications that have been tried [5].

5. AN APPROXIMATE CONSTITUTIVE EQUATION FOR FINITELY-EXTENSIBLE NONLINEAR ELASTIC (FENE) DUMBBELLS [DPL, §10.5]

A rather simple nonlinear spring force law is [9]:

$$F^{(c)} = \frac{H\underline{Q}}{1 - (Q^2/Q_0^2)} \qquad (Q \leq Q_0) \tag{10}$$

This kind of spring is Hookean at small values of the bead separation and cannot be stretched beyond a separation of Q_0. For this model there are two time constants: $\lambda_H = \zeta/4H$ and $\lambda_Q = \zeta Q_0^2/12kT$ (instead of the latter we use the ratio $b = 3\lambda_Q/\lambda_H = HQ_0^2/kT$). Comparisons of calculated rheological properties with experimental data [10,11] indicate that realistic values for the parameter b are in the range from 30 to 300. The stress-tensor expression is now (from Eq. (4)):

$$\underline{\underline{\tau}} = -\eta_s \underline{\underline{\dot{\gamma}}}_{(1)} - nH \left\langle \frac{\underline{Q}\underline{Q}}{1 - (Q^2/Q_0^2)} \right\rangle + nkT\underline{\underline{\delta}} \tag{11}$$

We would now like to eliminate the < ... > quantities as we did for Hookean dumbbells, but this obviously cannot be done. However, we can get an approximate constitutive equation if we replace Eq. (11) by:

$$\underline{\underline{\tau}} \doteq -\eta_s \underline{\underline{\dot{\gamma}}}_{(1)} - nH \frac{\langle \underline{Q}\underline{Q} \rangle}{1 - \langle Q^2/Q_0^2 \rangle} + (1 - \varepsilon b)nkT\underline{\underline{\delta}} \tag{12}$$

That is, we replace the average of the ratio by the ratio of averages, and then try to compensate for the error thus introduced by including an extra isotropic contribution containing the parameter ε. We then choose ε to be $2/[b(b + 2)]$ in order to make Eq. (12) be correct at equilibrium. Then one can eliminate from Eq. (5), Eq. (12), and the trace of Eq. (12) the two averages $\langle \underline{Q}\underline{Q} \rangle$ and $\langle Q^2/Q_0^2 \rangle$ to get the following approximate constitutive equation (almost the same equation was obtained by Tanner [12] by a different procedure):

$$Z\underline{\underline{\tau}}_p + \lambda_H \underline{\underline{\tau}}_{p(1)} - \lambda_H[\underline{\underline{\tau}}_p - (1 - \varepsilon b)nkT\underline{\underline{\delta}}]D \ln Z/Dt$$
$$= -(1 - \varepsilon b)nkT\lambda_H \underline{\underline{\dot{\gamma}}}_{(1)} \tag{13a}$$

$$Z = 1 + (3/b)[(1 - \varepsilon b) - \mathrm{tr}(\underline{\underline{\tau}}_p/3nkT)] \tag{13b}$$

In the limit that b becomes infinite (Hookean dumbbells), the quantity Z goes to unity, and the constitutive equation simplifies to Eq. (9); thus Z is seen to describe the influence of the nonlinear springs. Although Eq. (13) is an approximate constitutive equation, it leads to results which are in remarkably good agreement with those one gets by solving the diffusion equation numerically and then computing the components of the stress tensor [13].

The constitutive equation in Eq. (13) has been found to describe quite well the viscosity curves for several moderately concentrated polymer solutions [11], and it seems to describe a number of other rheological properties qualitatively.

It has been used to solve several flow problems [11,14]; once the flow problems have been solved to get the velocity field, the trace of Eq. (12) can be used to get the molecular stretching For this particular model then we can go through the entire sequence: molecular model → constitutive equation → solution to fluid dynamics problem → molecular stretching in a flow field. It is this sequence of activities that we would like to be able to carry out for more realistic molecular models.

6. MODEL OF INTERACTING KRAMERS CHAINS AS A MODEL FOR A POLYMER MELT

In the foregoing section it was seen that even for the very simple dilute solution model with nonlinear elastic dumbbells a constitutive equation could be obtained only by introducing a mathematical assumption. It is not surprising, therefore, that in developing a kinetic theory for a polymer melt, in which the polymer chains are always interacting with one another in a very complicated way, a number of simplifications have to be made. There are two kinds of theories for polymer melts: the older ones are "network" theories [15] in which the melt is envisioned as a temporary network formed by the transient junctions between the strands of the constituent molecules; the newer ones are the "mean field" theories [16,17], in which one modifies the dilute solution theory and looks just at the motion of a single molecule as it moves with its motion drastically curtailed by the other molecules in its immediate vicinity. In the network theories one has to make empirical statements about the rates of junction creation and destruction, whereas in the mean field theories one has to make empirical statements about the drag force on the beads and the Brownian motion forces--and both of these are now taken to be nonisotropic (the Brownian motion is presumed to be acting only along the polymer backbone, and the resultant back-and-forth wiggling is called "reptation").

If the polymer molecules are represented by Kramers chains, then the use of the nonisotropic hydrodynamic drag and the reptational Brownian motion lead after considerable effort to the following constitutive equation:

$$\underline{\underline{\tau}} = NnkT[\frac{1}{3}\underline{\underline{\delta}} - \int_{-\infty}^{t} \mu(t-t')\underline{\underline{A}}(t,t')\,dt' - \varepsilon\lambda\underline{\underline{\kappa}}:\int_{-\infty}^{t} \nu(t-t')\underline{\underline{\underline{\underline{B}}}}(t,t')\,dt'] \tag{14a}$$

$$\mu(s) = -(\lambda/2)d\nu/ds \tag{14b}$$

$$\nu(s) = \frac{16}{\pi^2\lambda}\Sigma_{\alpha,odd}\frac{1}{\alpha^2}e^{-\pi^2\alpha^2 s/\lambda} \tag{14c}$$

$$\underline{\underline{A}} = (1/4\pi)\int [1 + (\underline{\underline{\gamma}}^{[0]}:\underline{u}\underline{u})]^{-3/2}\underline{u}\underline{u}\,d\underline{u} \tag{14d}$$

$$\underline{\underline{\underline{\underline{B}}}} = (1/4\pi)\int [1 + (\underline{\underline{\gamma}}^{[0]}:\underline{u}\underline{u})]^{-3/2}\underline{u}\underline{u}\underline{u}\underline{u}\,d\underline{u} \tag{14e}$$

Here N is the number of beads in the Kramers bead-rod chain, $\lambda = N^{3+\beta}\zeta a^2/2kT$ is the time constant, a is the length of a rod in the chain, $\underline{\underline{\gamma}}^{[0]} = \{\underline{\underline{\Delta}}^{\dagger}\cdot\underline{\underline{\Delta}}\} - \underline{\underline{\delta}}$ is another finite strain tensor ($\underline{\underline{\Delta}}$ is the tensor inverse to $\underline{\underline{E}}$), $\underline{u}$ is a unit vector, and $d\underline{u}$ is an element of area on a unit sphere. The model contains a total of five constants: the number of beads in the chain N, the rod length a, the bead friction factor ζ, the "link tension coefficient" ε which is between zero and unity, and the "chain constraint exponent β which appears to be about 0.3 to 0.5; several of these parameters (ζ,β, and a) are lumped together in the time constant λ, which has to be determined from experiments.

In the final constitutive equation there are two integral terms, one containing a second order tensor $\underline{\underline{A}}$ and another containing a fourth order tensor $\underline{\underline{\underline{\underline{B}}}}$. Both integrals have the same structure: the integrand is a product of a term involving only the time constant of the fluid (μ or ν) and another term involving only the kinematics of the flow field ($\underline{\underline{A}}$ or $\underline{\underline{\underline{\underline{B}}}}$). Note that this same type of structure manifested itself in Eq. (8) for the elastic dumbbell. The term involving ε in Eq. (14) is traceable to the rigid connectors in the Kramers bead-rod chain model; it has been shown recently [19] that this model for a polymer melt will not show rod-climbing [27] unless $\varepsilon > 1/8$.

Extensive comparisons have been made between the rheological properties calculated for this model and the experimental measurements for monodisperse liquids made of chainlike polymers [18]. It is possible to choose values of N, λ, and ε to fit the experimental data for several different types of polymeric liquids (both undiluted polymers and very concentrated solutions). The constitutive equation given above has recently been used for making fluid dynamics calculations [20]. The theory has been extended to polydisperse systems by Curtiss [21]; this theory gives for the steady-state shear compliance $J_e^0 = \Psi_{1,0}/2\eta_0^2 \sim (\bar{M}_Z/\bar{M}_W)^4$, whereas one experimental data correlation gives an exponent of 3.7 [22].

7. USES OF THE KINETIC THEORY RESULTS

Most kinetic theories involve lengthy and tedious developments, and few people have the patience to work their way through them. However, as may be seen in the several examples cited above, the final results can often be cast in readily understandable form, and the constants that appear in the constitutive equation usually have simple physical meanings. The kinetic theory results that are now being obtained can be used in a number of ways:

a. Kinetic theory provides constitutive equations for numerical solution to flow problems. The numerical solutions of nontrivial flow problems require long and expensive computational work [23]. Realistic flow problems often involve very unusual combinations of stretching and shearing flows. The constitutive equations obtained from kinetic theory probably have a better chance of success then do empirical equations provided that sensible physical ideas are incorporated in the theory.

b. The constitutive equations generated by kinetic theory can often suggest useful empirical constitutive equations. In the kinetic theory development one is compelled to use rather simple molecular models and to make some mathematical and physical assumptions in the derivations. Nonetheless the constitutive equations produced by kinetic theory have suggested which kinds of kinematic tensors should be used, what kind of factorization may be profitable to explore in integral models, and what kinds of nonlinearities might appear in differential models.

c. Kinetic theory has also provided the basis for interrelating various rheological properties and also suggesting relations between rheological properties and optical properties (such as flow birefringence and light scattering) or transport properties (such as translational diffusivity).

d. Kinetic theory gives information about the molecular stretching and molecular orientation in flow fields, and this kind of information should be of interest in polymer processing.

e. By studying the rheological consequences of various kinds of modeling, one can begin to get information about the structure-property relationships. For example, the osculating multibead-rod model for rodlike macromolecules gives the interesting result that the second normal-stress difference is negative for large length-to-diameter ratios but positive for small ratios (see Note (c) in Table 1). Molecular theory also predicts that steady elongational viscosity for dilute solutions should be a monotone function of the elongation rate [1], whereas the same quantity for polymer melts should go through a maximum and then decrease with elongation rate [18].

f. Kinetic theory also gives information about some of the constants and functions that appear in various continuum mechanics expansions. As one example, we give in Table 1 the values of the constants in the retarded motion expansion (through third order) from several different molecular theories. For a wide range of models we find that the second- and third-order constants can be arranged in order of decreasing magnitude: $|b_2| > |b_{11}|$ (b_{11} changes sign for multibead rods), and $b_3 > b_{12} > b_{1:11}$. As another example, we point out the fact that for the multibead-rod model the kernel functions in a memory-integral expansion have been obtained [24].

g. Kinetic theory calculations provide important background material which will be necessary in order to check the techniques of the rapidly developing Brownian dynamics technique that will be useful for probing the details of molecular motions in polymeric liquids [25].

h. Kinetic theory can provide valuable information about the behavior of polymer blends. Most industrial polymers are polydisperse (they are mixtures of polymers of various molecular weights), and the kinetic theory of mixtures can show what the effect of polydispersity will be on various macroscopic properties.

i. In any field one of the main uses of theoretical developments is to suggest new experiments, and certainly the kinetic theory of polymers gives results from time to time that catalyze experimental activity. For example, according to the FENE dumbbell model there should be an overshoot in the mean-square end-to-end distance at the start up of a steady shear flow [11]; this suggests that one should do a light-scattering study of this flow system. The multibead-rod theory cited earlier suggests that there should be a change in the sign of the second normal-stress coefficient as one goes from short rods to long rods; perhaps this sign change could be observed in the tilted-trough experiment [26].

In conclusion, it should be clear that the field of polymer rheology is very much like a three-legged stool, with the three legs being represented by experiment, continuum mechanics, and kinetic theory. No one approach to the subject can supply all the information that is needed by the fluid dynamicist. Each approach makes a unique contribution and complements the other two. Kinetic theory results are only as good as the molecular models on which the calculations are based, and more often than not mathematical and physical approximations have to be made in the theoretical developments; physical insight and mathematical skill are both needed for doing kinetic theory research. The molecular approach to polymer science is a challenging subject with a bright and useful future.

TABLE 1

CONSTANTS IN THE RETARDED-MOTION EXPANSION FROM KINETIC THEORY CALCULATIONS[a] (MADE DIMENSIONLESS WITH a_i)

Order k	Constants	FENE Dumbbells (dilute solution)[b]	Multi-bead rods (dilute solution)[c]	Freely jointed bead-rod chain (melt)[d]
1	b_1/a_1	$\frac{\eta_s}{a_1} + \frac{b}{b+5}$	$\frac{\eta_s}{a_1} + [1 - \frac{2}{5}(1 - \frac{\lambda_2}{\lambda_1})]$	$\frac{1}{60} + \frac{\varepsilon}{90}$
2	b_2/a_2	$-\frac{b^2}{(b+5)(b+7)}$	$-\frac{3}{5}$	$-\frac{1}{600}$
	b_{11}/a_2	0	$-\frac{12}{35}(1 - \frac{\lambda_2}{\lambda_1})$	$-\frac{1}{1050} + \frac{\varepsilon}{1050}$
3	b_3/a_3	$\frac{b^3(2b+11)/(2b+7)}{(b+5)(b+7)(b+9)}$	$\frac{3}{5}$	$\frac{17}{100{,}800}$
	b_{12}/a_3	$\frac{4b^3/(2b+7)}{(b+5)(b+7)(b+9)}$	$\frac{12}{35}[1 + \frac{1}{2}(1 - \frac{\lambda_2}{\lambda_1})]$	$\frac{17}{117{,}600} - \frac{17\varepsilon}{352{,}800}$
	$b_{1:11}/a_3$	$\frac{3b^3/(2b+7)}{(b+5)^2(b+7)(b+9)}$	$\frac{3}{35}[1 + \frac{6}{5}(1 - \frac{\lambda_2}{\lambda_1})]$	$\frac{17}{705{,}600}$
	a_k	$nkT\lambda_H^{\;k}$	$nkT\lambda_1^{\;k}$	$NnkT\lambda^k$

Notes to TABLE 1:

[a] The coefficients tabulated here are for the retarded-motion expansion

$$\underline{\underline{\tau}} = -[b_1 \underline{\underline{\gamma}}_{(1)} + b_2 \underline{\underline{\gamma}}_{(2)} + b_{11}\{\underline{\underline{\gamma}}_{(1)} \cdot \underline{\underline{\gamma}}_{(1)}\} + b_3 \underline{\underline{\gamma}}_{(3)}$$
$$+ b_{12}\{\underline{\underline{\gamma}}_{(1)} \cdot \underline{\underline{\gamma}}_{(2)} + \underline{\underline{\gamma}}_{(2)} \cdot \underline{\underline{\gamma}}_{(1)}\} + b_{1:11}(\underline{\underline{\gamma}}_{(1)} : \underline{\underline{\gamma}}_{(1)}) \underline{\underline{\gamma}}_{(1)} + \ldots]$$

in which $\underline{\underline{\gamma}}_{(1)} = \nabla \underline{v} + (\nabla \underline{v})^{\dagger}$, and the higher-order rate-of-strain tensors are defined by

$$\underline{\underline{\gamma}}_{(n+1)} = \frac{D}{Dt} \underline{\underline{\gamma}}_{(n)} - \{(\nabla \underline{v})^{\dagger} \cdot \underline{\underline{\gamma}}_{(n)} + \underline{\underline{\gamma}}_{(n)} \cdot \nabla \underline{v}\}$$

Note that $b_1 \equiv \eta_0$, $b_2 \equiv -\Psi_{1,0}/2$, and $b_{11} \equiv \Psi_{2,0}$ where the subscript zero designates zero-shear-rate properties.

[b] The constants for the FENE dumbbell with no hydrodynamic interaction were obtained by R. C. Armstrong, 60, 724-728 (1974); the parameters λ_H and b are defined after Eq. (10).

[c] The constants for the multibead rod with Rotne-Prager-Yamakawa hydrodynamic interaction were obtained by X. J. Fan, using R. B. Bird and C. F. Curtiss, J. Non-Newtonian Fluid Mech., 14, 85-101 (1984). The time constants λ_1 and λ_2 are abbreviations for $\lambda_N(h, \xi^2)$ and $\lambda_N(2h, -\overline{\xi^2})$ in the Bird-Curtiss publication. For osculating beads the quantity $[1 - (\lambda_2/\lambda_1)]$ varies with the number of beads from -0.5000 for N=2 to -0.0284 for N=6, and from +0.0102 for N=7 to 0.3082 for N=70.

[d] The constants for an undiluted system of interacting Kramers bead-rod chains are taken from C. F. Curtiss and R. B. Bird, 74, 2016-2025, 2026-2033 (1981). The parameters N, λ, and ε are defined after Eq. (14).

REFERENCES

[1] R. B. Bird, O. Hassager, R. C. Armstrong, and C. F. Curtiss, Dynamics of Polymeric Liquids, Vol. 2, Kinetic Theory, Wiley, New York (1977).

[2] C. F. Curtiss and R. B. Bird, Physica 118A, 191-204 (1983).

[3] H. Giesekus, J. Non-Newtonian Fluid Mech., 11, 69-109 (1982); 12, 367-374 (1983); R. B. Bird and J. M. Wiest, Univ. of Wisconsin Rheology Research Center Report No. 90 (June 1984).

[4] R. I. Tanner, Trans. Soc. Rheol., 19, 557-582 (1975); X. J. Fan and R. B. Bird, Univ. of Wisconsin Rheology Research Center Report (Oct. 1984).

[5] G. G. Fuller and L. G. Leal, J. Non-Newtonian Fluid Mech., 8, 271-310 (1981).

[6] A. S. Lodge and Y. Wu, Rheol. Acta, 10, 539-553 (1971).

[7] P. H. van Wiechen and H. C. Booij, J. Engr. Math., 5, 89-98 (1971).

[8] H. King and D. F. James, J. Chem. Phys., 78, 4743-4748, 4749-4754 (1983); see also H. C. Booij, ibid. 80, 4571-4572 (1984).

[9] H. R. Warner, Jr., Ind. Eng. Chem. Fundamentals, 11, 379-387 (1972).

[10] R. L. Christiansen and R. B. Bird, J. Non-Newtonian Fluid Mech., 3, 161-177 (1977/1978).

[11] Y. Mochimaru, J. Non-Newtonian Fluid Mech., 9, 157-178, 179-194 (1981).

[12] R. I. Tanner, Trans. Soc. Rheol., 19, 37-65 (1975).

[13] X. J. Fan, J. Non-Newtonian Fluid Mech., 00, 000-000 (1985).

[14] Y. Mochimaru, J. Non-Newtonian Fluid Mech., 12, 135-152 (1983).

[15] A. S. Lodge, R. C. Armstrong, M. H. Wagner, and H. H. Winter, Pure and Appl. Chem., 54, 1349-1359 (1983).

[16] M. Doi and S. F. Edwards, J. Chem. Soc., Faraday Trans., II, 74, 1789-1801, 1802-1817, 1818-1832 (1978), 75, 38-54 (1979).

[17] C. F. Curtiss and R. B. Bird, J. Chem. Phys., 74, 2016-2025, 2026-2033 (1981); Physics Today, 37, No. 1,36-43 (1984).

[18] R. B. Bird, H. H. Saab, and C. F. Curtiss, J. Phys. Chem., 86, 1102-1105 (1982); R. B. Bird, H. H. Saab, and C. F. Curtiss, J. Chem. Phys., 77, 4747-4757 (1982); H. H. Saab, R. B. Bird, and C. F. Curtiss, ibid., 77, 4758-4766 (1982).

[19] O. Hassager, J. Rheol., 00, 000-000 (1985).

[20] D. S. Malkus and B. Bernstein, J. Non-Newtonian Fluid Mech., 16, 77-116 (1984).

[21] C. F. Curtiss (to appear in Chapter 19 of the Second Edition of Ref. [1]).

[22] M. Kotaka, Macromolecules, 17, 895-898 (1984).

[23] M. J. Crochet, A. R. Davies, and K. Walters, Numerical Simulation of Non-Newtonian Flow, Elsevier, Amsterdam (1984).

[24] R. B. Bird and C. F. Curtiss, J. Non-Newtonian Fluid Mech., 14, 85-101 (1984).

[25] P. J. Dotson, J. Chem. Phys., 79, 5730-5731 (1983); Ph.D. Thesis, University of Wisconsin (1984).

[26] M. Keentok, A. G. Georgescu, A. A. Sherwood, and R. I. Tanner, J. Non-Newtonian Fluid Mech., 6, 303-324 (1980).

[27] D. D. Joseph, G. S. Beavers, A. Cers, C. Dewald, A. Hoger, and P. T. Than, J. Rheol., 28, 325-346 (1984).

ACKNOWLEDGMENTS

The author wishes to thank the National Science Foundation (Grant No. CPE-8104705), the Vilas Trust Fund of the University of Wisconsin, and the John D. MacArthur Foundation for financial support during his work on kinetic theory of polymeric liquids. In addition the author wishes to acknowledge the assistance of Mr. X. J. Fan of Zhejiang University in the preparation of Table 1. Professor A. S. Lodge, Professor R. I. Tanner, Mr. J. M. Wiest, and Mr. H. H. Saab read the manuscript and suggested improvements. Finally generous thanks are due Professor C. F. Curtiss, with whom the author has collaborated extensively on research and teaching in the field of polymer kinetic theory.

ON SLOW-FLOW APPROXIMATIONS TO FLUIDS WITH FADING MEMORY

Bernard D. Coleman

Department of Mathematics
Carnegie-Mellon University
Pittsburgh, Pennsylvania

1. INTRODUCTION

According to the definition proposed by Walter Noll, an incompressible simple fluid is a material for which the mass density ρ never changes and the stress tensor $\mathbf{T}$ is determined, to within a pressure, by the history of the strain of past configurations relative to the present configuration.[1] The constitutive equation for the stress in such a material has the form[2]

---

[1] In his celebrated paper on the general theory of materials with memory [31], Noll introduced the concept of a simple material, developed a general mathematical theory of material symmetry for such substances, and defined a simple fluid to be a simple material for which the symmetry group (there called the "isotropy group") equals the unimodular group. That paper contains the equations which, for compressible fluids, correspond to the present equations (1.1)-(1.6). The minor modifications required of Noll's definition and constitutive equations when the fluid is incompressible were given in papers [14, 15, 19] in which Noll and I treated the dynamics of examples of the flows which I later [1] called the "viscometric flows". For further discussion see the Springer Tract, Viscometric Flows of Non-Newtonian Fluids, by Coleman, Markovitz, and Noll [10], and Volume III/3 of the Encyclopedia of Physics, The Non-Linear Field Theories of Mechanics, by Truesdell and Noll [35].

[2] Here $\mathbf{1}$ is the "unit tensor" for which $\mathbf{1u} = \mathbf{u}$ for each vector $\mathbf{u}$ in the translation space of the three-dimensional Euclidean space $\mathcal{E}$.

$$\mathbf{T}(t) + p(t)\mathbf{1} = \mathfrak{F}(\mathbf{F}^t), \tag{1.1}$$

where $\mathfrak{F}$ is a functional characterizing the material, and $\mathbf{F}^t$, called the <u>history</u> (up to t) <u>of the relative deformation gradient</u>, is a function defined on $[0,\infty)$ by the formula

$$\mathbf{F}^t(s) = \operatorname{grad}_{\mathbf{x}} \boldsymbol{\chi}_t(\mathbf{x}, t-s), \qquad s \in [0,\infty), \tag{1.2}$$

in which $\boldsymbol{\chi}_t(\mathbf{x}, t-s)$ is the location in a fixed Euclidean space $\mathcal{E}$ at time t - s of the material point that is at $\mathbf{x}$ in $\mathcal{E}$ at time t.

Using his principle of material frame-indifference, Noll [31] showed that the functional $\mathfrak{F}$ in (1.1) is determined by a functional $\mathfrak{H}$ such that, for each history $\mathbf{F}^*$ in the domain of $\mathfrak{F}$,

$$\mathfrak{F}(\mathbf{F}^*) = \mathfrak{H}(\mathbf{C}^*), \tag{1.3}$$

where $\mathbf{C}^* = \mathbf{F}^{*T}\mathbf{F}^*$,[3] and this functional $\mathfrak{H}$ is isotropic in the sense that, for each (constant) orthogonal tensor $\mathbf{Q}$,[4] it obeys the equation

$$\mathfrak{H}(\mathbf{Q}\mathbf{C}^*\mathbf{Q}^T) = \mathbf{Q}\mathfrak{H}(\mathbf{C}^*)\mathbf{Q}^T \tag{1.4}$$

as an identity in $\mathbf{C}^*$. By use of (1.3), the constitutive equation (1.1) can be written[5]

---

[3] I.e., $\mathbf{C}^*$ is the function on $[0,\infty)$ whose value at each $s \geq 0$ is $\mathbf{F}^*(s)^T\mathbf{F}^*(s)$, with $\mathbf{F}^*(s)^T$ the transpose of the tensor $\mathbf{F}^*(s)$.

[4] I.e., for each tensor $\mathbf{Q}$ obeying $\mathbf{Q}^T\mathbf{Q} = \mathbf{Q}\mathbf{Q}^T = \mathbf{1}$.

[5] For early applications of the relations (1.4)-(1.7) to tractable classes of flows see [1, 14, 15, 19, 20, 32] and the references given in [10, 35].

$$\mathbf{T}(t) + p(t)\mathbf{1} = \mathfrak{H}(\mathbf{C}^t), \tag{1.5}$$

where $\mathbf{C}^t$ is the function defined by

$$\begin{aligned}\mathbf{C}^t(s) &= \mathbf{F}^t(s)^T\mathbf{F}^t(s) \\ &= [\operatorname{grad}_{\mathbf{x}} \boldsymbol{\chi}_t(\mathbf{x},t-s)]^T[\operatorname{grad}_{\mathbf{x}} \boldsymbol{\chi}(\mathbf{x},t-s)], \qquad s \in [0,\infty).\end{aligned} \tag{1.6}$$

For each $s \geq 0$, the tensor $\mathbf{C}^t(s)$ is symmetric; it is called the <u>right Cauchy-Green tensor at time</u> $t-s$ <u>computed relative to the configuration at time</u> t. The condition that the density of the fluid never change is equivalent to the assertion that for each t,

$$\det \mathbf{C}^t(s) \equiv 1, \qquad s \in [0,\infty). \tag{1.7}$$

It follows from (1.4) that the value of $\mathfrak{H}$ at the constant function $\mathbf{1}^\dagger$ (for which $\mathbf{1}^\dagger(s) \equiv \mathbf{1}$) is a tensor of the form $-\nu\mathbf{1}$ [31], and as the pressure $p(t)$ in (1.5) is not determined by $\mathbf{C}^t$, the value of ν can be chosen to be zero so that

$$\mathfrak{H}(\mathbf{1}^\dagger) = \mathbf{0}. \tag{1.8}$$

In work published in 1960 [16], Noll and I considered the consequences of adding to the definition of a simple fluid an assumption of smoothness for the functional $\mathfrak{H}$ that in a very general way renders mathematical the concept of "gradually fading memory", i.e., the physical observation that, for materials called "viscoelastic", strains that occurred in the remote past have a sensible influence on the present stress, but this influence is weaker than that of recent strains of comparable magnitude. We obtained theorems justifying the intuitive idea that for very slow motions the equation (1.5) can be approximated by the constitutive equation for the stress in a Navier-Stokes fluid.[6] In other words, on assuming fading memory, one can prove that the constitutive equation

$$\underset{\sim}{T} + p\underset{\sim}{1} = 2\eta_0 \underset{\sim}{D} \tag{1.9}$$

with η_0 a material constant and with $\underset{\sim}{D}$ the symmetric part of the spatial gradient $\underset{\sim}{L}$ of the velocity $\underset{\sim}{v}$, i.e.,

$$\underset{\sim}{D} = \frac{1}{2}[\underset{\sim}{L}+\underset{\sim}{L}^T], \tag{1.10}$$

$$\underset{\sim}{L} = \underset{\sim}{L}(\underset{\sim}{x},t) = \text{grad}_{\underset{\sim}{x}} \underset{\sim}{v}(\underset{\sim}{x},t), \tag{1.11}$$

gives to the theory of perfect fluids a correction for viscoelastic effects that, in a precise sense, is complete to within order one in a retardation parameter α characterizing a change of time-scale.[7] In the same work, we observed that by carrying one step further the argument that yields this result, we could show that an equation of the form,

$$\underset{\sim}{T} + p\underset{\sim}{1} = \eta_0 \underset{\sim}{A}_1 + \beta \underset{\sim}{A}_1^2 + \gamma \underset{\sim}{A}_2, \tag{1.12}$$

gives to the classical relation (1.9) a correction for viscoelastic effects that is complete to within order two in the retardation parameter. In this last equation, called the constitutive equation of a *second-order fluid* or of a *fluid of second grade*, β and γ, like η_0, are material constants.

[6] *Navier-Stokes fluids*, also called (incompressible) *Newtonian fluids*, are the linearly viscous fluids familiar in classical hydrodynamics.

[7] A change of time-scale is effected by replacement of the elapsed time s in (1.6) by αs with $\alpha > 0$.

The tensors $\underset{\sim}{A}_N$, usually called Rivlin-Ericksen tensors,[8] are related to the velocity gradient $\underset{\sim}{L}$ by the recursion formulae,

$$\left.\begin{aligned} \underset{\sim}{A}_1 &= 2\underset{\sim}{D} = \underset{\sim}{L} + \underset{\sim}{L}^T, \\ \underset{\sim}{A}_N &= \dot{\underset{\sim}{A}}_{N-1} + \underset{\sim}{A}_{N-1}\underset{\sim}{L} + \underset{\sim}{L}^T\underset{\sim}{A}_{N-1}, \end{aligned}\right\} \qquad (1.13)$$

in which a superposed dot denotes a material time-derivative. The formulae (1.13) yield

$$\underset{\sim}{A}_2 = \dot{\underset{\sim}{A}}_1 + \underset{\sim}{A}_1\underset{\sim}{L} + \underset{\sim}{L}^T\underset{\sim}{A}_1 = \dot{\underset{\sim}{L}} + \dot{\underset{\sim}{L}}^T + \underset{\sim}{L}^2 + \underset{\sim}{L}^{T^2} + 2\underset{\sim}{L}^T\underset{\sim}{L}, \qquad (1.14)$$

$$\underset{\sim}{A}_1^2 = \underset{\sim}{L}^2 + \underset{\sim}{L}^{T^2} + \underset{\sim}{L}\underset{\sim}{L}^T + \underset{\sim}{L}^T\underset{\sim}{L}. \qquad (1.15)$$

Another formula for the tensors $\underset{\sim}{A}_N$ is [30]:

$$\underset{\sim}{A}_N = (-1)^N \frac{d^N}{ds^N}\underset{\sim}{C}^t(s)\Big|_{s=0} \qquad (1.16)$$

[8] Rivlin and Ericksen [33] defined the tensors $\underset{\sim}{A}_N$ by (1.13) and used them to develop a theory of isotropic materials of differential type. A Rivlin-Ericksen fluid is an incompressible material obeying a constitutive equation of the form $\underset{\sim}{T} + p\underset{\sim}{1} = \mathfrak{f}(\underset{\sim}{A}_1, \underset{\sim}{A}_2, \ldots, \underset{\sim}{A}_n)$ with $\mathfrak{f}$ a smooth, tensor-valued, isotropic function of n tensorial arguments. The Navier-Stokes fluid (1.9) and the second-order fluid (1.12) are special examples of Rivlin-Ericksen fluids, and it follows from (1.16) that each Rivlin-Ericksen fluid is a simple fluid. However, Rivlin-Ericksen fluids do not have gradually fading memory. (When a Rivlin-Ericksen fluid is held at rest with $\underset{\sim}{L} = \underset{\sim}{0}$ for an interval of time, no matter how brief, the stress in the fluid reduces to a pressure that is independent of the fluid's prior history.) If one were to start with the theory of such fluids (rather than with the theory of simple fluids with gradually fading memory), an elementary argument, involving little more than dimensional analysis and known representation theorems for linear and quadratic isotropic functions, could be invoked to show that for slow motions (1.9) and (1.12) give first- and second-order approximations to more general Rivlin-Ericksen fluids. (See, for example, Langlois' study [25] of the steady flow of Rivlin-Ericksen fluids between rotating spheres.) On the other hand, the theorems on retardation obtained in [16] show that the special Rivlin-Ericksen fluids, (1.9) and (1.12), although not themselves fluids with gradually fading memory, do approximate such truly viscoelastic fluids for highly retarded motions.

In this paper, in order to make clear the way Navier-Stokes and second-order fluids approximate fluids with memory and to show how these approximations are related to others, such as that behind the Boltzmann-Volterra theory of linear viscoelasticity, I shall summarize the hypotheses and some of the conclusions of the theory of fading memory. In the interest of brevity, I shall not attempt to describe the theory in its most modern and general form, or even at the level of generality of the original formulation of the theorems on retardation.[9]

2. FADING MEMORY

A positive, continuous, monotone-decreasing function h on $[0,\infty)$ here will be called an <u>influence function of order</u> r if $s^{r+\frac{1}{2}}h(s)$ is monotone decreasing for large s and

$$\lim_{s\to\infty} s^{r+\frac{1}{2}}h(s) = 0. \tag{2.1}$$

(For example, $(s+1)^{-\ell}$ is an influence function of order r if $\ell > r+\frac{1}{2}$, and $e^{-\ell s}$ with $\ell > 0$ is an influence function of all orders.) For a given influence function of positive order, the measurable symmetric-tensor-valued functions $\tilde{G}^*$ on $[0,\infty)$ for which the quantity,

$$\|\tilde{G}^*\| = \left[\int_0^\infty |\tilde{G}^*(s)|^2 h(s)^2 ds\right]^{1/2}, \tag{2.2}$$

[9]The generality sought here is approximately that of the following works on fluids with fading memory: [4, 7, 8, 9, 17, 18, 29].

is finite form a Hilbert space H with the inner product[10]

$$\langle \mathbf{G}_1^*, \mathbf{G}_2^* \rangle = \int_0^\infty \mathbf{G}_1^*(s)\cdot\mathbf{G}_2^*(s)\,h(s)^2\,ds. \tag{2.3}$$

Clearly, in the evaluation of the norm $\|\mathbf{G}^*\|$ of a function $\mathbf{G}^*$ in H, greater emphasis is placed on the values $\mathbf{G}^*(s)$ at small s than on the values at large s. The space H contains the constant function $\mathbf{1}^\dagger$, called the "rest history" because it equals the history $\mathbf{C}^t$ when material has been subjected to only rigid rotations. Two functions $\mathbf{G}_1^*$ and $\mathbf{G}_2^*$ in H are considered the same if $\|\mathbf{G}_1^* - \mathbf{G}_2^*\| = 0$, i.e., if $\mathbf{G}_1^*(s) = \mathbf{G}_2^*(s)$ for almost all s.

The *hypothesis of fading memory* is the assumption that (*i*) there is an influence function of order $r > 2$ such that the domain $\mathbb{D}$ of $\mathfrak{H}$ is the set of functions that lie in the space H corresponding to h and that have, for almost all s, values $\mathbf{C}^*(s)$ that are not only symmetric tensors but are also positive-definite and unimodular,[11] and (*ii*) $\mathfrak{H}$ is twice Fréchet differentiable at the point $\mathbf{1}^\dagger$ in $\mathbb{D} \subset H$ in the sense that there is a continuous linear functional $\delta\mathfrak{H}$ defined on H and a continuous, symmetric, bilinear functional $\delta^2\mathfrak{H}$ defined on $H \times H$ such that, for each $\mathbf{G}^*$ in H for which $\mathbf{1}^\dagger + \mathbf{G}^*$ is in $\mathbb{D}$,[12]

$$\mathfrak{H}(\mathbf{1}^\dagger + \mathbf{G}^*) = \mathfrak{H}(\mathbf{1}^\dagger) + \delta\mathfrak{H}(\mathbf{G}^*) + \tfrac{1}{2}\delta^2\mathfrak{H}(\mathbf{G}^*, \mathbf{G}^*) + o(\|\mathbf{G}^*\|^2). \tag{2.4}$$

Here, as usual, $o(\|\mathbf{G}^*\|^2)$ represents an unspecified function of $\mathbf{G}^*$ obeying

---

[10] For tensors $\mathbf{A}$ and $\mathbf{B}$, $\mathbf{A}\cdot\mathbf{B} = \mathrm{tr}(\mathbf{A}\mathbf{B}^T)$ and $|\mathbf{A}| = \sqrt{\mathbf{A}\cdot\mathbf{A}}$.

[11] A tensor $\mathbf{A}$ is called unimodular if $|\det \mathbf{A}| = 1$. If a symmetric tensor $\mathbf{C}$ is both unimodular and positive definite, then $\det \mathbf{C} = 1$.

[12] Recall that $\mathfrak{H}(\mathbf{1}^\dagger)$ has the form $-\nu\mathbf{1}$, and we can set ν equal to zero, if we wish, as is done in (1.8).

$$\lim_{\|\underset{\sim}{G}^*\|\to 0} \frac{o(\|\underset{\sim}{G}^*\|^2)}{\|\underset{\sim}{G}^*\|^2} = \underset{\sim}{0}. \tag{2.5}$$

It follows from (1.4) that, for each constant orthogonal tensor $\underset{\sim}{Q}$, the functionals, $\delta\mathcal{H}$ and $\delta^2\mathcal{H}$, obey the relations,

$$\delta\mathcal{H}(\underset{\sim}{Q}\underset{\sim}{G}^*\underset{\sim}{Q}^T) = \underset{\sim}{Q}\delta\mathcal{H}(\underset{\sim}{G}^*)\underset{\sim}{Q}^T, \tag{2.6a}$$

$$\delta^2\mathcal{H}(\underset{\sim}{Q}\underset{\sim}{G}^*\underset{\sim}{Q}^T, \underset{\sim}{Q}\underset{\sim}{G}^*\underset{\sim}{Q}^T) = \underset{\sim}{Q}\delta^2\mathcal{H}(\underset{\sim}{G}^*, \underset{\sim}{G}^*)\underset{\sim}{Q}^T, \tag{2.6b}$$

as identities in $\underset{\sim}{G}^*$.

As every continuous linear functional on a Hilbert space may be written as an inner product, $\delta\mathcal{H}$, the first Fréchet derivative of $\mathcal{H}$ at $\underset{\sim}{1}$†, has a representation of the form,

$$\delta\mathcal{H}(\underset{\sim}{G}^*) = \int_0^\infty \underset{\sim}{\Gamma}(s)\{\underset{\sim}{G}^*(s)\}h(s)^2 ds; \tag{2.7}$$

here $\underset{\sim}{\Gamma}(s)\{\cdot\}$ is, for each $s \geq 0$, a linear transformation of the space of symmetric tensors into itself, and

$$\int_0^\infty |\underset{\sim}{\Gamma}(s)|^2 h(s)^2 ds < \infty. \tag{2.8}$$

It follows from (2.6a) that $\underset{\sim}{\Gamma}$ is determined by a single scalar-valued function m, and (2.7) can be simplified:

$$\delta\mathcal{H}(\underset{\sim}{G}^*) = \int_0^\infty m(s)\underset{\sim}{G}^*(s) ds. \tag{2.9}$$

For the material function m we have, by (2.8),

$$\int_0^\infty m(s)^2 h(s)^{-2} ds < \infty. \tag{2.10}$$

Because the term $\frac{1}{2}\delta^2\mathcal{H}(\underset{\sim}{G}^*, \underset{\sim}{G}^*)$ in (2.4) is $O(\|\underset{\sim}{G}^*\|^2)$, the hypothesis of fading memory yields the following approximation formula for the basic constitutive relation (1.5):

$$\underset{\sim}{T}(t) + p(t)\underset{\sim}{1} = \int_0^\infty m(s)\underset{\sim}{G}^t(s)\,ds + o(\|\underset{\sim}{G}^t\|^2), \tag{2.11}$$

where the function $\underset{\sim}{G}^t$, defined by

$$\underset{\sim}{G}^t(s) = \underset{\sim}{C}^t(s) - \underset{\sim}{1} = \underset{\sim}{F}^t(s)^T\underset{\sim}{F}^t(s) - \underset{\sim}{1}, \tag{2.12}$$

may be called the <u>relative-strain history</u>. The equation obtained by striking out the term $o(\|\underset{\sim}{G}^t\|^2)$ in (2.11) is called <u>the constitutive equation of finite linear viscoelasticity for an incompressible fluid</u>. In [17] Noll and I observed that for fluids with fading memory an equation of the form,

$$\underset{\sim}{T}(t) + p(t)\underset{\sim}{1} = \int_0^\infty m(s)\underset{\sim}{J}^t(s)\,ds + o(\|\underset{\sim}{J}^t\|^2), \tag{2.13}$$

remains valid not only when $\underset{\sim}{J}^t(s) = \underset{\sim}{G}^t(s)$ but also when $\underset{\sim}{J}^t(s)$ is any finite measure of strain that differs from $\underset{\sim}{G}^t(s)$ by terms $o(|\underset{\sim}{G}^t(s)|^2)$. For example, (2.13) holds when $\underset{\sim}{J}^t(s)$ is the measure of strain,

$$\underset{\sim}{J}^t(s) = \underset{\sim}{1} - \underset{\sim}{C}^t(s)^{-1} \tag{2.14}$$

occurring naturally in a molecular model put forward by Lodge [26]. Experimental confirmation of the relation (2.13) with $\underset{\sim}{J}^t(s) = \underset{\sim}{C}^t(s) - \underset{\sim}{1}$, with $\underset{\sim}{J}^t(s) = \underset{\sim}{1} - \underset{\sim}{C}^t(s)^{-1}$, or with $\underset{\sim}{J}^t = 2[(\underset{\sim}{C}^t(s))^{1/2} - \underset{\sim}{1}]$, does not confirm the correctness of a specific molecular hypothesis, but does show that $\mathcal{H}$ has a certain very general property of smoothness.[13]

[13] In view of comments made during this Symposium, it does not appear inappropriate to observe here that, although equation (2.11), without the term $o(\|\underset{\sim}{G}^t\|^2)$, describes a particular BKZ fluid, the derivation of that equation for the limit $\|\underset{\sim}{G}^t\| \to 0$ does <u>not</u> require the assumption that when higher-order terms in $\|\underset{\sim}{G}^t\|$ are retained the resulting constitutive equation must be of the BKZ type. Not all generalizations of finite linear viscoelasticity are BKZ fluids. In particular, the expression obtained by placing (2.18) in (2.15) does not describe a BKZ fluid. I believe that other examples and references to the theory of BKZ fluids are given in the paper of Louis Zapas in these Proceedings.

In view of (1.5) and (2.11), the assumption (2.4) yields

$$\underset{\sim}{T}(t) + p(t)\underset{\sim}{1} = \int_0^\infty m(s)\underset{\sim}{G}^t(s)\,ds + \tfrac{1}{2}\delta^2\mathfrak{H}(\underset{\sim}{G}^t,\underset{\sim}{G}^t) + o(\|\underset{\sim}{G}^t\|^2). \quad (2.15)$$

The hypothesis of fading memory, as stated above, does <u>not</u> imply that the bilinear continuous functional $\delta^2\mathfrak{H}$ can be represented as a sum of iterated integrals; if, however, one assumes that $\delta^2\mathfrak{H}$ is of Hilbert-Schmidt type, then $\delta^2\mathfrak{H}$ does have the representation,

$$\delta^2\mathfrak{H}(\underset{\sim}{G}^*,\underset{\sim}{G}^*) = \int_0^\infty\int_0^\infty \underset{\sim}{\Gamma}(s_1,s_2)\{\underset{\sim}{G}^*(s_1),\underset{\sim}{G}^*(s_2)\}h(s_1)^2h(s_2)^2ds_1ds_2, \quad (2.16)$$

where $\underset{\sim}{\Gamma}(s_1,s_2)\{\cdot\,,\cdot\}$ is, for each pair (s_1,s_2), a bilinear mapping of pairs of symmetric tensors into symmetric tensors, and

$$\int_0^\infty\int_0^\infty |\underset{\sim}{\Gamma}(s_1,s_2)|^2h(s_1)^2h(s_2)^2ds_1ds_2 < \infty. \quad (2.17)$$

The identity (2.6b) implies that $\delta^2\mathfrak{H}$ has the form [17],

$$\begin{aligned}\tfrac{1}{2}\delta^2\mathfrak{H}(\underset{\sim}{G}^*,\underset{\sim}{G}^*) &= \int_0^\infty\int_0^\infty \underline{a}(s_1,s_2)\underset{\sim}{G}^*(s_1)\underset{\sim}{G}^*(s_2)\,ds_1ds_2 \\ &\quad + \int_0^\infty\int_0^\infty b(s_1,s_2)[\mathrm{tr}\,\underset{\sim}{G}^*(s_1)]\underset{\sim}{G}^*(s_2)\,ds_1ds_2,\end{aligned} \quad (2.18)$$

where $\underline{a}$ is symmetric in the sense that $\underline{a}(s_1,s_2) = \underline{a}(s_2,s_1)$, and

$$\int_0^\infty\int_0^\infty \underline{a}(s_1,s_2)^2h(s_1)^{-2}h(s_2)^{-2}ds_1ds_2 < \infty. \quad (2.19)$$

The formula (2.19) need not be employed here; the principal results of the theory of fading memory require no properties of the second Fréchet derivative $\delta^2\mathfrak{H}(\cdot\,,\cdot)$ other than bilinearity and continuity on $\mathbb{H}\times\mathbb{H}$.

3. ON INFINITESIMAL DEFORMATIONS

Intuition suggests that $\|\underset{\sim}{G}^t\|$ in (2.11) should be small if all deformations from a fixed reference configuration $\mathcal{R}$ have been "infinitesimal". To make this idea precise, let

$$\underset{\sim}{H}(\tau) = \underset{\sim}{F}_{\mathcal{R}}(\tau) - \underset{\sim}{1} \tag{3.1}$$

and

$$\varepsilon = \sup_{-\infty<\tau\leq t} |\underset{\sim}{H}(\tau)|, \tag{3.2}$$

where $\underset{\sim}{F}_{\mathcal{R}}(\tau)$ is the deformation gradient at time τ for the material point under consideration when $\mathcal{R}$ is taken as the reference configuration. One says that the _deformations from $\mathcal{R}$ have been infinitesimal_ (up to time t) _if ε is small compared to unity_. The symmetric part of the _displacement gradient tensor_ $\underset{\sim}{H}$ defined in (3.1) is the _infinitesimal strain tensor_ $\underset{\sim}{E}$ (relative to $\mathcal{R}$):

$$\underset{\sim}{E} = \frac{1}{2}(\underset{\sim}{H}+\underset{\sim}{H}^T). \tag{3.3}$$

It follows from (1.6) and (2.12) that, for each s,

$$\underset{\sim}{G}^t(s) = 2[\underset{\sim}{E}(t-s)-\underset{\sim}{E}(t)] + o(\varepsilon^2), \tag{3.4}$$

and hence (2.2) yields

$$\|\underset{\sim}{G}^t\| = O(\varepsilon). \tag{3.5}$$

Thus it is indeed true that if ε is small, i.e., if the deformations from some fixed configuration $\mathcal{R}$ have been infinitesimal up to time t, then $\|\underset{\sim}{G}^t\|$ is small.[14]

Placement of (3.4) and (3.5) in (2.11) yields[15]

$$\underset{\sim}{T}(t) + p(t)\underset{\sim}{1} = 2\underline{\underline{G}}(0)\underset{\sim}{E}(t) + 2\int_0^{\infty} \frac{d\underline{\underline{G}}(s)}{ds} \underset{\sim}{E}(t-s)\,ds + o(\varepsilon^2), \tag{3.6}$$

---

[14] Footnote is on following page.

[15] The present derivation of (3.6) is that given in [17].

where $\underset{=}{G}$ is a scalar-valued function determined as follows by the function m in (2.9):

$$\underset{=}{G}(s) = -\int_s^\infty m(\zeta)\,d\zeta. \tag{3.7}$$

The terms exhibited explicitly in (3.6) are $O(\varepsilon)$ and have the forms found in the Boltzmann-Volterra theory of viscoelasticity, specialized, of course, to incompressible fluids. Thus the theory obtained by crossing out the term $O(\varepsilon^2)$ is the classical *linear theory of infinitesimal viscoelasticity for incompressible fluids*.

Rheologists call the material function $\underset{=}{G}$ the *stress relaxation function for shear* or, simply, the *shear relaxation modulus*; for, in an experiment in which $\underset{\sim}{E}$ has the time-dependence,

$$\underset{\sim}{E}(t) = \begin{cases} \underset{\sim}{0}, & \text{for } t < 0, \\ \underset{\sim}{E}_o, & \text{for } t \geq 0, \end{cases} \tag{3.8}$$

i.e., in a "relaxation test", equation (3.6) yields, for $t > 0$,

$$\underset{\sim}{T}(t) + p(t)\underset{\sim}{1} = 2\underset{=}{G}(t)\underset{\sim}{E}_o + O(\varepsilon^2). \tag{3.9}$$

---

[14] In a *steady simple shearing flow* for which the velocity vector has the components $v^x = 0$, $v^y = v(x)$, $v^z = 0$ in a Cartesian coordinate system, the matrix of the Cartesian components of $\underset{\sim}{G}^t(s)$ has the form $\begin{bmatrix} \kappa^2 s^2 & -\kappa s & 0 \\ -\kappa s & 0 & 0 \\ 0 & 0 & 0 \end{bmatrix}$ with $\kappa = dv(x)/dx$, and $\underset{\sim}{G}^t$ is independent of t. (Cf. [14, 10].) In such a motion ε is *not* small; in fact for each choice of $\mathfrak{R}$, $\varepsilon = \infty$; however, $\|\underset{\sim}{G}^t\| = O(\kappa)$, and hence $\|\underset{\sim}{G}^t\|$ is small when the rate of shear, κ, is small. Thus small ε is a sufficient, but not a necessary condition to have $\|\underset{\sim}{G}^t\|$ small; $\|\underset{\sim}{G}^t\|$ is small also in motions that are "slow" in the sense of the retardation theorem to be stated in §4 .

It follows from (2.10) and (3.7) that

$$\lim_{s\to\infty} \underset{=}{G}(s) = 0. \tag{3.10}$$

In fact, using (2.1) and our assumption that h is an influence function of order $r > 2$, one can show easily that there is a $\delta > 0$ such that, in the limit of large s,[16]

$$\underset{=}{G}(s) = O\left(\frac{1}{s^{2+\delta}}\right), \tag{3.11}$$

which yields

$$\lim_{s\to\infty} s^2 \underset{=}{G}(s) = 0. \tag{3.12}$$

For each $\omega \geq 0$, the real numbers $\underset{=}{G}'(\omega)$ and $\underset{=}{G}''(\omega)$, called, respectively, the _shear storage modulus_ and the _shear loss modulus_, are defined by

$$\left.\begin{aligned} \underset{=}{G}'(\omega) &= \omega\int_0^\infty \underset{=}{G}(s)\sin\omega s\, ds,\\ \underset{=}{G}''(\omega) &= \omega\int_0^\infty \underset{=}{G}(s)\cos\omega s\, ds. \end{aligned}\right\} \tag{3.13}$$

When $\underset{\sim}{E}$ is a sinusoidal function of time with circular frequency ω, so that, for all t,

$$\underset{\sim}{E}(t) = \underset{\sim}{E}_o \sin\omega t, \tag{3.14}$$

(3.6) yields

$$\underset{\sim}{T}(t) + p(t)\underset{\sim}{1} = 2\underset{\sim}{E}_o[\underset{=}{G}'(\omega)\sin\omega t + \underset{=}{G}''(\omega)\cos\omega t] + O(\varepsilon^2). \tag{3.15}$$

---

[16] See [4] §11.

4. ORIGIN AND PROPERTIES OF SECOND-ORDER FLUIDS

For the norm $\|\underset{\sim}{G}^t\|$ defined in (2.2) to be small, i.e., for the relations (2.11) and (2.15) to be useful, it is not necessary that there be a reference configuration $\mathcal{R}$ for which ε in (3.2) is small. Indeed, it follows from our assumptions about the influence function h in (2.2) that for $\|\underset{\sim}{G}^t\|$ to be small it is sufficient that the deformation relative to the present configuration be small in the recent past. Such is the case in motions that are "slow".

If $\underset{\sim}{G}^*$ is a function on $[0,\infty)$ and α is in $(0,1]$, the function ${}_{(\alpha)}G^*$, defined by

$$ {}_{(\alpha)}\underset{\sim}{G}^*(s) = \underset{\sim}{G}^*(\alpha s), \tag{4.1}$$

is called the <u>retardation of $\underset{\sim}{G}^*$ by the factor</u> α, or simply, the α-<u>retardation</u> of $\underset{\sim}{G}^*$. If, after a suitable alteration of $\underset{\sim}{G}^*$ on a set of Lebesgue measure zero, the limits $\overset{(\ell)}{\underset{\sim}{G}}{}^*$, defined recursively by

$$ \overset{(0)}{\underset{\sim}{G}}{}^* = \lim_{s\to 0} \underset{\sim}{G}^*(s), \qquad \overset{(\ell)}{\underset{\sim}{G}}{}^* = \lim_{s\to 0} \frac{\ell!}{s^\ell}\left(\underset{\sim}{G}^*(s) - \sum_{k=0}^{\ell-1} \frac{s^k}{k!} \overset{(k)}{\underset{\sim}{G}}{}^*\right), \tag{4.2}$$

exist for $\ell = 0,1,\ldots,n$, then we say that $\underset{\sim}{G}^*$ has n <u>generalized derivatives at zero</u>. Thus, $\underset{\sim}{G}^*$ has n generalized derivatives at zero if it differs on at most a set of measure zero from a function that has a limit at the point 0 in $[0,\infty)$ and has n ordinary derivatives at 0.

Each function $\underset{\sim}{G}^*$ that has n generalized derivatives at zero determines a function $\underset{n}{\Pi}\underset{\sim}{G}^*$, called the <u>n'th-order Taylor transformation of</u> $\underset{\sim}{G}^*$, defined as

$$ \underset{n}{\Pi}\underset{\sim}{G}^*(s) = \sum_{\ell=0}^{n} \frac{s^\ell \overset{(\ell)}{\underset{\sim}{G}}{}^*}{\ell!}. \tag{4.3}$$

It turns out that if $\underset{\sim}{G}^*$, in addition to having n generalized derivatives at zero, is also a member of the space $\mathbb{H}$, i.e., has $\|\underset{\sim}{G}^*\| < \infty$, and the order r of the influence function in (2.2) exceeds n, then $\underset{n}{\Pi}\underset{\sim}{G}^*$ and ${}_{(\alpha)}\underset{\sim}{G}^*$ are in $\mathbb{H}$, and, in the limit as $\alpha \to 0$, the n'th order Taylor transformation of the

α-retardation of $\underset{\sim}{G}^*$, $\underset{n}{\Pi}\,{}_{(\alpha)}\underset{\sim}{G}^*$, differs from ${}_{(\alpha)}\underset{\sim}{G}^*$ by a function whose norm in H goes to zero very rapidly with α, in fact as $o(\alpha^n)$. This is the content of the retardation theorem.[17]

If h is an influence function of order greater than n and $\underset{\sim}{G}^*$ in H has n generalized derivatives at zero, then for each α in (0,1], ${}_{(\alpha)}\underset{\sim}{G}^*$ and $\underset{n}{\Pi}\,{}_{(\alpha)}\underset{\sim}{G}^*$ are in H, and, as $\alpha \to 0$,

$$\lim_{\alpha\to 0} \frac{1}{\alpha^n} \| {}_{(\alpha)}\underset{\sim}{G}^* - \underset{n}{\Pi}\,{}_{(\alpha)}\underset{\sim}{G}^* \| = 0. \tag{4.4}$$

When $\underset{\sim}{G}^t$ is a relative-strain history, as defined in (2.12),

$$\underset{\sim}{G}^t(0) = \underset{\sim}{0}. \tag{4.5}$$

Let us assume now that $\underset{\sim}{G}^t$ is continuous at zero and that $\overset{(n)}{\underset{\sim}{G}}{}^t$ exists for n = 1 and 2. Then by (1.16), (4.2), and (4.5)

$$\overset{(0)}{\underset{\sim}{G}}{}^t = 0, \qquad \overset{(1)}{\underset{\sim}{G}}{}^t = \underset{\sim}{A}_1, \qquad \overset{(2)}{\underset{\sim}{G}}{}^t = \underset{\sim}{A}_2 \tag{4.6}$$

with $\underset{\sim}{A}_1$ and $\underset{\sim}{A}_2$ Rivlin-Ericksen tensors. The Rivlin-Ericksen tensors ${}_{(\alpha)}\underset{\sim}{A}_N$ corresponding to the α-retardation, ${}_{(\alpha)}\underset{\sim}{G}^t$, of $\underset{\sim}{G}^t$ are related as follows to those corresponding to $\underset{\sim}{G}^t$:

$${}_{(\alpha)}\underset{\sim}{A}_N = \alpha^N \underset{\sim}{A}_N. \tag{4.7}$$

As we assumed in item ($\underset{\sim}{i}$) of the hypothesis of fading memory that the order of the influence function h is greater than 2, it here follows from (4.4)-(4.7) that

$$\| {}_{(\alpha)}\underset{\sim}{G}^t \| = O(\alpha). \tag{4.8}$$

Thus, slow flows do give rise to strain histories $\underset{\sim}{G}^t$ with small norm.[18]

---

[17] [16], Thm. 1.

[18] Footnote is on following page.

It follows from (2.11), (4.6), and the conclusion (4.4) of the retardation theorem that we have the following approximation formula for the stress ${}_{(\alpha)}\underset{\sim}{T}$ corresponding to ${}_{(\alpha)}\underset{\sim}{G}^t$:

$$ {}_{(\alpha)}\underset{\sim}{T} + p\underset{\sim}{1} = \eta_0\ {}_{(\alpha)}\underset{\sim}{A}_1 + O(\alpha^2); \tag{4.9} $$

η_0 is a material constant given by

$$ \eta_0 = -\int_0^\infty s\, m(s)\, ds. \tag{4.10} $$

As $\underset{\sim}{A}_1 = 2\underset{\sim}{D}$ and the term ${}_{(\alpha)}\underset{\sim}{A}_1$ is $O(\alpha)$, the relation (4.9) justifies the intuitive idea that for motions sufficiently slow the constitutive equation $\underset{\sim}{T} + p\underset{\sim}{1} = \mathfrak{H}(\underset{\sim}{C}^t)$ for the stress in a simple fluid with gradually fading memory is well approximated by the constitutive equation $\underset{\sim}{T} + p\underset{\sim}{1} = 2\eta_0\underset{\sim}{D}$ of a Navier-Stokes fluid.

From (2.15), (2.6b), and (4.4) with n = 2, one may calculate $\mathfrak{H}({}_{(\alpha)}\underset{\sim}{C}^t) = \mathfrak{H}(\underset{\sim}{1}^\dagger + {}_{(\alpha)}\underset{\sim}{G}^t)$ to within an error $o(\alpha^2)$; one finds that[19]

$$ {}_{(\alpha)}\underset{\sim}{T} + p\underset{\sim}{1} = \eta_0\ {}_{(\alpha)}\underset{\sim}{A}_1 + \beta\ {}_{(\alpha)}\underset{\sim}{A}_1^2 + \gamma\ {}_{(\alpha)}\underset{\sim}{A}_2 + o(\alpha^2). \tag{4.11} $$

---

[18] If one assumes only that $\underset{\sim}{G}^t$ is continuous at zero, then (4.4) yields, in place of (4.8), the weaker conclusion that $\lim_{\alpha\to 0} \| {}_{(\alpha)}\underset{\sim}{G}^t \| = 0$. For the validity of (4.8) it suffices that $\underset{\sim}{G}^t$ have just one generalized derivative at zero and that the order r of h exceed one.

[19] Noll and I obtained the results (4.9), (4.11), and some generalizations of them in our paper [16] on the retardation theorem. Shortly thereafter, Markovitz and I [8, 29] made the observation that γ and β in (4.11) are related to m and $\underset{-}{a}$ as shown in equations (4.12) and (4.14). Extensions of these results are given in [18], [4], and [9].

The material constant γ is, like η_0, determined by the first functional derivative $\delta\mathfrak{H}$ of $\mathfrak{H}$; for γ we have, in place of (4.10):

$$\gamma = \frac{1}{2}\int_0^\infty s^2 m(s)\,ds. \tag{4.12}$$

The constant β is determined by the quadratic functional $\delta^2\mathfrak{H}$ through the formula,

$$2\beta \underset{\sim}{A}_1^2 = \delta^2\mathfrak{H}(\iota\underset{\sim}{A}_1, \iota\underset{\sim}{A}_1), \tag{4.13}$$

which holds for every choice of the traceless symmetric tensor $\underset{\sim}{A}_1$; here ι is the identity function on $[0,\infty)$: $\iota(s) = s$. In the special case in which $\delta^2\mathfrak{H}$ has a representation as an iterated integral, (2.18) may be employed to write (4.13) in the form,

$$\beta = \int_0^\infty\int_0^\infty s_1 s_2 \underline{a}(s_1, s_2)\,ds_1 ds_2. \tag{4.14}$$

In (4.11) the term $\beta\, {}_{(\alpha)}\underset{\sim}{A}_1^2 + \gamma\, {}_{(\alpha)}\underset{\sim}{A}_2$ is $O(\alpha^2)$, and as the "error term" is, as shown, $o(\alpha^2)$, the derivation of (4.11) shows that the constitutive equation of a second-order fluid,

$$\underset{\sim}{T} + p\underset{\sim}{1} = \eta_0\underset{\sim}{A}_1 + \beta\underset{\sim}{A}_1^2 + \gamma\underset{\sim}{A}_2, \tag{4.15}$$

does give to the classical relation $\underset{\sim}{T} + p\underset{\sim}{1} = 2\eta_0\underset{\sim}{D}$, (as claimed in §1) a correction for viscoelastic effects that is complete to within order two in the retardation parameter α.

In view of (3.7) and (3.11), the relation (4.10) for η_0 can be written in terms of the shear relaxation modulus $\underline{\underline{G}}$ appearing in the linear theory of infinitesimal deformations,

$$\eta_0 = \int_0^\infty \underline{\underline{G}}(s)\,ds, \tag{4.16}$$

and in terms of the shear loss modulus $\underline{\underline{G}}''$ defined in (3.13),

$$\eta_0 = \lim_{\omega\to 0} \frac{G''(\omega)}{\omega}. \tag{4.17}$$

Relations equivalent to (4.16) and (4.17) have been known for some time. However, when Hershel Markovitz and I [8, 29] noticed that (3.7), (3.11), and (3.13), permit one to write the relation (4.12) for γ in terms of $\underline{\underline{G}}$ as

$$\gamma = -\int_0^\infty s\underline{\underline{G}}(s)\,ds, \tag{4.18}$$

and in terms of the shear storage modulus $\underline{\underline{G}}'$ as

$$\gamma = -\lim_{\omega \to 0} \frac{\underline{\underline{G}}'(\omega)}{\omega^2}, \tag{4.19}$$

the result was new, and led to some surprises.

First, it follows from (4.18) that a normal stress effect occuring in steady simple shearing flows and called the <u>first normal stress difference</u>, although intrinsically non-linear because it is an even function of the rate of shear and is not exhibited by Navier-Stokes fluids, is, nonetheless, determined to within a remainder $o(\kappa^2)$, if one knows the function $\underline{\underline{G}}$ (or, equivalently, $\underline{\underline{G}}'$) that governs material behavior in the linear theory of infinitesimal viscoelastic deformations.[20]

---

[20] Specifically, in steady simple shearing flow (see Footnote 14) for which the velocity field has the Cartesian components $v^x = 0$, $v^y = v(x)$, $v^z = 0$, a Navier-Stokes fluid yields $T^{xx} - T^{yy} = 0$, but for a general simple fluid with memory [16, 10]: $T^{xx} - T^{yy} = c\kappa^2 + o(\kappa^2)$ with c a constant and $\kappa = dv(x)/dx$. (The "first normal stress difference" is the quantity $T^{yy} - T^{xx}$.) Upon retardation, κ is proportional to α and (4.15) yields, for a second-order fluid, $T^{xx} - T^{yy} = 2\gamma\kappa^2$; thus,

$$T^{xx} - T^{yy} = 2\gamma\kappa^2 + o(\kappa^2),$$

with γ as in (4.18). In earlier work [18], Noll and I had shown that in an oscillatory shearing motion of small amplitude, i.e., with ε of equation (3.2) small, the same difference of normal stresses, $T^{xx} - T^{yy}$, is determined by $\underline{\underline{G}}$ to within an error $o(\varepsilon^2)$, and so also is the difference in radial thrusts on the inner and outer cylinders in a periodic Couette flow. Markovitz and I treated such effects further in [9] and [30].

Second, as all rheological studies of truly viscoelastic materials yield $\underset{=}{G}(s) > 0$, (4.18) and (4.19) imply

$$\gamma < 0, \tag{4.20}$$

and this inequality was soon shown to imply a striking lack of stability of solutions, $\underset{\sim}{v}$, of the differential equations obtained by placing the constitutive equation of a second-order fluid, (4.15), in the law of balance of linear momentum,

$$\operatorname{div} \underset{\sim}{T} = \rho \dot{\underset{\sim}{v}}. \tag{4.21}$$

I cannot review here all results on the instability of flows of second-order fluids, but I shall try to give some idea of the nature of the subject by stating pertinent conclusions from the first study of the consequences of (4.20).

A (non-steady) <u>simple shearing</u> flow is one for which there is a Cartesian coordinate system, x, y, z, in which the components of $\underset{\sim}{v}$ have the form

$$v^x = 0, \qquad v^y = v(x,t), \qquad v^z = 0. \tag{4.22}$$

For such a flow, if there are no applied body forces, and if the pressure has no gradient in the direction of flow, then use of the classical constitutive equation of a Navier-Stokes fluid (1.9) in (4.21) yields, as is well known, an equation for v with the form of the classical diffusion equation, i.e.,

$$\eta_0 \partial_x^2 v = \rho \partial_t v. \tag{4.23}$$

Under the same conditions, the constitutive equation of a second-order fluid yields [16]

$$\eta_0 \partial_x^2 v + \gamma \partial_x \partial_t \partial_x v = \rho \partial_t v. \tag{4.24}$$

Let us consider now the case in which the fluid lies between plates at x = 0 and x = H, with H > 0, and the fluid adheres to the plates. As one of the plates may be assumed at rest, one boundary condition is v(0,t) = 0. At the other plate (x=H), one may prescribe either the velocity v(H,t) or the shearing force per unit area, which, by (4.15), is $\eta_0\partial_x v(H,t) + \gamma\partial_t\partial_x v(H,t)$. Thus, two types of boundary data are studied:

$$\text{Type I:} \quad v(0,t) = 0, \qquad v(H,t) = v^*(t), \tag{4.25a}$$

$$\text{Type II:} \quad v(0,t) = 0, \qquad \left[\eta_0\partial_x v(x,t) + \gamma\partial_t\partial_x v(x,t)\right]_{x=H} = f^*(t), \tag{4.25b}$$

with v^* or f^* a specified function. For initial data one takes

$$v(x,0) = \bar{v}(x), \tag{4.26}$$

with $\bar{v}$ given on $[0,H]$.

Equation (4.25) with these initial and boundary data was studied by Ting [34] under the assumption that ρ, η_0, and γ are all positive. Ting's conclusion that solutions of (4.25) are stable rested heavily on his assumption that ρ, η_0, and γ are of one sign. There is no doubt that[21]

$$\rho > 0, \qquad \eta_0 > 0. \tag{4.27}$$

The relation (4.18) and its implication (4.20) were found while Ting's work was in progress, and experiments on polymer solutions confirmed that γ is negative.[22] An elementary calculation showed that Ting's conclusions of stability of solutions of (4.24) on the section of the (x,t)-plane by

$$0 \leq x \leq H, \quad 0 \leq t < \infty \tag{4.28}$$

does not hold for $\gamma < 0$. In fact, the following conclusion

[21]The positivity of η_0 is, of course, compatible with the relation (4.16) and the positivity of $\underline{\underline{G}}$.

[22]See the early work of Markovitz [27] and Markovitz and Brown [28] and the recent investigation of Joseph, Beavers, Cers, Dewald, Hoger, and Than [24].

may be drawn [5]: <u>For</u> $\rho > 0$, $\eta_0 > 0$, <u>and</u> $\gamma < 0$, <u>and with given boundary data of</u> Type I <u>or</u> Type II, <u>every bounded solution</u> $v_1(x,t)$ <u>of</u> (4.24) <u>is unstable against arbitrarily small perturbations of the initial data in the sense that for each</u> $\varepsilon > 0$ <u>there is an unbounded solution</u> $v_2(x,t)$ <u>obeying the same boundary data with</u> $|v_1(x,0)-v_2(x,0)| < \varepsilon$. To see this one should observe that if v_1 obeys (4.24) then so also does

$$v_2(x,t) = v_1(x,t) + \varepsilon \exp\left(\frac{\eta_0 b^2 t/(-\gamma)}{b^2-1}\right) \sin\left(bx\sqrt{\rho/(-\gamma)}\right), \qquad (4.29)$$

for $b^2 \neq 1$, and this function v_2 is unbounded when $b^2 > 1$. If the given boundary data is of Type I, we may note that v^* in (4.25a) is the same for v_1 and v_2 if and only if

$$b^2 = \frac{n^2\pi^2(-\gamma)}{H^2\rho}, \qquad n = 1,2,\ldots\ . \qquad (4.30)$$

In such a case, v_2 is unbounded if

$$n^2 > \frac{H^2\rho}{-\gamma\pi^2}, \qquad (4.31)$$

and n certainly can be chosen so that this holds. Similarly, f^* in (4.25b) is the same for v_1 and v_2 if and only if

$$b^2 = \frac{(2n-1)^2\pi^2(-\gamma)}{4H^2\rho}, \qquad n = 1,2,\ldots\ , \qquad (4.32)$$

and hence, in the case of given data of Type II, v_2 is unbounded if

$$\left(n-\frac{1}{2}\right)^2 > \frac{H^2\rho}{-\gamma\pi^2}, \qquad (4.33)$$

and this condition is easily fulfilled.

In the same work in which we made these observations, Duffin, Mizel, and I [5] obtained a uniqueness theorem showing that the lack of stability of solutions of (4.24) is <u>not</u> the result of an underspecification of either initial data or boundary data: <u>When</u> (4.27) <u>and</u> (4.20) <u>hold, if</u>

boundary data of Type I or Type II and initial data $\bar{v}(x)$ *are given, then, for every* $\tau > 0$, (4.24) *has at most one solution* $\underset{\sim}{v}(x,t)$ *on the set* $[0,H]\times[0,\tau]$.

As I said above, I shall not review the large and growing literature about the dynamical behavior of second-order fluids with $\gamma < 0$. (For example, in the paper just cited [5], that deals only with shearing flows of the form (4.22), there are results which I have not had a chance to discuss showing a general paucity of solutions in narrow gaps, i.e., on sets of the form (4.28) with H small.[23]) But I should mention that, since the publication of this work in 1965, results have been obtained demonstrating, for second-order fluids with $\gamma < 0$, anomalous behavior of flows that are not of the form (4.22) and that are not such that the three non-linear equations (4.21) reduce to one linear equation. In 1979, Fosdick and Rajagopal [23] showed that for a fluid obeying (4.15) with $\rho > 0$, $\eta > 0$, $\gamma < 0$, and $\beta + \gamma \neq 0$, an isochoric velocity field $\underset{\sim}{v}$ that is a solution of (4.21) in a sufficiently small region Ω, with $\underset{\sim}{v} = \underset{\sim}{0}$ on $\partial\Omega \times [0,\infty)$ and with $\underset{\sim}{v}(\underset{\sim}{x},0)$ smooth but non-trivial, must describe a flow that never subsides, i.e., a flow for which

$$\lim_{t\to\infty} \max_{\underset{\sim}{x}\in\Omega} |\underset{\sim}{A}_1(\underset{\sim}{x},t)| > 0, \tag{4.34}$$

and if the viscosity η_0 is large enough, the flow must be unbounded in the sense that $\int_\Omega |\underset{\sim}{A}_1(\underset{\sim}{x},t))|^2 dv$ becomes arbitrarily large with increasing t. (For the same boundary condition, $\underset{\sim}{v} = \underset{\sim}{0}$ on $\partial\Omega\times[0,\infty)$, Dunn and Fosdick [22] had obtained, in 1974, theorems showing the instability of equilibrium of a fluid obeying (4.15) with $\eta_0 > 0$, $\gamma < 0$, and $\beta + \gamma = 0$.)

Navier-Stokes theory can be viewed either as an approximation to the theory of fluids with fading memory, valid for sufficiently slow flows or, as it is usually considered, a special theory of interest in its own right.

---

[23] See §§4 and 5 of [5] and the further discussion in [13].

The second viewpoint is acceptable for Navier-Stokes fluids, because the theory of such fluids has not yet been shown to give rise to unphysical behavior when applied outside the range of slow flows. The theorems on retardation justify the status of second-order fluids as higher approximations, for slow flows, to the theory of fluids with fading memory, and in so doing lead to $\gamma < 0$, in accord with experiment; but the theorems on lack of stability of flows of second-order fluids show that it is not useful to regard the constitutive equation of a second-order fluid, with material parameters agreeing in sign with those resulting from the process of retardation, as a constitutive assumption to be employed in fast flows. Indeed, for a fluid with fading memory, the instabilities found in the theory of the corresponding second-order fluid are unlikely to be significant; for such instabilities, as they make $|\underset{\sim}{A}_1|$ large,[24] will take the flow out of the range in which the "error term", $o(\alpha^2)$, in (4.11) can be safely neglected.

That Navier-Stokes theory is free from some of the difficulties that arise when one applies the theory of second-order fluids outside of the asymptotic limit for which it has been derived is an interesting fact but not one that can be used to discredit the theory of fluids with fading memory or careful use of the retardation theorem and its consequences (4.11) and (4.16)-(4.20). As Truesdell and Noll have remarked, there are other examples in which an approximation of higher order to a sound theory introduces difficulties not found in lower-order approximations ([35], p. 513):

[24]Recall equation (4.7).

> The mere fact that a second approximation is in some ways worse than the first is not without precedent. Consider, for example, approximation of a real function by a power series. If the function itself is monotone increasing from $-\infty$ to $+\infty$, the first approximation at a typical point, being a straight line, preserves this property, while every second approximation, being a parabola, fails to share it. E.g., infinitesimal elasticity predicts behavior that is not *a priori* ridiculous, however incorrect it may be, in large pure strains, but the second-order theory if interpreted outside the range of small strains leads to outright impossibility: either that large extension requires pressure, or that large compression requires tension. The price of closer approximation in a narrow range may be poorer approximation in a great one.

5. A CONDITION SUFFICIENT FOR FAILURE OF THE SECOND-ORDER APPROXIMATION

A *substantially stagnant flow*[25] is one for which, as a particle moves along its path, the history of the right relative Cauchy-Green tensor at the particle either remains the same function or is rotated in time so that for each t there is an orthogonal tensor $\mathbf{Q}(t)$ relating the history up to time t, $\mathbf{C}^t$, to the history up to time 0, $\mathbf{C}^0$, by the formula

$$\mathbf{C}^t(s) = \mathbf{Q}(t)\mathbf{C}^0(s)\mathbf{Q}(t)^T. \tag{5.1}$$

My own early study of these flows grew out of an interest in the restrictions that the laws of thermodynamics place on material behavior. In 1961, before I had in hand a procedure

[25]This kinematical concept was introduced in [1]; such flows are called also *motions with constant stretch history*, cf. [32].

for finding conditions on constitutive relations necessary and sufficient to insure that the rate of production of entropy never be negative, I noticed that if, in a substantially stagnant flow of an incompressible simple fluid with memory, one thermodynamic variable, such as the specific internal energy or the temperature, is constant along a particle path, then the specific entropy, the specific Helmholtz free energy, and the stress power $\underset{\sim}{T} \cdot \underset{\sim}{D}$ are also constant along the path. Using this observation, it was not difficult to show that it is a consequence of the laws of thermodynamics that[26], _in an isothermal substantially stagnant flow of an incompressible simple fluid with memory_,

$$\underset{\sim}{T} \cdot \underset{\sim}{D} \geq 0. \tag{5.2}$$

Let us here continue to follow the standard practice of assuming that temperature is constant in space and time, so that when I refer below to substantially stagnant flows it will be understood that they are isothermal.

While the research on the stability of solutions of equation (4.24) was in progress, I mentioned to Walter Noll that, as yet, no one had found the full implications of the inequality (5.2) for second-order fluids.[27] Noll then showed that, _if an incompressible fluid obeys the defining equation of a second-order fluid_ (4.15) _in all isochoric flows_ (no matter how fast, i.e., no matter how large $|\underset{\sim}{D}|$), _then to have_ $\underset{\sim}{T} \cdot \underset{\sim}{D} \geq 0$ in _all isochoric_

[26] See [1]. The _principle of positive internal production of entropy_ employed in [1] was shown to be a consequence of balance of energy and the Clausius-Duhem inequality in [2].

[27] As observed in [1], because each steady simple shearing flow is substantially stagnant, it follows from (5.2) that the viscosity function η of a general simple fluid cannot be negative and therefore it was known that (5.2) yields $\eta_0 \geq 0$; at the time of my discussions with Noll, however, it was not known whether (5.2) places restrictions on β and γ.

substantially stagnant flows (again, no matter how fast), it is necessary and sufficient that both of the following relations hold:

$$\eta_0 \geq 0 \tag{5.3}$$

and

$$\beta + \gamma = 0. \tag{5.4}$$

Duffin, Mizel, and I quoted Noll's result in a footnote to our paper [5]. We observed there that while (5.3) obviously is obeyed, the relation (5.4) is not in accord with experiments on viscoelastic fluids, such as polymer solutions.[28] Indeed, because the constitutive relation (4.15) does not hold in arbitrarily fast motions of an incompressible simple fluid with fading memory, the derivation of (5.4) is not meaningful for such a fluid, and one should not expect to have $\beta = -\gamma$ in (4.11).[29]

By the late 1960's methods had been developed[30] for finding the complete set of restrictions that the second law of thermodynamics, in the form of the Clausius-Duhem inequality, places on a given set of constitutive relations. It was clear that the second law requires that, in isothermal motions (for which the temperature is not only temporally constant, but also spatially uniform), the rate of internal

[28] This was true in 1964 and remains true today; see the recent measurements reported in [24].

[29] Within the theory of the thermodynamics of materials with memory [2, 3], a stronger statement can be made: the Clausius-Duhem inequality does not imply that, for an incompressible simple fluid with memory, $\delta\mathcal{H}$ and $\delta^2\mathcal{H}$ be such that (4.12) and (4.13) yield $\beta = -\gamma$.

[30] Cf., [2, 3, 6, 11, 12].

dissipation, ξ, defined by

$$\xi = \mathbf{T}\cdot\mathbf{D} - \dot{\psi} \tag{5.5}$$

with ψ the Helmholtz free energy per unit volume, never be negative.[31]

In 1974, Dunn and Fosdick published a paper [22] on second-order fluids in which they considered fluids for which in all isochoric and isothermal[32] motions the tensor $\mathbf{T}$ obeys (4.15) and ψ is given by a smooth function of the velocity gradient $\mathbf{L}$ and its material time-derivative $\dot{\mathbf{L}}$. For such a fluid, it follows from the principle of material frame-indifference that

$$\psi = \hat{\psi}(\mathbf{A}_1,\mathbf{A}_2) \tag{5.6}$$

with $\hat{\psi}$ isotropic in the sense that, for each orthogonal tensor $\mathbf{Q}$,

$$\hat{\psi}(\mathbf{Q}\mathbf{A}_1\mathbf{Q}^T, \mathbf{Q}\mathbf{A}_2\mathbf{Q}^T) = \hat{\psi}(\mathbf{A}_1,\mathbf{A}_2). \tag{5.7}$$

Dunn and Fosdick found that to have $\xi \geq 0$ in every isochoric flow of a fluid that obeys (4.15) and (5.6) in *all* such flows (again, no matter how fast) it is necessary and sufficient that the material constants η_0, β, and γ obey (5.3) and (5.4) *and* that ψ not only be independent of $\mathbf{A}_2$ but be given by an equation of the form

$$\psi = \frac{\gamma}{4}\mathbf{A}_1\cdot\mathbf{A}_1 + c. \tag{5.8}$$

[31] In [2] the term internal dissipation is applied to σ, the rate of production of entropy (per unit mass) by effects other than the transfer of heat through flux or radiation. The present ξ is related to σ by the formula $\xi = \rho\theta\sigma$ in which θ is the absolute temperature.

[32] I here specialize the assumptions and principal results of reference [22] to the important case of isothermal motion.

Since $\underset{\sim}{A}_1 = 2\underset{\sim}{D}$ and the constant c can be set equal to zero, (5.8) can be written

$$\psi = \frac{\gamma}{4}|\underset{\sim}{A}_1|^2 = \gamma|\underset{\sim}{D}|^2. \tag{5.9}$$

As we have seen, the relation (5.4) is not observed in experiments and is not a theorem in the thermodynamics of fluids with fading memory. Equation (5.9), on the other hand, has not been ruled out by experiment.

In the thermodynamics of materials with fading memory [2, 3], the Helmholtz free energy ψ of an incompressible simple fluid subject to only isothermal processes is given by a functional $\mathfrak{p}$ of the relative strain history $\underset{\sim}{G}^t$:

$$\psi(t) = \mathfrak{p}(\underset{\sim}{G}^t). \tag{5.10}$$

In a forthcoming paper I shall show that there is a broad class of incompressible fluids with fading memory for which the retardation theorem and results from the thermodynamics of materials with memory imply that if ${}_{(\alpha)}\underset{\sim}{G}^t$ is the α-retardation of a relative strain history $\underset{\sim}{G}^t$ that is continuous at zero and such that $\overset{(1)}{\underset{\sim}{G}}{}^t$ and $\overset{(2)}{\underset{\sim}{G}}{}^t$ exist, then

$$\mathfrak{p}(\underset{\sim}{G}^t) = \frac{1}{4}\gamma|{}_{(\alpha)}\underset{\sim}{A}_1|^2 + o(\alpha^2). \tag{5.11}$$

In other words, it appears that (5.9) can be accepted as an approximation to ψ valid for slow flows to the same order in the retardation parameter as (4.15). Very recently, Thomas Wicks and I have observed that <u>if</u> equation (5.9) is accepted in this way, <u>then</u> (4.15) and (5.9), together, yield [after some algebra and use of (1.13)-(1.15)] the following remarkably simple formula for the rate of internal dissipation ξ defined in (5.5):

$$\begin{aligned} \xi &= 2\eta \operatorname{tr}(\underset{\sim}{D}^2) + 4(\beta+\gamma)\operatorname{tr}(\underset{\sim}{D}^3) \\ &= \frac{1}{2}\eta \operatorname{tr}(\underset{\sim}{A}_1^2) + \frac{1}{2}(\beta+\gamma)\operatorname{tr}(\underset{\sim}{A}_1^3). \end{aligned} \tag{5.12}$$

Of course, this formula can be valid for a fluid with fading memory only to within an error $o(\alpha^3)$ in the retardation parameter α. That is: for a fluid with memory, (5.12) is to be interpreted as saying that, when $\tilde{G}^t$ is sufficiently smooth near zero, ${}_{(\alpha)}\xi$, the value of ξ corresponding to ${}_{(\alpha)}\tilde{G}^t$, obeys, for small α, the asymptotic relation

$$2\,{}_{(\alpha)}\xi = \eta_0\,\mathrm{tr}({}_{(\alpha)}\tilde{A}_1^2) + (\beta+\gamma)\,\mathrm{tr}({}_{(\alpha)}\tilde{A}_1^3) + o(\alpha^3). \tag{5.13}$$

It is interesting that here only the first Rivlin-Ericksen tensor is needed to determine the internal dissipation to an approximation of order three in α.

Now, as $\beta \neq -\gamma$ for real viscoelastic fluids, there are values of $\tilde{D}$ for which (5.12) yields $\xi < 0$, i.e., for which the rate of production of entropy, computed using (4.15) and (5.9), turns out negative. When such is the case, the error term, $o(\alpha^2)$, in (4.11) and (5.11) cannot be neglected, for the rate of production of entropy is never negative in practice. This observation, upon which Thomas Wicks and I shall elaborate in a future publication, appears to provide a tool for showing that a particular flow of a fluid with fading memory is too fast for the constitutive equation of the corresponding second-order fluid to be taken as an approximation. Of course, a negative value for the right hand side of (5.12) is only a sufficient, not a necessary condition for a flow to be out of the range of usefulness of the theory of second-order fluids.

REFERENCES

1. Coleman, B. D., Kinematical concepts with applications in the mechanics and thermodynamics of incompressible viscoelastic fluids, Arch. Rational Mech. Anal. 9 (1962), 273-300.
2. Coleman, B. D., Thermodynamics of materials with memory, Arch. Rational Mech. Anal. 17 (1964), 1-46.
3. Coleman, B. D., On thermodynamics, strain impulses, and viscoelasticity, Arch. Rational Mech. Anal. 17 (1964) 230-254.
4. Coleman, B. D., E. H. Dill, and R. A. Toupin, A phenomenological theory of streaming birefringence, Arch. Rational Mech. Anal. 39, (1970), 358-399.

5. Coleman, B. D., R. J. Duffin, and V. J. Mizel, Instability, uniqueness, and nonexistence theorems for the equation $u_t = u_{xx} - u_{xtx}$ on a strip, Arch. Rational Mech. Anal. 19 (1965), 100-116.

6. Coleman, B. D., and M. E. Gurtin, Thermodynamics with internal state variables, J. Chem. Phys. 47 (1967), 597-613.

7. Coleman, B. D., and M. E. Gurtin, On the stability against shear waves of steady flows of non-linear viscoelastic fluids, J. Fluid Mech. 33 (1968), 165-171.

8. Coleman, B. D., and H. Markovitz, Normal stress effects in second-order fluids. J. Appl. Phys. 35 (1964), 1-9.

9. Coleman, B. D., and H. Markovitz, Asymptotic relations between shear stresses and normal stresses in general incompressible fluids, J. Polymer Science: Polymer Physics Edition 12 (1974), 2195-2207.

10. Coleman, B. D., H. Markovitz, and W. Noll, Viscometric Flows of Non-Newtonian Fluids, Theory and Experiment, Springer-Verlag, Berlin, etc., 1966.

11. Coleman, B. D., and V. J. Mizel, Thermodynamics and departures from Fourier's law of heat conduction, Arch. Rational Mech. Anal. 13 (1963), 245-261.

12. Coleman, B. D., and V. J. Mizel, Existence of caloric equations of state in thermodynamics, J. Chem. Phys. 40 (1964), 1116-1125.

13. Coleman, B. D. and V. J. Mizel, Breakdown of laminar shearing flows for second-order fluids in channels of critical width, Z.A.M.M. 46 (1966), 445-448.

14. Coleman, B. D., and W. Noll, On certain steady flows of general fluids, Arch. Rational Mech. Anal. 3 (1959), 289-303.

15. Coleman, B. D., and W. Noll, Helical flow of general fluids, J. Appl. Phys. 30 (1959), 1508-1512.

16. Coleman, B. D., and W. Noll, An approximation theorem for functionals, with applications in continuum mechanics, Arch. Rational Mech. Anal. 6 (1960), 355-370.

17. Coleman, B. D., and W. Noll, Foundations of linear viscoelasticity, Rev. Mod. Phys. 33 (1961), 239-249; errata, ibid. 36, 1103.

18. Coleman, B. D., and W. Noll, Normal stresses in second-order viscoelasticity, Trans. Soc. Rheol. 5 (1961), 41-46.
19. Coleman, B. D., and W. Noll, Recent results in the continuum theory of viscoelastic fluids, Ann. N.Y. Acad. Sci. 89 (1961), 672-714.
20. Coleman, B. D., and W. Noll, Steady extension of incompressible simple fluids, Phys. Fluids 5 (1962), 840-843.
21. Coleman, B. D., and W. Noll, The thermodynamics of elastic materials with heat conduction and viscosity, Arch. Rational Mech. Anal. 13 (1963), 167-178.
22. Dunn, J. E., and R. L. Fosdick, Thermodynamics, stability, and boundedness of fluids of complexity 2 and fluids of second grade, Arch. Rational Mech. Anal. 56 (1974), 191-252.
23. Fosdick, R. L., and K. R. Rajagopal, Anomalous features in the model of "second order fluids", Arch. Rational Mech. Anal. 70 (1979), 145-152.
24. Joseph, D. D., G. S. Beavers, A. Cers, C. Dewald, A. Hoger, and P. T. Than, Climbing constants for various liquids, J. Rheology 28 (4) (1984), 325-345.
25. Langlois, W. E., Steady flow of a slightly visco-elastic fluid between rotating spheres, Quart. Appl. Math. 21 (1963), 61-71.
26. Lodge, A. S., A network theory of flow and stress in concentrated polymer solutions, Trans. Faraday Soc. 52 (1956), 120-130.
27. Markovitz, H., Normal stress measurements on polymer solutions, Proc. 4th Intl. Congress Rheol. 1 (1963), 189-212.
28. Markovitz, H., and D. R. Brown, Parallel plate and cone-plate normal stress measurements on polyisobutylene-cetane solutions, Trans. Soc. Rheol. 7 (1963), 137-154.
29. Markovitz, H., and B. D. Coleman, Incompressible second-order fluids. Adv. Appl. Mech. 8 (1964) 69-101.
30. Markovitz, H., and B. D. Coleman, Nonsteady helical flows of second-order fluids, Phys. Fluids 7, (1964) 833-841.

31. Noll, W., A mathematical theory of the mechanical behavior of continuous media, Arch. Rational Mech. Anal. 2 (1958/59), 197-226.

32. Noll, W. Motions with constant stretch history, Arch. Rational Mech. Anal. 11 (1962), 97-105.

33. Rivlin, R. S., and J. L. Ericksen, Stress-deformation relations for isotropic materials, J. Rational Mech. Anal. 4 (1955), 323-425.

34. Ting, T. W., Certain non-steady flows of second-order fluids, Arch. Rational Mech. Anal. 14 (1963) 1-26.

35. Truesdell, C., and W. Noll, *The Non-Linear Field Theories of Mechanics*, Vol. III/3 of the Encyclopedia of Physics, Springer-Verlag, Berlin, etc., 1965.

I am grateful to T. Wicks, H. Markovitz, and K. R. Rajagopal for helpful discussions about the subjects covered here. During the months that I was writing this paper, my research was supported in part by the National Science Foundation under Grant MCS-82-02647 and in part by Sandia National Laboratories under Contract 21-2953.

A COMPARISON OF MOLECULAR AND NETWORK-CONSTITUTIVE THEORIES FOR POLYMER FLUIDS

Hanswalter Giesekus

Department of Chemical Engineering
University of Dortmund
Dortmund, West Germany

1. INTRODUCTION

As is well known there exist two essentially different approaches for constructing constitutive theories of polymer fluids. In the first approach the polymer molecules are idealized by chain-like structures consisting of "beads" joined by either rigid or elastic connections termed "rods" and "springs" respectively. This approach, in the following somewhat non-specifically called the *molecular* approach, was originally only used to describe the behaviour of dilute polymer solutions but it has since been generalized to model also concentrated polymer solutions and even undiluted polymer systems.

However, this last type of polymer system had been described much earlier by an approach derived from the theory of rubber elasticity based on the model of a cross-linked polymer network. In order to describe polymer fluids it was only necessary to substitute the permanent junctions between the molecular chains (due to cross-links in the original model) by transient junctions arising due to different bonding mechanisms but usually summed up under the designation "entanglements". In the following this network approach is compared with the afore-mentioned molecular approach to determine their relative advantages and disadvantages for modelling the viscoelastic flow properties of polymer fluids.

2. COMPARISON OF THE SIMPLEST REPRESENTATIVES OF MOLECULAR AND TRANSIENT NETWORK MODELS

A dilute suspension of Hookean dumbbells (figure 1) is assumed to be the simplest non-trivial representative of molecular models[1]. The number of dumbbells per unit volume is n, the distance vector between the centres of the beads $\underline{R}$ and the drag coefficient of each bead with respect to the motion relative to the surrounding fluid ζ. A distribution function $\psi(\underline{R},t)$ exists denoting the number of dumbbells in the infinitesimal cubic volume between the Cartesian components R_i and $R_i + dR_i$ ($i = 1,2,3$) at time t, so that

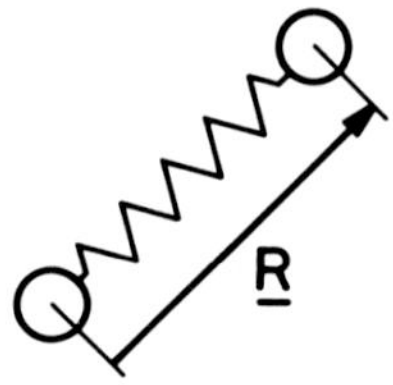

Fig. 1. Elastic dumbbell molecule

$$\int \Psi(\underline{R},t)\,d\underline{R} = n \tag{1}$$

where $d\underline{R} = dR_1 dR_2 dR_3$ and the integration is extended over the full $\underline{R}$-space.

This distribution function obeys the so-called diffusion equation

$$\frac{\partial \Psi}{\partial t} + \frac{\partial}{\partial \underline{R}} \cdot (\dot{\underline{R}}\Psi) = 0 \tag{2}$$

due to the fact that no dumbbells are generated or annihilated.

The average macroscopic motion at the position $\underline{x}$ at time t is given by $\underline{v}(\underline{x},t)$ and its gradient

$$\left[\frac{\partial}{\partial \underline{x}}\, \underline{v}(\underline{x},t)\right]^{\dagger} = \underline{\underline{L}} = \underline{\underline{D}} + \underline{\underline{W}} , \tag{3}$$

is assumed to be constant within a microscopic neighbourhood, where $\underline{\underline{D}}$ and $\underline{\underline{W}}$ designate its symmetric and antisymmetric parts i.e the rate of deformation and the rate of rotation tensors, and the dagger ($\dagger$) signifies the transpose of a tensor.
Using equation (3) the derivative $\dot{\underline{R}}$ can be decomposed into a force-free term and a term proportional to the forces acting on the beads:

$$\dot{\underline{R}} = \underline{\underline{L}} \cdot \underline{R} + \frac{2}{\zeta}\, \underline{F} . \tag{4}$$

---
[1] For more details see, for example, the monograph by Bird et al.[1].

Here $\underline{F}$ is the force acting on that bead to which the vector $\underline{R}$ points. Substituting this expression in equation (2) results in

$$\frac{\partial\Psi}{\partial t} + \frac{\partial}{\partial \underline{R}}\cdot(\underline{\underline{L}}\cdot\underline{\underline{R}}\Psi) = -\frac{2}{\zeta}\frac{\partial}{\partial\underline{R}}\cdot(\underline{F}\Psi) \ . \tag{5}$$

The force $\underline{F}$ can be described as the negative gradient of the differential Helmholtz free energy $\Phi(R) + kT \ln \Psi$, with $\Phi(R)$ being the spring potential depending only on the magnitude $R = |\underline{R}|$, and $kT \ln \Psi$ the Brownian motion term calculated from momentum transfer in the near-equilibrium state. In the special case of Hookean dumbbells one has

$$\Phi(R) = \frac{H}{2}\ \underline{R}\cdot\underline{R} \tag{6}$$

with H as the "spring constant". This results in

$$\underline{F} = -\frac{\partial}{\partial\underline{R}}\left(\frac{H}{2}\ \underline{R}\cdot\underline{R} + kT \ln\Psi\right) \ . \tag{7}$$

Substituting this in equation (5), the basic differential equation for the distribution function $\Psi(\underline{R},t)$ is obtained:

$$\frac{\partial\Psi}{\partial t} + \frac{\partial}{\partial\underline{R}}\cdot(\underline{\underline{L}}\cdot\underline{R}\Psi) = \frac{2}{\zeta}\left[\frac{H}{2}\ \frac{\partial}{\partial\underline{R}}\cdot\frac{\partial}{\partial\underline{R}}\ (\underline{R}\cdot\underline{R}\Psi) + kT\ \frac{\partial}{\partial\underline{R}}\cdot\frac{\partial\Psi}{\partial\underline{R}}\right] \ . \tag{8}$$

However, this equation does not need to be solved because after multiplication with the tensorial product $\underline{R}\underline{R}$, integration over the full space of $\underline{R}$, and some straightforward calculations a very much simpler equation is obtained:

$$\frac{\mathcal{D}}{\mathcal{D}T}\left\langle \underline{R}\underline{R}\right\rangle = -\frac{4}{\zeta}\left[H\left\langle \underline{R}\underline{R}\right\rangle - kT\ \underline{\underline{1}}\right] , \tag{9}$$

where the arrow-shaped brackets designate ensemble averages and $\mathcal{D}/\mathcal{D}t$ denotes the so-called "Upper Oldroyd Derivative" defined for a tensor $\underline{\underline{T}}$ by

$$\frac{\mathcal{D}\underline{\underline{T}}}{\mathcal{D}t} = \frac{D\underline{\underline{T}}}{Dt} - (\underline{\underline{L}}\cdot\underline{\underline{T}} + \underline{\underline{T}}\cdot\underline{\underline{L}}^{\dagger}) \ . \tag{10}$$

For the equilibrium state ($\underline{\underline{L}} \equiv 0$) the tensor $\left\langle \underline{R}\underline{R}\right\rangle$ becomes isotropic

$$\left\langle \overset{o}{\underline{R}}\overset{o}{\underline{R}}\right\rangle = \frac{1}{3}\left\langle \overset{o}{\underline{R}}\cdot\overset{o}{\underline{R}}\right\rangle \underline{\underline{1}} = \frac{kT}{H}\ \underline{\underline{1}} \ . \tag{11}$$

The last part of this expression follows from equation (9), the left-hand side of which has to vanish in the case of equilibrium.

Let the mapping of $\left\langle \overset{o}{\underline{R}}\overset{o}{\underline{R}}\right\rangle$ into $\left\langle \underline{R}\underline{R}\right\rangle$ be described by a configuration tensor $\underline{\underline{b}}$ as follows:

$$\langle \underline{R}\underline{R} \rangle = \underline{\underline{b}} \cdot \langle \overset{o}{\underline{R}}\overset{o}{\underline{R}} \rangle = \frac{1}{3} \langle \overset{o}{\underline{R}} \cdot \overset{o}{\underline{R}} \rangle \underline{\underline{b}} \ . \tag{12}$$

Using this expression, equation (9) can be written as

$$\frac{\mathcal{D}\underline{\underline{b}}}{\mathcal{D}t} + \frac{1}{\lambda} (\underline{\underline{b}} - \underline{\underline{1}}) = 0, \qquad \lambda = \frac{\zeta}{4H} \ . \tag{13}$$

Furthermore, it can easily be derived that the tensor of the excess stresses, resulting from the drag of the dumbbells, is given by

$$\underline{\underline{S}} = - n \langle \underline{R}\underline{F} \rangle = G(\underline{\underline{b}} - \underline{\underline{1}}) \tag{14}$$

with

$$G = n \frac{H}{3} \langle \overset{o}{\underline{R}} \cdot \overset{o}{\underline{R}} \rangle = n \, kT \ . \tag{15}$$

Introducing this relation into equation (13) and noting that $\mathcal{D}\underline{\underline{1}}/\mathcal{D}t = - 2 \, \underline{\underline{D}}$, the constitutive equation

$$\underline{\underline{S}} + \lambda \frac{\mathcal{D}\underline{\underline{S}}}{\mathcal{D}t} = 2\eta\underline{\underline{D}} \ , \qquad \text{with} \quad \eta = G\lambda \tag{16}$$

is obtained, which corresponds to the "Upper Convected Maxwell Fluid".

This result has been well known for about 40 years. Its derivation has been sketched here only for the sake of detailed comparison with the corresponding result of the network approach, which is presented below.

The simplest transient network model is the one first introduced by Green and Tobolsky [2] [2]. As is indicated in figure 2 there are "junctions" connected by elastic "segments" (or "strands"). These are idealized as Hookean springs with their length and orientation given by $\underline{R}'$ and they are distributed independently according to the distribution function $\Psi'(\underline{R}',t)$ obeying the condition

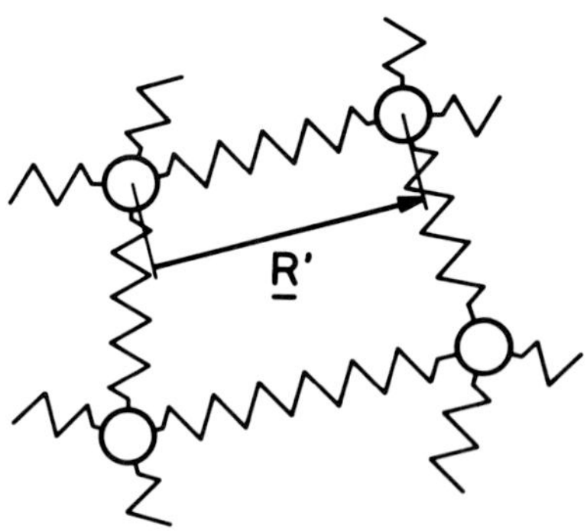

Fig. 2. Elastic network

---

[2] More details are given, for example, by Lodge et al.[3].

$$\int \Psi'(\underline{R}',t)d\underline{R}' = n'(t) \ . \tag{17}$$

$n'(t)$ is the number of segments per unit volume and in contrast to the number of dumbbells is generally considered to be time-dependent. Now, the basic assumption is that junctions are produced and dissociated leading to generation and loss rates of segments k and ℓ respectively. As a result $\Psi'(\underline{R}',t)$ changes not only because of a flux $\dot{\underline{R}}'\Psi'$ but also because of "sources" and "sinks":

$$\frac{\partial \Psi'}{\partial t} + \frac{\partial}{\partial \underline{R}'} \cdot (\dot{\underline{R}}'\Psi') = k - \ell \ . \tag{18}$$

In contrast to the motion of a dumbbell it is assumed that the junctions move "affinely", that is they follow the average motion of the continuum

$$\dot{\underline{R}}' = \underline{\underline{L}} \cdot \underline{R}' \ , \tag{19}$$

and that ℓ is proportional to Ψ', whereas k is proportional to the isotropic equilibrium distribution $\overset{0}{\Psi}{}'$:

$$\ell = \frac{1}{\lambda'} \Psi' \ , \qquad k = \frac{1}{\lambda'} \overset{0}{\Psi}{}' \ . \tag{20}$$

The equality of both constants of proportionality is a consequence of the fact that in the equilibrium state the rate of generation has to be equal of the rate of loss.

In this way the basic differential equation

$$\frac{\partial \Psi'}{\partial t} + \frac{\partial}{\partial \underline{R}'} \cdot (\underline{\underline{L}} \cdot \underline{R}'\Psi') = -\frac{1}{\lambda'} [\Psi' - \overset{0}{\Psi}{}'] \tag{21}$$

is obtained. Its structure is quite different from the corresponding equation (8) for the dumbbell suspension and it does not even allow for a definite solution without some additional assumption about the form of $\overset{0}{\Psi}{}'$. It is therefore all the more surprising that multiplication of this equation with the tensorial product $\underline{R}'\underline{R}'$, integration over the $\underline{R}'$-space and introduction of the segment-configuration tensor $\underline{\underline{b}}'$, defined completely analogously to the dumbbell-configuration tensor $\underline{\underline{b}}$ in equation (12), results in the equation

$$\frac{\mathcal{D}\underline{\underline{b}}'}{\mathcal{D}t} + \frac{1}{\lambda'}(\underline{\underline{b}}' - \underline{\underline{1}}) = 0 \ , \tag{22}$$

which has exactly the same structure as equation (13).

Up to this point no assumptions about forces and stresses have been introduced. If, however, the segments are assumed to act as Hookean dumbbells under the influence of spring and Brownian forces, relations analogous to the excess stresses of the dumbbell suspension, equations (14) and (15), are obtained:

$$\underline{\underline{S}} = G'(\underline{\underline{b}}' - \underline{\underline{1}}) \quad , \quad G' = n'kT \ . \tag{23}$$

Correspondingly the constitutive equation of an "Upper Convected Maxwell Fluid" is obtained once again.

This result, although also well known for about 30 years[3], has always been perceived as surprising. There are apparently two entirely different mechanisms, on the one hand the steady motions of beads relative to the surrounding continuum as a result of internal forces, and on the other hand the affine motion of network junctions but with a randomly occurring loss and regenation of segments. Recalling that the dumbbell springs are introduced to idealize reactions of more or less entropic nature in coiled polymer molecules which endeavour to occupy their most probable configuration, the dissimilarity seems somewhat less marked. However, the distinction remains that the dumbbell model covers that limiting case in which the irregular Brownian motion occurs on such a fine scale that it can be smoothed out in the ensemble resulting in a continuous motion, whereas in the network model this motion is in contrast assumed to occur in such big jumps that on the average each jump completely restores the equilibrium state of the respective segment[4].

With this in mind we can interpret the results of our comparison as follows: Despite the essential differences in the structure of the basic differential equations (8) and

[3] It was first discussed by Lodge [4] in a slightly different form.

[4] Here a slightly different interpretation of the process of loss and regenation of junctions is used in which at least the identity of the segments is retained. This presupposes that the frequency of both processes is equal, independent of the state of deformation.

(21), which reflect the differences in the individual processes, the procedure of ensemble averaging, by which the configuration and stress tensors are obtained, removes these differences completely, as is seen from equations (13) and (22). Of course, in the case of a dilute dumbbell suspension the stress due to the viscosity of the surrounding fluid has to be added and this has no counterpart in the case of an elastic network structure. However, our analysis encourages the application of molecular models, of which the dumbbell model is a simple example, to describe not only dilute solutions but also moderately dilute or even undiluted polymer fluids as well. The recognition emerges that molecular and network theories could indeed be able to model the same material from two more or less equivalent points of view. In the molecular models the structural units ("beads") and their motion are the basic notions, a certain number of which are always trapped in entanglements or being released. In constrast, in the network theories the entanglements and their development are the basic notions, but, of course, these are nothing more than the interactions of molecular units.

However, there are also some essential distinctions between the theories: Polymer molecules as a whole as well as their monomeric units are invariable quantities, whereas junctions are not. Molecular models, at least in their present form, have to smooth interaction of the structural units with all elements of the surrounding polymer molecules in terms of a drag coefficient or the like. Network models, in contrast, restrict attention to pairwise interactions between two elements of different molecules. However, before being able to make a more detailed comparison, the two models discussed above have to be generalized.

3. MULTI-MODE MOLECULAR AND NETWORK MODELS

It is easy to understand that the above models are too simple to describe any finer details of the flow behaviour of either dilute or concentrated polymer fluids. One of the most obvious generalizations is to proceed from models with only one relaxation time λ to a multitude (or "spectrum") λ_j

of relaxation times. In the case of a dumbbell suspension this means that different types of dumbbells with concentrations n_j are present with either different spring constants or drag coefficients, of which every type is subjected to equations of the type presented above:

$$\underline{\underline{S}} = \sum_j \underline{\underline{S}}_j \ , \quad \underline{\underline{S}}_j = G_j(\underline{\underline{b}}_j - \underline{\underline{1}}), \ \frac{\mathcal{D}\underline{\underline{b}}_j}{\mathcal{D}t} + \frac{1}{\lambda_j}(\underline{\underline{b}}_j - \underline{\underline{1}}) = 0 \ . \quad (24)$$

Equivalent relations hold for a network with segments of different "complexity", leading to a "Lodge Rubberlike Liquid" [5].

The above equations (24) are easily integrated and summed to give an explicit integral expression for $\underline{\underline{S}}$. This fluid is then called the Walters Fluid B' in analogy to a slight generalization of the Upper Convected Maxwell Fluid called the Oldroyd Fluid B. Of course, in this model the summing may be replaced by intergration over a continuous relaxation spectrum.

Furthermore, in the class of molecular models there also exists another generalization leading to a discrete relaxation spectrum, namely the transition from a Hookean dumbbell to a Rouse chain molecule [6]. This is the chain-like addition of, say, N + 1 beads connected by N Hookean springs with free draining links (figure 3). The derivation of the constitutive equation is more complicated in this case, see for example Bird et al. [1]. However, the result

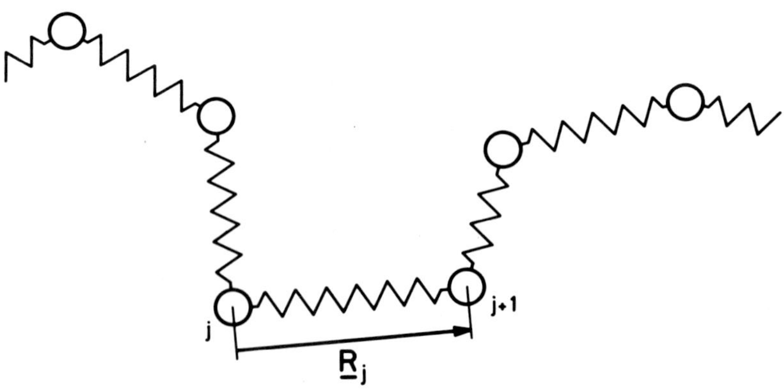

Fig. 3. Elastic chain molecule

can be formulated in terms of equations (24) with the specialization that with $j = 1, \ldots, N$ the moduli $G_j = G = nkT$ and the relaxation times obey a somewhat complicated relation, which for $j << N$ can be simplified to give

$$\lambda_j = \left(\frac{N+1}{j\pi}\right)^2 \frac{\zeta}{2H} \,. \tag{25}$$

In this case the $\underline{\underline{b}}_j$ are, however, not directly related to the vectors $\underline{R}_j$ from figure 3 but to the so-called normal coordinates $\underline{Q}_j$ obtained from the set $\underline{R}_j$ by a principal-axes transformation.

This generalization has no counterpart in the scope of network models. Segments cannot be modelled by Rouse chains, for this would mean that the beads between the junction points could contribute to the stresses by modes of motion that are not related to the motion of the end points themselves.

4. MOLECULAR MODELS WITH HYDRODYNAMIC INTERACTION AND NETWORK MODELS WITH NON-AFFINE MOTION

Linear-viscoelastic behaviour can be described with the above models by a suitable fit of the relaxation spectrum, whereas the predictions of the Rouse model describe experimental results over only a quite limited range. However, the most obvious shortcoming of these models is that they predict the shear viscosity and the first normal-stress coefficient in steady shear flows to be independent of the shear rate and the second normal-stress coefficient to be zero. Additionally, they predict infinite values of the extensional viscosity for finite rates of extensions in all types of steady extensional flow.

Most of these unrealistic predictions can in the scope of molecular models at least qualitatively be corrected by replacing the Hookean springs by non-linear springs that become stiffer with increasing stretching. In particular, this has been clearly shown with the FENE (finitely extendable non-linear elastic) dumbbell model of Warner [7]. However, even with this modification the second normal-stress coefficient continues to be zero. To overcome this, the influence of hydrodynamic interaction between the beads has to be taken into account. In principle this has been done for the bead-spring and bead-rod models in the theory of Kirkwood and Riseman [8] by the introduction of the so-called Oseen tensor. This theory is, however, very unwieldly for application to real problems and only a few simple models have been evaluated with full consideration of hydrodynamic interaction, e.g. rigid dumbbells [9], shishkebab structures [10, 11], and (along the same lines) spheroids [9].

As a full consideration of hydrodynamic interaction was not practicable in more complicated models, an equilibrium-averaged hydrodynamic interaction was introduced. The most famous application is Zimm's modification of the Rouse model [12], which indeed allows for some more realistic predictions in dilute polymer solutions. However, this preaveraging procedure leads back to a constitutive equation of the type discussed above with only a modified relaxation spectrum. Thus it does not allow for a shear-dependent viscosity or first normal-stress coefficient, nor a non-zero second normal-stress coefficient or finite limiting value of the extensional viscosity.

A more successful approach, though not based on real molecular ideas, was proposed by Gordon and Schowalter [13]. Starting from the phenomenological theory of anisotropic fluids established by Ericksen, they arrived at a modification of the Hookean dumbbell model by substituting the average velocity gradient $\underline{\underline{L}}$, cf. equation (3), by a reduced gradient

$$\hat{\underline{\underline{L}}} = \underline{\underline{L}} - \xi\underline{\underline{D}} \tag{26}$$

in the diffusion equation (5). This results in a substitution of the upper Oldroyd derivative in the subsequent equations, in particular equations (13) and (16), by a combined Oldroyd derivative[5]

[5] This is sometimes also called the Gordon-Schowalter derivative, but it was earlier introduced (though in an implicit manner) by Oldroyd, cf. [14].

$$\frac{\mathbb{D}\underline{\underline{T}}}{\mathbb{D}t} = \frac{\mathcal{D}\underline{\underline{T}}}{\mathcal{D}t} + \xi(\underline{\underline{D}}\cdot\underline{\underline{T}} + \underline{\underline{T}}\cdot\underline{\underline{D}}), \quad 0 \leq \xi \leq 1 \quad . \tag{27}$$

Indeed, this model predicts shear-dependent viscosity, first and even second normal-stress coefficients, the negative ratio of the latter being given by $\xi/2$.

Although Gordon and Schowalter do not propose their generalization as an effect of hydrodynamic interaction, their substitution of $\underline{\underline{L}}$ by $\hat{\underline{\underline{L}}}$ can be interpreted as an attempt to include flow hampering resulting from hydrodynamic interaction. The fluid surrounding the two beads of the dumbbell no longer moves with the velocity difference $\underline{\underline{L}}\cdot\underline{R}$ but with a reduced difference $\hat{\underline{\underline{L}}}\cdot\underline{R}$.

This substitution, introduced by the above authors in the theory of dilute polymer solutions, has been adopted in the network theories independently by Segalman [15] and by Phan-Thien and Tanner [16, 17]. The reasoning is here somewhat different, namely to describe the effect of "non-affine motion" or "effective slip" of the junctions. The correspondence to the molecular theories is as follows: Polymer molecules, such as dumbbells, move trivially in a non-affine manner, as it is just the non-affine part of the motion which is connected with the internal forces. Non-trivial non-affinity refers therefore to the motion of the fluid surrounding the beads and just means substitution of the tensor $\underline{\underline{L}}$, defining affine motion, by a modified tensor $\hat{\underline{\underline{L}}}$.

In dilute solutions it is easy to understand that flow hampering occurs, although the special supposition (26) may be questionable. It is likewise obvious that in concentrated systems the motion of either the material surrounding a bead or network junctions is not strictly affine, but a systematic (i.e. not only random) deviation demands the assumption of some micro-inhomogeneity. An example is sketched for an assembly of only partly interpenetrating polymer coils in figure 4. If penetration only occurs in the periphery of the coils, then in the central regions (designated by the hatching) the

Fig. 4. Partly penetrating polymer coils

velocity gradient is certainly reduced. Because this gradient must on average be equal to the macroscopic value $\underline{\underline{L}}$ it must correspondingly be increased in the peripheral regions. If, however, the density of polymer beads or junctions there is smaller than in the central regions a correspondingly weighted average $\hat{\underline{\underline{L}}}$ should be smaller than $\underline{\underline{L}}$ [6].

By taking into account respectively flow hampering and non-affine motion of the junctions by means of the substitution (27), these constitutive equations provide more realistic predictions than the original ones but they also show some peculiarities in more complicated flow situations, cf. for example Larson [18], and with respect to flow stability. This leads to the question of whether the concept of flow hampering could be introduced in a modified way allowing a clearer physical interpretation. This author [14] has recently proposed that instead of equation (26) the following reduced velocity gradient be substituted

$$\hat{\underline{\underline{L}}} = \underline{\underline{L}} - \xi^* \underline{\underline{D}}_{(p)} , \qquad (28)$$

where $\underline{\underline{D}}_{(p)}$ is the irreversible portion of the velocity gradient (which can be shown to necessarily be a symmetric tensor) defined according to Leonov [19]:

$$\frac{\mathcal{D}\underline{\underline{b}}}{\mathcal{D}t} = -\left(\underline{\underline{D}}_{(p)} \cdot \underline{\underline{b}} + \underline{\underline{b}} \cdot \underline{\underline{D}}_{(p)}\right) . \qquad (29)$$

Indeed, this leads to a more general objective time derivative $\mathbb{D}^*/\mathbb{D}t$ by which the force-free motion of a particle within a deformed molecular structure is described in a much more plausible manner than by $\mathbb{D}/\mathbb{D}t$, the former containing this derivative as a limiting case for a non-deformable structure, i.e. $\underline{\underline{b}} \equiv \underline{\underline{1}}$, only. However, as can be derived immediately from equation (29), in the case of application to the configuration tensor $\underline{\underline{b}}$ itself this derivative degenerates to the upper Oldroyd derivative

---

6 This simple reflection suggests the hypothesis that deviation from affine motion should be larger in dilute and only moderately concentrated solutions (in particular with poor solvents) than in undiluted systems.

apart from a "shielding factor" $(1 - \xi^*)$,

$$\frac{\mathbb{D}^* \underline{\underline{b}}}{\mathbb{D}t} = (1 - \xi^*) \frac{\mathcal{D}\underline{\underline{b}}}{\mathcal{D}t} , \tag{30}$$

this shielding factor causing a modification of only the relaxation time $\lambda^* = (1 - \xi^*)\lambda$.

Of course, this would be a disappointing result if no alternatives existed to obtain a constitutive model covering particle interaction by the use of a more plausible physical hypothesis. However, such an hypothesis exists a least for molecular models and is realized by the concept of configuration-dependent tensorial mobility. This means that the constant scalar drag coefficient ζ, introduced in equation (4), is substituted by a tensorial drag coefficient $\underline{\underline{\zeta}}$ which degenerates to an isotropic tensor only in the case of the equilibrium configuration

$$\overset{o}{\underline{\underline{\zeta}}} = \overset{o}{\zeta}\, \underline{\underline{1}} . \tag{31}$$

This substitution can be achieved in at least two different ways. The first starts from the idea that a polymer molecule finds it easier to move in its own direction than perpendicular to it; metaphorically one can say that it prefers to move like a reptile. This results in $\underline{\underline{\zeta}}$ depending on the configuration of the respective molecule itself. In the simplest case of a chain-like molecule the drag of the j-th bead has to depend on the two neighbouring configuration vectors $\underline{R}_j$ and $\underline{R}_{j+1}$. This concept has been used by Curtiss and Bird [20] for the model of a free-draining bead-rod chain, whereas an earlier approach to a "reptational theory" is connected with the names of De Gennes [21] and Doi and Edwards [22]. Actually these models predict all the above mentioned properties, including the existence of a second normal-stress difference, at least in a qualitative way. It seems, however, to be better suited for undiluted systems of polymer molecules with a simple chemical structure than for melts of more complex (e.g. branched) polymers or moderately concentrated polymer solutions.

In the latter case a modification of the above concept, which has been proposed by this author [23 - 26], seems to be more appropriate. This makes allowance for the fact that the polymer structure *as a whole* is made anisotropic by the deformational motion and ζ thus depends essentially on the *mean* configuration as characterized by the configuration tensor $\underline{\underline{b}}$. With this assumption it is very easy to generalize the derivation of the simple dumbbell model. After introduction of the relative mobility tensor

$$\underline{\underline{\zeta}}^{-1} = \overset{o}{\zeta}^{-1} \underline{\underline{\beta}} \quad \text{with} \quad \overset{o}{\underline{\underline{\beta}}} = \underline{\underline{1}} \tag{32}$$

one obtains instead of equation (13):

$$\frac{\mathcal{D}\underline{\underline{b}}}{\mathcal{D}t} + \frac{1}{2\overset{o}{\lambda}} [\underline{\underline{\beta}}\cdot(\underline{\underline{b}} - \underline{\underline{1}}) + (\underline{\underline{b}} - \underline{\underline{1}})\cdot\underline{\underline{\beta}}^{\dagger}] = 0, \; \lambda = \overset{o}{\zeta}/4H. \tag{33}$$

If $\underline{\underline{\beta}}$ is assumed to be a function of $\underline{\underline{b}}$ only at the same time, this tensor must be symmetric and also commutable with $\underline{\underline{b}}$, so that a simpler equation

$$\frac{\mathcal{D}\underline{\underline{b}}}{\mathcal{D}t} + \frac{1}{\overset{o}{\lambda}} \underline{\underline{\beta}}\cdot(\underline{\underline{b}} - \underline{\underline{1}}) = 0 \tag{34}$$

results.

Using the linear substitution

$$\underline{\underline{\beta}} - \underline{\underline{1}} = \alpha(\underline{\underline{b}} - \underline{\underline{1}}), \quad 0 \le \alpha \le 1 , \tag{35}$$

this constitutive equation has been applied to a number of simple steady and transient flow situations, and it has been found that the respective predictions show qualitative agreement with most experimental observations. It predicts a shear rate dependent viscosity and first and second normal-stress coefficients (of which the initial ratio is given by $-\alpha/2$) in steady shear flow as well as finite limiting values of extensional viscosity in steady extensional flow. In addition, so-called stress-overshoot in the start-up flow is predicted for simple shear but not for simple extension [25, 26]. It should be mentioned that just recently Volkov and Vinogradov [27] have arrived at exactly the same model using somewhat different reasoning.

This one-mode model can be generalized to a multi-mode model in a straightforward manner. Here the only problem that arises is whether the tensors $\underline{\underline{\beta}}_j$ should be assumed to depend solely on the respective $\underline{\underline{b}}_j$ or on the full set of configuration tensors $\underline{\underline{b}}_k$. The first does not complicate the model significantly, whereas the second, for which however more plausible arguments exist, leads to a coupling of the whole system of differential equations resulting in an enormous increase in the work required for evaluation.

It is evident that in network theories that are only based on the mechanism of destruction and regeneration of junctions there is no counterpart for anisotropic drag because no drag coefficient at all is introduced. However, in some recent concepts a generalization is found which also allows for quasisteady motions of junctions. The original meaning of "entanglement" provokes such a concept (figure 5), whereas a sudden linking and unlinking is based on the idea of formation and breakage of van-der-Waals bonds.

Of course, with this interpretation all ideas developed in the context of molecular theories can be transferred to the network theories, including that of configuration-dependent mobility, see for example Larson [18]. However, when

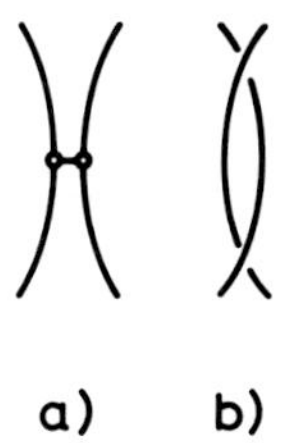

Fig. 5. Different types of transient junction:
a) rigid connection, which can only be broken,
b) entanglement allowing continuous sliding motion

using this generalized concept the notion of non-affine motion becomes ambiguous. Motion of junctions due to forces acting in the segments is trivially "non-affine" motion in the same sense as in the motion of beads; non-trivial "non-affine" motion should be understood to always presuppose micro-inhomogeneity, as discussed above.

5. NETWORK MODELS WITH TIME DEPENDENT JUNCTION DENSITY AND MOLECULAR MODELS WITH FUNCTIONAL-TYPE CONFIGURATION-DEPENDENT MOBILITY

The network models discussed so far were specialized to possess constant rates of loss and regeneration of junctions. However, it is an obvious generalization to assume these to be dependent on mean properties of the network as, for example, the segment configuration or the stress are. In the limiting case of equilibrium, of course, both rates must be equal. This generalization requires substitution of equations (20) by the expressions

$$\ell = h\Psi', \quad k = g(\overset{o}{\Psi}' - \Psi') - (p - h)\Psi' , \tag{36}$$

this nomenclature having been introduced by Mewis and Denn [28], where h,g and p are functions of $\underline{\underline{b}}'$ and $\overset{o}{p} = 0$. With these expressions the diffusion equation becomes

$$\frac{\partial\Psi'}{\partial t} + \frac{\partial}{\partial \underline{R}'} \cdot (\underline{\underline{L}} \cdot \underline{R}'\Psi') = - [g(\Psi' - \overset{o}{\Psi}') + p\Psi'] . \tag{37}$$

Integration over the full $\underline{R}'$ space results, because of equation (17), in

$$\dot{n}'(t) = - \{g[n'(t) - \overset{o}{n}'] + pn'(t)\} , \tag{38}$$

where $\overset{o}{n}'$ denotes the equilibrium density of junctions. If a relative density is introduced by the definition $x(t) = n'(t)/\overset{o}{n}'$, this can be written as

$$\dot{x} = -[g(x - 1) + px] . \tag{39}$$

If equation (37) is multiplied with $\underline{R}'\underline{R}'$, integrated and equation (39) is introduced, an evolution equation is obtained which looks quite similar to equation (22), namely

$$\frac{\mathcal{D}\underline{\underline{b}}'}{\mathcal{D}t} + \frac{g}{x} (\underline{\underline{b}}' - \underline{\underline{1}}) = 0 . \tag{40}$$

The only difference is that the constant relaxation time λ' is substituted by an expression x/g, which depends on the deformation history. This can be understood to be the "*configuration relaxation time*". It must, however, now be distinguished from the "*stress relaxation time*", as a substitution of equations (23) by

$$G' = n' kT = x \overset{o}{n}'kT = x\overset{o}{G}' \tag{41}$$

results in

$$[\underline{\underline{S}}/G'] + \frac{x}{g} \frac{\mathcal{D}}{\mathcal{D}t} [\underline{\underline{S}}/G'] = 2 \frac{x}{g} \underline{\underline{D}} \tag{42}$$

which with the use of equation (39) is transformed to give

$$\underline{\underline{S}} + \frac{1}{g+p} \frac{\mathcal{D}\underline{\underline{S}}}{\mathcal{D}t} = 2 \frac{G'}{g+p} \underline{\underline{D}} . \tag{43}$$

This constitutive equation looks quite similar to an upper convected Maxwell equation but with a time dependent "stress relaxation time" $\lambda' = (g+p)^{-1}$ and a time dependent viscosity $\eta' = G'/(g+p)$.

This model can be generalized without any essential change to a multi-mode model. $\mathcal{D}/\mathcal{D}t$ may also be substituted by the combined Oldroyd derivative $\mathbb{D}/\mathbb{D}t$ with the only change being that $\eta = (1-\xi)\, G'/(g+p)$.

A simplification is obtained by introducing the "pseudo-steady state" assumption $\dot{x} \approx o$, from which it follows that

$$x = \frac{g}{g+p}, \quad n' = \frac{g}{g+p}\, \overset{o}{n}' \; . \tag{44}$$

Because g and p are functions of $\underline{\underline{b}}$ or $\underline{\underline{S}}$ this results in $n' = n'(\underline{\underline{b}})$ or $n'(\underline{\underline{S}})$, where this notation means that these scalar quantities depend on the invariants of either $\underline{\underline{b}}$ or $\underline{\underline{S}}$. As, at the same time, $\lambda = (g+p)^{-1}$ is a function of $\underline{\underline{b}}$ or $\underline{\underline{S}}$, the above result (43) can be written (using the second alternative) as

$$\underline{\underline{S}} + \lambda(\underline{\underline{S}})\, \frac{\mathcal{D}(\underline{\underline{S}})}{\mathcal{D}t} = 2\,\eta(\underline{\underline{S}})\, \underline{\underline{D}} \quad , \tag{45}$$

which is of the type proposed by Kaye [29]. It should be noted that this equation predicts a disappearing second normal-stress difference. This shortcoming is, however, removed if $\mathcal{D}/\mathcal{D}t$ is substituted by $\mathbb{D}/\mathbb{D}t$, as is done in the Phan-Thien-Tanner models [16, 17]. Together with some specializations of the dependence of λ and η on $\underline{\underline{S}}$, this results in

$$\sigma(\mathrm{Tr}\, \underline{\underline{S}})\, \underline{\underline{S}} + \overset{o}{\lambda} \frac{\mathbb{D}\underline{\underline{S}}}{\mathbb{D}t} = 2\overset{o}{\eta}\underline{\underline{D}} \; . \tag{46}$$

Here, for convenience, only the one-mode form is given, whereas these authors formulate their model as a multi-mode model.

There are, however, also models which do not apply the "pseudo-steady state" approach. Best known of these is the Marrucci model [30, 31] which (if for the sake of simplicity it is once again written in the one-mode form) corresponds exactly to equation (42) with the configuration relaxation time given by the empirical relation

$$x/g = \overset{o}{\lambda}{}' \; x^{1.4} \tag{47}$$

and the evolution equation for the relative density of junctions by a semi-empirical equation

$$\dot{x} = \frac{1}{\overset{o}{\lambda}{}'} \left[(1-x)\; x^{-1.4} - a\; x^{-0.4} \sqrt{\frac{1}{2}\,\mathrm{Tr}\,(\underline{\underline{S}}/G')} \right] . \tag{48}$$

The corresponding expression for the excess rate of junction loss is also quite complicated

$$p = \frac{x^{-1.4}}{\overset{o}{\lambda}{}'} \left[a\sqrt{\frac{1}{2}\,\mathrm{Tr}\,(\underline{\underline{S}}/G')} - \frac{(1-x)^2}{x} \right] . \tag{49}$$

It should be remembered that because $\underline{\underline{S}}/G' = \underline{\underline{b}}' - \underline{\underline{1}}$ the network configuration tensor is the basic variable in these equations; this point seems to have been overlooked in the discussion in Mewis and Denn [28].

The Marrucci model not only predicts most phenomena in steady flows quite satisfactorily but also gives a fairly good description of start-up phenomena. More precisely, it predicts stress-overshoot in start-up flows not only in shear but also in extension. However, as can be seen from inspection of equation (42), it does not predict a non-zero second normal-stress difference.

This model is the first type of constitutive model discussed here in which the stress depends not only on the present configuration but also on the configuration history because the number of segments varies under the deformational motion. Thermodynamically this means that the number of degrees of freedom which contribute to the macroscopic stresses is no longer constant but decreases with increasing configuration and stress anisotropy respectively.

Returning to molecular models once again it is difficult to imagine time dependent moduli because here the number of beads and springs or rods is supposed to be constant.
At first sight the mobility tensor should also only depend on the present configuration. However, it must be borne in mind that the real state of configuration is by no means fully described by the configuration tensor $\underline{\underline{b}}$. This is particularly so in the extremely crude approximation of the one-mode model, because this quantity only contains the

second-order moment of the distribution function but no information at all on, for example, fourth-order moments or correlations between neighbouring configuration vectors $\underline{R}_j$ and $\underline{R}_{j+k}$, which may also influence the kinetics of the motion of beads. As a result it seems more realistic to consider that $\underline{\underline{b}}$ no longer characterizes the complete momentary configuration but rather only the basic configuration associated with the large structures which convey the stresses. At the same time it can be considered to represent a quasi-equilibrium state to which the instantaneous configuration of the structures tends under the counteracting effects of distortion by the bulk deformational motion and relaxation. This means that the relative mobility $\underline{\underline{\beta}}$ should no longer be considered to be a function of the present value of the configuration tensor $\underline{\underline{b}}$ but as a functional of its history.

To obtain a practicable model it seems most appropriate to test a relaxation-type dependence between the anisotropic parts of $\underline{\underline{\beta}}$ and $\underline{\underline{b}}$ which contains the linear relation (35) as the limiting case:

$$(\underline{\underline{\beta}}-\underline{\underline{1}}) + \overset{o}{\varkappa}\,\frac{\mathcal{D}(\underline{\underline{\beta}}-1)}{\mathcal{D}t} = \alpha(\underline{\underline{b}}-\underline{\underline{1}}) \ . \tag{50}$$

Here $\overset{o}{\varkappa}$ designates the retardation of the mobility behind the present configuration of the basic structure. The choice of the upper Oldroyd derivative is of course arbitrary but is stimulated by the favoured role that it plays throughout this theory[7]. Since $\underline{\underline{\beta}}$ and $\underline{\underline{b}}$ are no longer commutable, the evolution equation has to be chosen in the form of equation (33) instead of equation (34) though $\underline{\underline{\beta}}$ should be understood to remain symmetric. This generalized theory contains only one additional constant which is appropriately made dimensionless with the relaxation time $\overset{o}{\lambda}$ to give $L = \overset{o}{\varkappa}/\overset{o}{\lambda}$. Indeed, as has been discussed quite recently [32, 33], this model allows prediction of all the well-known properties of concentrated polymer systems including steady extensional flow with a maximum viscosity at a certain rate of uniaxial

[7] Moreover it can be shown that substitution by the corotational derivative leads to unacceptable predictions [32].

extension and stress overshoot in the respective start-up curve. In the start-up curves for shear flow the amount of stress-overshoot is intensified and is no longer solely dependent on the anisotropy coefficient α.

Of course, equation (50) is not believed to give the best possible fit to experimental results but is chosen above all because of its simplicity. Undoubtedly by inclusion of terms with the invariants of $\underline{\underline{b}}$ (in analogy to the Phan-Thien-Tanner or Marruci equations) or by generalization to more than one mode better approximation would be obtained. However, the manipulation would then become more and more cumbersome and eventually it would no longer be at all clear whether the newly introduced constants had a real physical meaning or were only fitting coefficients.

6. CONCLUSIONS

The above comparison of molecular and network theories has been undertaken with many simplifications and the omission of numerous important ideas. The aim was to establish the similarities as well as the differences between the two types of models. First it was shown that there are far-reaching analogies between the two types which allow the establishment of a unified approach. To every material point or "particle" of the macro-continuum a set of interpenetrating (statistical) micro-continua can be associated of which the instantaneous configuration is described by configuration tensors and possibly additional scalar density-type quantities. These all depend on the deformation history in a macroscopically infinitesimal surrounding of the respective particle. The stress tensor or equivalently the material functions contained in the constitutive model (moduli, viscosities or relaxation time spectra) are understood to be either functions or functionals of the quantities characterizing the configuration. Notwithstanding the functional character of the constitutive model, a relatively small number of constants are sufficient to characterize the respective material. It was not mentioned above, but all constitutive models must of course also obey the basic theorems of continuum mechanics and thermodynamics.

This is strongly emphasized in the continuum theories of Leonov [19], Dashner and VanArsdale [34] and, quite recently, Kwon and Shen [35].

In contrast, the second intention was to elucidate the most important differences between the two classes of theories. The crucial distinction is the fact that the objects of the molecular theories are vectors connecting structural units or their representatives (beads) which retain their identity, whereas those of network theories are vectors connecting junctions which consist of pairs of selected structural units with limited lifetimes and a density that depends on the state of deformation. However, the most important aspect is the interdependence (not mentioned earlier) of the creation and loss of junctions of different complexity as a consequence of stoichiometric constraints. As a result each broken junction means at the same time the loss of two shorter segments and the creation of a new one and vice versa. Another point is the problem of loose ends, the influence of which, if considered at all, has to be taken into account by the formalism of the molecular models, cf. Jongschaap et al. [36].

In addition, the shorter relaxation modes, which are determined by the cooperative motions of short segments or side groups, are not described by the loss and regeneration process. On the other hand, if network theories are generalized to include steady motion of junctions besides impulsive breakage and regeneration, the flexibility for describing different interaction mechanisms is considerably increased.

The shortcomings of the molecular theories in their present state are due to the need to completely smooth out the influence of interaction of the neighbouring particles by, in its to date most general form, a configuration-dependent mobility. Here, only the two extreme situations were analysed, the mobility being solely dependent on either the configuration of the respective part of the molecule under consideration or the mean configuration of all molecules, whereas the real situation is certainly intermediate between these two.

Finally, consideration should be given to the deformation-dependent modulus, which is essential in the most elaborate network theories and implies the assumption of variance of the degrees of freedom contributing to the stress. A question worthy of further reflection seems to be under which conditions is the assumption of a deformation-dependent modulus adequate for describing the transfer of forces, or is the variation of the drag resistance with a constant modulus physically more satisfactory.

If this author were asked which type of model he prefers and considers to have the better future, he would not hesitate to opt for the molecular theories. However, this would not be the opinion of an independent referee but rather that of an involved advocate with the intention of challenging the supporters of the network theories to show the superiority of their favourite.

REFERENCES

1. Bird, R.B., O. Hassager, R.C. Armstrong, and C.F. Curtiss, *Dynamics of Polymeric Liquids. Vol.* II. *Kinetic Theory.* John Wiley & Sons (New York, 1977).

2. Green, M.S., and A.V. Tobolsky, J. Chem. Phys. 14, 80 (1946).

3. Lodge, A.S., R.C. Armstrong, M.H. Wagner, and H.H. Winter, Pure & Appl.Chem. 54, 1349-59 (1982).

4. Lodge, A.S., in: S. Onogi (ed.), *Proc.* V. *Intern. Congr. Rheology Kyoto 1968, Vol. 4, pp. 169-178,* University of Tokyo Press (Tokyo, 1970).

5. Lodge, A.S. Trans. Faraday Soc. 50, 120 (1956).

6. Rouse, Jr., P.E., J. Chem. Phys. 21, 1272-80 (1953).

7. Warner, H.R., Ind. Eng. Chem. Fundamentals 11, 379-87 (1972).

8. Kirkwood, J.G., and J. Riseman, J. Chem. Phys. 16, 565-73 (1948); 22, 1626-27 (1954); J. Riseman and J.G. Kirkwood, in: F.R. Eirich (ed.), *Rheology, Vol.* I, *Chapt. 16, pp. 662-66,* Academic Press (New York, 1956).

9. Giesekus, H., Rheol. Acta 2, 50-62 (1962).

10. Kotaka, T., J. Chem. Phys. 30, 1566-67 (1959).

11. Bird, R.B., and C.F. Curtiss, J. Non-Newtonian Fluid Mech. 14, 85-101 (1984).

12. Zimm, B.H., J. Chem. Phys. 24, 269-78 (1956).

13. Gordon, R.J., and W.R. Schowalter, Trans. Soc. Rheology 16, 79-97 (1972).

14. Giesekus H., J. Non-Newtonian Fluid Mech. 14, 47-65 (1984).

15. Johnson, Jr., M.W., and D. Segalman, J. Non-Newtonian Fluid Mech. 2, 255-70 (1977).

16. Phan-Thien, N., and R.I. Tanner, J. Non-Newtonian Fluid Mech. 2, 353-65 (1977).

17. Phan-Thien, N., J. Rheology 22, 259-83 (1978).

18. Larson, R.G., J. Non-Newtonian Fluid Mech. 13, 279-308 (1983).

19. Leonov, A.I., Rheol. Acta 15, 85-98 (1976).

20. Curtiss, C.F., and R.B. Bird, J. Chem. Phys. 74, 2016-25, 2026-33 (1981).

21. De Gennes, P.G., J. Chem. Phys. 55, 572-78 (1971).

22. Doi, M., and S.F. Edwards, J. Chem. Soc. Faraday Trans. II, 74, 1789-1801, 1802-17, 1819-32 (1978); 75, 38-54 (1979).

23. Giesekus, H., Rheol. Acta 5, 29-35 (1966).

24. Giesekus, H,, Rheol. Acta 21, 366-75 (1982).

25. Giesekus, H., J. Non-Newtonian Fluid Mech. 11, 69-109 (1982).

26. Giesekus, H., J. Non-Newtonian Fluid Mech. 12, 367-74 (1983).

27. Volkov, V.S., and G.V. Vinogradov, Rheol. Acta 23, 231-37 (1984); cf. also H. Giesekus, Rheol. Acta 23, 564 (1984).

28. Mewis, J., and M.M. Denn, J. Non-Newtonian Fluid Mech. 12, 69-83 (1983).

29. Kaye, A., Brit. J. Appl. Phys. 17, 803-06 (1966).

30. Marrucci, G., G. Titomanlio, and G.C. Sarti, Rheol. Acta 12, 269-75 (1973).

31. Acierno, D., F.P. La Mantia, G. Marrucci, and G. Titomanlio, J. Non-Newtonian Fluid Mech. 1, 125-46, 147-57 (1976).

32. Giesekus, H., Constitutive Equations for Polymer Fluids Based on the Concept of Configuration-Dependent Molecular Mobility: A Generalized Mean-Configuration Model, J. Non-Newtonian Fluid Mech. (in the press).

33. Giesekus, H., in: B. Mena, A. Garcia-Rejón, C. Rangel-Nafaile (eds.), *Advances in Rheology: 1. Theory, pp. 39-63,* Universidad Nacional Autónoma de México (México, 1984).

34. Dashner, P.A., and W.E. VanArsdale, J. Non-Newtonian Fluid Mech. 8, 59-67 (1981).

35. Kwon, T.H., and S.F. Shen, Rheol. Acta 23, 217-30 (1984).

36. Jongschaap, R.J.J., H. Kamphuis, and D.K. Doeksen, Rheol. Acta 22, 539-49 (1983).

The author is indebted to Dr. M. Hibberd for his help in improving the English.

ON USING RUBBER AS A GUIDE FOR UNDERSTANDING POLYMERIC LIQUID BEHAVIOR

Arthur S. Lodge

Department of Engineering Mechanics,
Engineering Experiment Station and
Mathematics Research Center,
University of Wisconsin-Madison
Madison, Wisconsin

1. INTRODUCTION

Polymeric materials, composed of very long, flexible molecules, can exhibit a variety of pronounced rheological properties not exhibited by materials composed solely of small molecules. It is desirable to explain these properties in terms of molecular structure, and to discover appropriate constitutive equations to describe the material behavior in terms of a macroscopic continuum supposedly composed of a set {P} of fictitious point particles P which at each instant t fill a connected region of space {Q}. In a complete theory, the constitutive equations would be derived from molecular structure by suitable processes of averaging over physically infinitesimal material elements.*

At the continuum level, the variety of properties can be ascribed to a combination of finite strain and long memory effects, for whose description the fundamental variables at any given particle P can fairly safely be assumed to be the covariant body metric tensor $\boldsymbol{\gamma}(P,t)$ and the contravariant body stress tensor $\boldsymbol{\pi}(P,t)$, both tensors being symmetric and of second rank. (For some materials of importance, such as liquid crystals, additional vector

*Underlined terms in the text are defined in the appendix.

variables have been invoked, but we shall not consider these here). It is natural to consider constitutive equations involving single or multiple time integrals of the two body tensor fields, with suitable kernels to represent memory. All such body tensor equations in which $\boldsymbol{\pi}$ and $\boldsymbol{\gamma}$ are the only time-dependent tensor variables are automatically a priori admissible as possible equations for describing intrinsic properties of a viscoelastic continuum which are, by hypothesis, in a suitable sense independent of orientation of the material relative to directions fixed in space. The earliest most general acceptable formulation of constitutive equations for viscoelastic materials was given by Oldroyd [1] in 1950 in a paper of wide-ranging significance.

Other choices for the fundamental tensor variables can be made as a matter of convenience according to context without affecting the physical content of the resulting theory. For example, in discussing the stress-birefringence properties, it is natural to use the <u>Cauchy stress tensor</u> $\mathbf{p}(Q,t)$ and the <u>refractive index tensor</u> $\mathbf{n}(Q,t)$, both of which are <u>Cartesian space tensors</u>, not body tensors.

For the general formulation and discussion of constitutive equations, body tensors are simpler to use than space tensors: suppose, for example, we stretch a piece of rubber (regarded for the moment as a perfectly elastic solid) and then rotate it rigidly (with respect to directions fixed in space); the stress and the body stress tensor $\boldsymbol{\pi}$ do not change during the rotation, but the Cauchy stress tensor $\mathbf{p}$ does change. Use of $\mathbf{p}$ in this context is tantamount to choosing a variable to describe a constant - a strange, if legal, procedure. The reason is that space tensors unavoidably involve reference to space-fixed directions which are irrelevant for the task of describing intrinsic material properties. This well-known complicating feature of space tensors is sometimes dealt with in elasticity theory by using a two-state stress tensor (such as the Kirchhoff-Piola tensor), but the price paid is the introduction of a second ("reference") state in the description of stress, which is a one-state object. The body stress tensor $\boldsymbol{\pi}$ does not suffer from this inconvenience. Body tensors and space tensors belong to disjoint sets, but there is a <u>time-dependent correspondence</u> between them; for example, $\boldsymbol{\pi}(P,t) \xrightarrow{t} \mathbf{p}(Q,t)$ at time t, where Q is the place in space occupied by particle P at time t. The extra complication introduced by the use of the Cauchy stress tensor $\mathbf{p}$ is well illustrated by the correspondence between the time derivatives:

$$\frac{\partial}{\partial t}\boldsymbol{\pi}(P,t) \xrightarrow{t} \left(\frac{\partial}{\partial t} + \mathbf{v}\cdot\nabla\right)\mathbf{p}(Q,t) - \mathbf{p}\cdot\nabla\mathbf{v} - (\widetilde{\mathbf{p}\cdot\nabla\mathbf{v}}) \qquad (1.1)$$

$\boldsymbol{\nabla}$ here denotes the spatial gradient operator, and $\mathbf{v}$ denotes the velocity of P in space. The dot $\cdot$ denotes <u>contraction</u>, and $\sim$ denotes the <u>transpose</u>.

Most of the current work is based on the not unreasonable assumption that, for any infinitesimal material element, containing a particle P, say, the stress at any time t is determined by the history of temperature T and metric $\boldsymbol{\gamma}$ in that element alone, so that we write:

$$\boldsymbol{\pi}(P,t) = H\{\boldsymbol{\gamma}(P,t'),t',\boldsymbol{\kappa}_n(T(P,t'),P)\,|-\infty < t' \leq t; \quad n = 1,2,\ldots,N\}\,, \qquad (1.2)$$

where H denotes a "constitutive functional" (whose form varies from one material to another) and $\{\boldsymbol{\kappa}_n\}$ denotes a set of fourth, or higher, rank "material constant" body tensors dependent on temperature but not explicitly on time. The same particle P occurs throughout. In the first instance, the main interest centers on materials that are homogeneous at time t in the sense that, if the temperature is independent of time and location in the body,

$$\boldsymbol{\nabla}_t\boldsymbol{\kappa}_n = 0 \qquad (n = 1,2,\ldots,N) \qquad (1.3)$$

where $\boldsymbol{\nabla}_t$ denotes the covariant derivative operator formed from the body metric tensor $\boldsymbol{\gamma}(P,t)$. According to one definition, the material (1.2) is isotropic if all the material constant tensors $\boldsymbol{\kappa}_n$ can be expressed solely in terms of the <u>right covariant mixed unit tensor</u> $\boldsymbol{\delta}$ and scalars.

2. PERFECTLY ELASTIC SOLIDS

According to one definition, a material is a perfectly elastic solid if, for an arbitrary infinitesimal material element at P, (a) $\boldsymbol{\gamma}$ and T form a complete set of thermodynamic equilibrium state variables and (b) there is, at zero stress and any given T, a unique metric tensor $\boldsymbol{\gamma}_0(P,T)$, say, the "zero stress metric". Since the work done per unit mass by contact forces minus (the work done by body forces plus the change in kinetic energy) can be expressed in the form $\boldsymbol{\pi} : d\boldsymbol{\gamma}/(2\rho)$ (where ρ denotes the density) for an arbitrary infinitesimal strain $\boldsymbol{\gamma} \to \boldsymbol{\gamma} + d\boldsymbol{\gamma}$, it follows from the first and second laws of thermodynamics that

$$\pi^{ij} = \frac{1}{\rho}\left\{\frac{\partial}{\partial\gamma_{ij}} + \frac{\partial}{\partial\gamma_{ji}}\right\}F \, .$$

where F denotes the Helmholtz free energy per unit mass, and π^{ij} and γ_{ij} denote components of $\boldsymbol{\pi}$ and $\boldsymbol{\gamma}$ referred to an arbitrary <u>body coordinate system</u> $B : P \to \xi$. This particularly simple form for the general constitutive equation for anisotropic and isotropic perfectly elastic solids was given by Brillouin [3] in 1925 in terms of <u>convected components</u> of space tensor fields, which are numerically equal to certain corresponding components of body tensor fields.

Regarding (2.1) as a special case of (1.2), it can be seen that the zero-stress body metric tensor $\boldsymbol{\gamma}_0(P,T)$ must be included in the set $\{\boldsymbol{\kappa}_n\}$ of material constant tensors. For the isotropic case, F must be expressible solely in terms of T and scalars formed from $\boldsymbol{\gamma}_0 := \boldsymbol{\gamma}_0(P,T)$, $\boldsymbol{\gamma} := \boldsymbol{\gamma}(P,T)$, and δ. One such set of independent scalars is

$$I_1 = \boldsymbol{\gamma}_0^{-1} : \boldsymbol{\gamma}, \quad I_2 = \boldsymbol{\gamma}^{-1} : \boldsymbol{\gamma}_0, \quad I_3 = |\boldsymbol{\gamma}_0^{-1}\cdot\boldsymbol{\gamma}| \, , \qquad (2.2)$$

where $|\ |$ denotes a determinant, and the arguments P, T have been omitted. It follows from (2.1) and (2.2) that

$$\boldsymbol{\pi}\cdot\boldsymbol{\gamma} = 2\rho\left\{\boldsymbol{\gamma}_0^{-1}\cdot\boldsymbol{\gamma}\,\frac{\partial}{\partial I_1} - \boldsymbol{\gamma}^{-1}\cdot\boldsymbol{\gamma}_0\,\frac{\partial}{\partial I_2} + \boldsymbol{\delta}\,\frac{\partial}{\partial I_3}\right\}F \qquad (2.3)$$

where the Helmholtz free energy F is now expressed as a function of the strain invariants I_κ, the temperature, and scalar material constants (not shown).

If the solid is incompressible (an idealization convenient for many purposes), then (at constant T) $I_3 = 1$ and in place of (2.3) one has

$$\boldsymbol{\pi}\cdot\boldsymbol{\gamma} = \{\boldsymbol{\gamma}_0^{-1}\cdot\boldsymbol{\gamma}\frac{\partial}{\partial I_1} - \boldsymbol{\gamma}^{-1}\cdot\boldsymbol{\gamma}_0\frac{\partial}{\partial I_2}\}W(I_1,I_2,T) - p\boldsymbol{\delta} \,. \qquad (2.4)$$

where $W(I_1,I_2,T) = 2\rho F(I_1,I_2,1,T)$ and p, a Lagrange multiplier associated with the constraint $I_3 = 1$, is a scalar function of P determined, not by the metric and temperature history only, but by the complete system of constitutive equations, stress equations of motion, boundary values, and the constant volume condition.

A property of isotropic, perfectly elastic, solids (important for the present purposes) is that, at any P, the <u>principal axes</u> of stress and strain coincide; it follows, in particular, that for a <u>shear strain</u> $\boldsymbol{\gamma}_0 \to \boldsymbol{\gamma}$ of <u>magnitude</u> s,

$$N_1/(s\sigma) = 1 \,, \qquad (2.5)$$

where N_1 and σ denote the <u>first normal stress difference</u> and <u>shear stress</u>, respectively. These results follow from (2.3) or (2.4).

At the molecular level, the characteristic rheological properties of polymeric solids and liquids are to be understood in terms of the thermal motion of the long, flexible, polymer molecules coupled with the interactions between these molecules. It is the mathematical difficulties associated with the interactions that make condensed phase molecular theory so difficult to develop; reasonable treatments of single polymer molecules are available.

According to the molecular network theory for lightly cross-linked elastomers [4], polymer molecules are joined to one another at a few points by permanent chemical bonds so that a 3-dimensional polymer molecule network extends throughout the sample; except at the cross-links, interaction forces are weak and serve only to determine the sample volume (or density), which is not calculated by the theory. Thermal motion of the molecular network generates the non-isotropic contribution to stress. Other more or less questionable assumptions are introduced in order to make the analysis tractable. As is the case for most molecular theories in polymer rheology, the material and the strains considered are taken to be uniform in the first instance, so that one really obtains from the molecular model a constitutive equation appropriate to an infinitesimal material element (of the macroscopic continuum). Application to non-uniform strains involves an additional assumption to the effect that strain gradients have no effect on the non-isotropic contribution to stress at any given particle.

In body tensors [5], the results of the network theory can be expressed in the form

$$F = F_1(T,\rho) + \boldsymbol{\mu} : \boldsymbol{\gamma} \ , \tag{2.6}$$

where F_1 is a contribution from volume-determining forces not given by the theory and $\boldsymbol{\mu}$ denotes a second-rank constant contravariant "structure tensor" which can be calculated if the temperature and network connectivity are given. From (2.6), one can obtain the constitutive equation

$$\boldsymbol{\pi} = -\rho^2 \frac{\partial F_1}{\partial \rho} \boldsymbol{\gamma}^{-1} + 2\rho\boldsymbol{\mu}(T) \ , \tag{2.7}$$

Putting $\boldsymbol{\pi} = \mathbf{O}$, it follows that, at given T, there is a unique zero-stress metric tensor

$$\boldsymbol{\gamma}_0 = \frac{1}{2} q_0 \boldsymbol{\mu}^{-1} \ , \tag{2.8}$$

where q_0 denotes the zero stress value of

$$q := -\rho \frac{\partial}{\partial \rho} F_1(T,\rho) \ . \tag{2.9}$$

It follows that the material is a perfectly elastic solid; and because the constitutive equation can be rewritten in the form

$$\boldsymbol{\pi}\cdot\boldsymbol{\gamma} = -\rho q \boldsymbol{\delta} + \mu \boldsymbol{\gamma}_0^{-1}\cdot\boldsymbol{\gamma} \ , \tag{2.10}$$

where

$$\mu = \frac{2}{3}\,\rho\boldsymbol{\gamma}_0 : \boldsymbol{\mu}\,, \tag{2.11}$$

the solid is isotropic, because (2.11) is a particular case of (2.3) with F independent of I_2. The network theory thus favors the contravariant zero-stress metric term in $\boldsymbol{\gamma}_0^{-1}$ in the constitutive equation - a feature attributable to thermal motion of sufficiently long and free polymer "strands" (i.e., parts of polymer molecules between successive cross-links) which one might expect to find also in polymeric liquids (not cross-linked).

This brief introduction to the network theory of rubber elasticity does not do justice to the extensive and still growing literature which has appeared even since the publication of Treloar's classic monograph [6]. This literature naturally focusses on the original theory's admitted weaknesses (e.g., the unrealistic form of stress-strain relations predicted by (2.10) for many cross-linked elastomers) but in so doing it could mislead the casual reader into underestimating the remarkable breadth of success enjoyed by this theory: the description of large elastic recovery; the predicted linearity of the relation between stress and refractive index tensors even at large strains; the predicted order of magnitude of the shear modulus and its dependence on temperature and on the concentration of cross-links; and even the prediction of the correct form of tension-elongation relation for polymers cross-linked in a special way [7].

3. POLYMERIC LIQUIDS

In polymeric materials having no permanently cross-linked network, individual polymer molecules are free to move past one another and there is no mechanism for giving the material a unique zero-stress metric: the material behaves like a liquid instead of a solid. The consequent increase in complexity of possible consitutive equations - of the general form (1.2), for example - is so great that one probably cannot hope for a simple closed form of any generality analogous to (2.3) or (2.4) for the isotropic, perfectly elastic, solid. The variety of forms of constitutive equations published to date - involving single and multiple time integrals with a variety of forms for the integrands - bears this out.

It is natural to seek some systematic approach to the problem of finding the "correct" constitutive equation for a given polymeric liquid or class of polymeric liquids. (If "correct" is too much to hope for, perhaps we could hope for a sequence of possible equations which converges to the "correct" equation). Expansion schemes, based on an assumed Fréchet differentiability of the constitutive functional H in (1.2), have been used, but the need to impose restrictions on some parameter such as strain or strain rate carries the danger that the large-strain features which characterize the variety of polymer rheological properties will be lost or emasculated.

As an alternative to a completely systematic approach, I believe that it is fruitful to use the comparatively well understood field of rubberlike solid elasticity as a guide: molten (or undiluted) polymers have several striking properties in common with lightly cross-linked elastomers: large, rapid elastic recovery; a linear relation between stress and refractive index tensors extending well into the finite strain region [8]; and the same order of magnitude of shear modulus (for solids) and plateau modulus G' for liquids [9].

In my opinion, the large elastic recovery property could be one of the most crucial in polymer rheology as a guide to realistic molecular explanations and to the choice of constitutive equations. It has received comparatively little attention in the literature. A rubber band can be stretched to about five times its zero-stress length and, on release, will return to within a few percent of its original length. A filament of a molten polymer (such as low-density polyethylene, "LDPE") on being elongated steadily and then cut can decrease in length by a factor of 10 [10]. If the filament is instead held at constant length after initial elongation, its tension will slowly decrease to zero, and so the material can reasonably be regarded as a liquid. LDPE thus affords us a striking example of a liquid which can exhibit even greater elastic recovery than that commonly demonstrable in cross-linked elastomers. Another significant example is afforded by "raw" rubber (natural unmilled, uncrosslinked rubber) [11], which also exhibits large, rapid recovery although it has no permanent cross-links (as attested by the fact that it is completely soluble in suitable solvents: a cross-linked elastomer would swell but not dissolve).

The success of the network theory in describing large recovery is attributable to three features [12]: (a) the existence of molecular structures (long flexible polymer molecules) which can sustain large changes of shape without damage; (b) a mechanism (thermal motion of comparatively free strands) which yields large changes of shape when external forces on the sample are removed; and (c) the interconnection of such structures (at a comparatively few points into a three-dimensional network extending throughout the sample) adequate to ensure the transmission of strand recovery to the macroscopic sample as a whole.

In this context, the importance of large recovery must be stressed: small recovery can (e.g., in crystals) be explained in terms of small displacements of an atomic lattice - a completely different mechanism which is incapable of producing large recoveries. Since different microscopic mechanisms can explain small recovery, it cannot be a crucial property in the present sense.

An important unanswered question concerns (c): to what extent is the (permanent or temporary) cross-link type of local interaction essential for the mechanism of large rapid elastic recovery in the macroscopic sample? On the one hand, the temporary-junction network theories [13-16], based as they are on the network theory for rubber elasticity, are clearly capable of explaining and describing large rapid recovery. On the other hand, for certain "one-polymer-molecule models" [17, 18], the situation appears to be less clear: these models focus on a single polymer molecule and idealize its environment (either as a "tube" [17] or as if it were a viscous Newtonian liquid [18]); ad hoc assumptions are made about how the environment acts on the polymer molecule (for application to problems in which external forces act on the whole sample to produce a flow history) but the published papers do not as yet seem to have made it clear how the polymer molecule acts on its environment (which would seem to be essential if elastic recovery of the sample as a whole is to be explained). According to one paper [19], each polymer molecule retracts (under certain circumstances) within its tube with the polymer molecule mass center remaining fixed in space - a mechanism which clearly can give no contribution to the macroscopic recovery of the sample as a whole, because, if every polymer molecule's mass center is fixed, then the macroscopic sample is also fixed; whether subsequent "tube" motion contributes to recovery is unclear because the laws of tube motion do not seem to have been stated; I am not sure, however, whether I am doing justice to the "tube" papers partly because, on a first reading, at least, they appear to involve self-contradictory concepts (embedded unit vectors [17]; uniform diameter curved tubes which undergo affine displacement without change of diameter [19]; and "Rouse molecules" which deform affinely [20]); and invoke the working of Maxwell Demons without giving their laws of motion. Non-uniform tube models have been considered by others [21].

4. THE ONE-STEP SHEAR EXPERIMENT

The large recovery exhibited by molten LDPE makes it plausible to suppose that the stress at any instant is determined (to within an additive isotropic stress) by the thermal motion of a temporary-junction polymer molecule network. In contrast to the case of a permanently cross-linked network, the network junctions here must be continually created and lost - presumably due to thermal motion (the only mechanism when the liquid metric tensor is independent of time) possibly supplemented by some mechanism attributable to the strain or stress history in the sample itself. This mechanism could also, conceivably, affect the mean lifetimes of junctions.

Clearly, a complete theory based on this conjecture would require detailed assumptions about junction loss and creation. Is there any experimental test of the conjecture which could be made without embarking upon this difficult task?

According to (1.2), the current stress is determined by an infinite set of metric tensors. For a particular history, however, we may hope to reduce this set effectively to 2: we consider a single bounded jump $\boldsymbol{\gamma}_- \to \boldsymbol{\gamma}_+$ at time 0:

$$\gamma(P,t) = \boldsymbol{\gamma}_- + \varepsilon(t)(\boldsymbol{\gamma}_+ - \boldsymbol{\gamma}_-) \quad \text{(for all } t) \; ; \qquad (4.1)$$

$\varepsilon(t) = 0$ $(t < 0)$ or 1 $(t > 0)$. Can we hope that the stress tensor at times $t > 0$ will be determined by the single strain $\boldsymbol{\gamma}_- \to \boldsymbol{\gamma}_+$ alone? We can if the network model is valid and there is no change of network connectivity during the jump strain; for the material would then, from $t = 0-$ to $0+$, be indistinguishable from a permanently cross-linked solid and so one would expect the stress at $t = 0+$ to be given by an equation of the form (2.3) (appropriate to an isotropic solid) with γ_- as the zero-stress metric and $\boldsymbol{\gamma}_+$ as the current metric. It would follow, in particular, that the principal axes of stress at $t = 0+$ coincide with the principal axes of the strain $\boldsymbol{\gamma}_- \to \boldsymbol{\gamma}_+$. Moreover, for $t > 0+$, the principal values of stress would be expected to decrease due to loss of junctions with creation of new junctions on strands which give little or no nonisotropic contribution to the stress; it seems reasonable to expect, however, that this process will not change the principal directions of stress. If so, then:

for $t > 0+$, the principal directions of stress coincide with the principal directions of the strain $\boldsymbol{\gamma}_- \to \boldsymbol{\gamma}_+$. (4.2)

The condition $t > 0+$ (rather than $t = 0+$ alone) makes it feasible to conduct experimental tests. (4.2) is motivated by the network model but is not restricted to cases involving a particular form of stress-strain relation such as (2.10): only the isotropy (2.3) is required. The validity of (4.2) does not imply that the network model is valid.

Experimental tests of (4.2) have been made for the case in which the step strain is a shear, of magnitude s, say; (4.2) reduces to

$$N_1(s,t)/\{s\sigma(s,t)\} = 1 \qquad (t > 0), \tag{4.3}$$

a relation proposed by Lodge and Meissner [22]. N_1 and σ can be determined from the thrust F and torque M exerted by the sample when sheared between a cone and plate. For samples of high viscosity, such as molten polymers, there are numerous sources of systematic error [23], and the measurements are difficult to perform reliably. Errors due to sample volume change have been investigated recently in a new apparatus [47]. By taking sufficient care, all known sources of error can now be dealt with, at least for materials having no yield stress. Interpretation of thrust data when the yield stress is not zero is fraught with difficulty [48]. Published data, though limited, is consistent with (4.3) for LDPE [24], high density polyethylene and polystyrene [25, 26], polybutadiene [27], and for a concentrated solution of polystyrene [28], [29]. Some of these tests were made by using the birefringence extinction angle and shear stress to determine N_1, and thus depend on the coaxiality of stress and refractive index tensors which has been well established in other experiments [8]; these tests have the advantage that they are not subject to some of the experimental difficulties involved in cone-plate thrust measurements.

These results are consistent with Weissenberg's early proposal [49] that, in a flowing elastic liquid, the stress at any instant is an isotropic function of a finite-strain tensor. Weissenberg suggested that this be identified with the "recoverable strain" tensor, but did not make clear whether this recovery should be instantaneous, delayed, or ultimate.

The one-step shear strain experiment can be used [23] to classify constitutive equations according to the values predicted for the left-hand side of (4.3): Class I for the value 1; II for a constant other than 1; III for a function of s alone; IV for a function of s and t; and V if the ratio is indeterminate (usually because of singularities in N_1 and σ at $t = 0$). Simple examples in all these classes can be constructed [23], and several published constitutive equations, not being in Class I, are thus incompatible with the data obtained for the polymeric materials listed above [23].

The advantage of this procedure is that classes of possible constitutive equations can be eliminated for a given material by means of a single experiment without the need for curve fitting or computation which is required for more comprehensive tests involving comparison with results of a variety of measurements. An objection to the procedure can be made on the grounds that step strains cannot be imposed in practice: at best, one can impose a "ramp" strain in which the strain grows at a large constant (but bounded) rate during a short time interval Δt and then remains constant. To approach the step strain envisaged in the proposed test, one would have to increase the rate of strain growth; even if this could be done in practice, however, an upper limit on the strain rate would be imposed by the need to avoid significant inertial effects and by the need to remain within the "rubberlike" response zone of the polymer under investigation: too high a rate would take one into the hard, "glassy", response zone which is not suitable for the present test because the network theory motivation is inapplicable and because the material would be likely to fracture before a large strain could be attained. Large shear strains are essential for the present test. A possible resolution of this difficulty is to repeat the test with more than one rise time and to consider values of $N_1/(s\sigma)$ for $t > 10\ \Delta t$, say; if the values are insensitive to a change in the rise time, one can have some confidence that an acceptable compromise has been found. One can also make theoretical corrections for the finite rise time [24], but to do so one must assume some form of constitutive equation; and the calculations involved detract from the simplicity claimed above for the test. In spite of these difficulties, the test appears to be practically useful when adequate precautions are taken; attainment of Class I behavior (i.e., (4.3)) is even regarded as an indication of satisfactory apparatus performance [30].

5. SOME CLASS I EQUATIONS

We shall use the abbreviations $\boldsymbol{\pi}$, $\boldsymbol{\gamma}$, $\boldsymbol{\gamma}'$, $\boldsymbol{\gamma}''$ for $\boldsymbol{\pi}(P,t)$, $\boldsymbol{\gamma}(P,t)$, $\boldsymbol{\gamma}(P,t')$, $\boldsymbol{\gamma}(P,t'')$, respectively, and we shall consider incompressible liquids only.

The "rubberlike liquid" constitutive equation

$$\boldsymbol{\pi} + p\boldsymbol{\gamma}^{-1} = \int_{-\infty}^{t} m(t - t')\boldsymbol{\gamma}'^{-1}dt' \tag{5.1}$$

has been derived [14] from the temporary junction network model for the case in which the junction creation and loss rates are constant for a given system at a given temperature. The "memory function" $m(t - t')$ is a monotonic decreasing function, usually expressed as a sum

$$m(\tau) = \sum_{i=1}^{M} a_i \exp(-\tau/\tau_i) \tag{5.2}$$

with non-negative coefficients a_i and positive exponents τ_i. (5.1) can be obtained from the incompressible version of the rubberlike solid equation (2.10) by replacing the zero-stress metric $\boldsymbol{\gamma}_0$ by $\boldsymbol{\gamma}'$, the coefficient μ by $m(t - t')dt'$ and integrating over all t' up to the current time t. It is convenient to refer to this process as one of "liquidating" a solid.

For a LDPE melt, constants in (5.2) have been chosen to fit stress-strain data obtained in elongation at low elongation rates [31]; (5.1) was then used to calculate similar data for higher elongation rates, for recovery after such elongation, for growth of N_1 and σ following a step shear rate, and for free recovery after shear flow [32]; agreement with measurements was found to be good for moderate strains from rest (e.g., up to about two shear units) but (5.1) seriously overestimated stress and recovery at higher strains [2]. In addition, (5.1) gives zero for the value of N_2, the second normal stress difference;

measurements for molten LDPE (and for other polymers) give results in the region $N_2 \approx -N_1/10$ [8, 33].

"Liquidation" of the general isotropic perfectly elastic solid equation (2.4) leads to the so-called K-BKZ equation [34], [35]:

$$\pi \cdot \gamma + p\boldsymbol{\delta} = \int_{-\infty}^{t} \{\boldsymbol{\gamma}'^{-1} \cdot \boldsymbol{\gamma} \frac{\partial}{\partial I_1} - \boldsymbol{\gamma}^{-1} \cdot \boldsymbol{\gamma}' \frac{\partial}{\partial I_2}\} W(I_1, I_2, t-t')dt' \qquad (5.3)$$

where

$$I_1 = I_1(t', t) = \boldsymbol{\gamma}'^{-1} : \boldsymbol{\gamma} ,$$
$$I_2 = I_2(t', t) = \boldsymbol{\gamma}^{-1} : \boldsymbol{\gamma}' . \qquad (5.4)$$

Equations of this form have been used to give a good description of LDPE melt data for a variety of experiments [36-39] over considerable ranges of strain and strain rate, provided that at least one principal strain (measured from rest to the current time t) increases monotonically. For other experiments, such as multistep shear sequences with changes of sign of shear [40] and for free recovery following elongational flow [41], significant disagreement has been found - with the K-BKZ equation overestimating recoverable strain by factors up to 3.3.

In order to tackle the latter type of disagreement, Wagner and Stephenson [41] added an extra strain functional factor "Min{ }" in the integrand, yielding an equation:

$$\boldsymbol{\pi} + p\boldsymbol{\gamma}^{-1} = \int_{-\infty}^{t} m(t - t')\boldsymbol{\gamma}'^{-1} \underset{t''=t'}{\overset{t''=t}{\mathrm{Min}}} \{h[I_1(t'', t'), I_2(t'', t')]\}dt' , \qquad (5.5)$$

where

$$h(I_1, I_2) = f_1 \exp(-n_1\sqrt{I}) + (1 - f_1)\exp(-n_2\sqrt{I}) , \qquad (5.6)$$

$$I = \alpha I_1 + (1 - \alpha)I_2 . \qquad (5.7)$$

There are four constants f_1, n_1, n_2, α which are chosen empirically to fit nonlinear experiments; the memory function m is taken to be of the form (5.7) with its constants chosen to fit linear data. The equation (5.5) is thereby completely determined; it gives a quantitatively successful description of data for an additional nine nonlinear tests, all on the same LDPE. To the best of my knowledge, this investigation represents the most extensive published comparison between predictions of a single constitutive equation and data from a variety of experiments all performed on one polymeric material.

In order to test thoroughly the applicability of any proposed constitutive equation, it is essential to use as great a variety of rheological experiments as possible; in particular, one needs both shear flow and shear free flow experiments, for these form disjoint sets: to see this most simply, the term

$$\int_{-\infty}^{t} M(t - t')\boldsymbol{\gamma}'^{-1}\mathrm{Trace}\{(\boldsymbol{\lambda}'\cdot\dot{\boldsymbol{\lambda}} - \dot{\boldsymbol{\lambda}}\cdot\boldsymbol{\lambda}')^2\}dt' , \qquad (5.8)$$

where

$$\dot{\boldsymbol{\lambda}} = \boldsymbol{\gamma}^{-1}\cdot\partial\boldsymbol{\gamma}/\partial t, \quad \boldsymbol{\lambda}' = \boldsymbol{\gamma}^{-1}\cdot\boldsymbol{\gamma}' , \qquad (5.9)$$

can readily be shown to vanish in all shear-free flows and to be non-zero in steady shear flow. The term (5.8) could be added to any contravariant form of constitutive equation, so as to give an added contribution to the stress tensor, say; this suffices to ensure that, if the original equation failed to distinguish between shear and shear-free flows, the final equation would distinguish between them.

The most extensive development of experimental techniques for shear-free flows is due to Meissner [10, 42]. These require the samples under test to have such high viscosities that in practice the techniques are applicable only to molten or undiluted polymers of sufficiently high molecular weight. It follows that, at present, adequately extensive testing of constitutive equations is restricted to these materials; dilute and moderately concentrated polymer solutions are unfortunately outside the scope of shear-free flow testing, at least under sufficiently well-controlled conditions over large ranges of strain.

Meissner's experiments on LDPE at constant elongation rate up to elongation ratios of 1100 show that the tensile stress never attains a constant value [43] and even passes through a maximum at elongation ratios of about 150; the reality of this maximum (in a filament that has at these strains a rather small cross-section) is demonstrated by the existence of a similar maximum in the magnitude of free recovery [44].

It follows that, for this polymer under the conditions investigated, elongational viscosity does not exist. This result is given, in particular, by the "rubberlike liquid" (5.1) for elongation rates greater than a certain critical rate. To compare constitutive predictions with data for these experiments, one should calculate not elongational viscosity but tensile stress or, more precisely, a difference of principal stresses, as a function of time for a step-function elongation rate.

Recent data of Meissner for polyisobutylene subjected to shear-free flows other than simple elongation [44] appear to disagree with the predictions of the Wagner-Stephenson form of equation (5.5), but it may be possible to choose an alternative to (5.7) which would repair the damage. (5.5), of course, gives $N_2 = 0$ (not a serious demerit for commercial melts); addition of a term in γ' to (5.5) could remedy this defect, but would also affect certain shear-free flow predictions.

For one-step shear strain experiments, the functional "Min{ }" reduces to { } and (5.5) is then indistinguishable from a K-BKZ equation. It follows, in particular, that (like all K-BKZ equations) the Wagner-Stephenson equation (5.5) is of Class I. It is easy to show [23] that Class I includes all constitutive equations in which the current stress tensor is expressible as a sum of terms each of which is a multiple time integral with integrands formed from body metric tensors evaluated at various times $t', t'', \ldots$. Their time derivatives should be excluded except to the extent that integration by parts could convert the terms to forms not containing time derivatives.

6. APPENDIX: DEFINITIONS

We give here a brief summary of definitions of terms in the above text. A complete treatment has been given elsewhere [2]. In common use, the term vector has two different meanings: an element of a linear space, and an element of a linear space associated with a geometric manifold in a certain way (described below). We shall use the term algebraic vector for the former and vector for the latter. A similar distinction will be drawn for tensors.

Continuum mechanics deals with the motion of one geometric manifold - the body - moving through another - space. Each may be regarded as a set of undefined elements - points - to which certain properties are assigned by hypothesis. The essential feature of our formalism is that, as far as possible, the two manifolds should be treated on an equal footing (in the sense made clear below). The more widely-used formalisms single out the space manifold for special treatment: they use space vectors and tensors for all purposes, and do not use body vectors and tensors for any purpose. It is simpler to use both.

The body is a set $\{P\}$ of point particles which, at any instant t, fill a connected region of space $\{Q\}$, a set of point places, so that there is a time-dependent 1-1 correspondence $\{P\} \overset{t}{\longleftrightarrow} \{Q\}$ which we may refer to as the state t or the configuration t.

A body coordinate system $B : \{P\} \to \mathbf{R}^3$ is a time-independent correspondence $P \to \xi$ between particles and ordered sets of three real numbers. ξ is an abbreviation for ξ^1, ξ^2, ξ^3. A space coordinate system $S : \{Q\} \to \mathbf{R}^3$ is a time-independent correspondence between places and ordered sets x^1, x^2, x^3 of three real numbers; x is an abbreviation for x^1, x^2, x^3. B and S are congruent at time t (written $B \overset{t}{\equiv} S$) if $\xi^i = x^i$ when $P \overset{t}{\longleftrightarrow} Q$. We consider sets $\{B\}$, $\{S\}$ of body and space coordinate systems which include congruent pairs. For each pair B, $\bar{B}$ there is a nonsingular 3×3 coordinate transformation matrix

$$\Lambda_P(B,\bar{B}) = [\partial\bar{\xi}^r/\partial\xi^c]_P \ . \qquad (6.1)$$

r, c are row and column labels; this matrix is independent of time and, in general, is dependent on P. Similarly, for each pair of space coordinate systems S, $\bar{S}$, there is a 3×3 matrix

$$L_Q(S,\bar{S}) = [\partial \bar{x}^r / \partial x^c]_Q \tag{6.2}$$

$\bar{\xi}$ denotes the coordinates of P in $\bar{B}$, and $\bar{x}$ those of Q in $\bar{S}$.

A <u>contravariant body vector</u> at P is a function $\boldsymbol{\theta} : \{B\} \to A^{3\times 1}$ whose arguments are body coordinate systems B, $\bar{B}$,... and whose values are real 3×1 column matrices θ, $\bar{\theta}$,... such that each pair are related by the <u>contravariant vector transformation law</u>

$$\bar{\theta} = \Lambda_P(B,\bar{B})\theta \qquad (\text{all } B, \bar{B}) \; . \tag{6.3}$$

For brevity, we write $\theta = \boldsymbol{\theta}B$, $\bar{\theta} = \boldsymbol{\theta}\bar{B}$,... A <u>covariant body vector</u> at P is a similar function $\boldsymbol{\psi} : \{B\} \to A^{3\times 1}$ with a <u>covariant vector transformation law</u>:

$$\bar{\psi} = \tilde{\Lambda}_P^{-1}(B,\bar{B})\psi \; . \tag{6.4}$$

Covariant vectors at P and contravariant vectors at P form disjoint linear spaces. A vector <u>field</u> is a set of vectors of the same kind (co- or contravariant), one for each particle of the body. Vectors at different particles belong to disjoint linear spaces. At P, $\boldsymbol{\theta}$ and $\boldsymbol{\psi}$ may be <u>contracted</u> to form a scalar $\boldsymbol{\theta}\cdot\boldsymbol{\psi}$ whose value $\tilde{\theta}\psi = \tilde{\bar{\theta}}\bar{\psi} = \ldots$ is independent of B, $\bar{B}$,... .

<u>Second rank body tensors</u> are similar functions whose values are 3×3 matrices; four kinds are distinguished by their transformation laws:

$$\begin{aligned} &\text{contravariant:} \quad \bar{\Theta} = \Lambda\,\Theta\,\tilde{\Lambda}; \\ &\text{covariant:} \quad \bar{\Psi} = \tilde{\Lambda}^{-1}\Psi\Lambda^{-1} \\ &\text{right covariant mixed:} \quad \bar{\Phi} = \Lambda\Phi\Lambda^{-1} \\ &\text{left covariant mixed:} \quad \bar{\Xi} = \tilde{\Lambda}^{-1}\,\Xi\,\tilde{\Lambda} \; . \end{aligned} \tag{6.5}$$

The <u>transpose</u> and <u>reciprocal</u> of a second rank tensor are formed by taking the transpose and reciprocal of their representative matrices. Applying these operations to the transformation laws (6.5), it is seen that the transpose of a covariant tensor is covariant, the reciprocal is contravariant, etc. Contraction of tensors is defined as for vectors. The trace $\mathrm{Tr}(\boldsymbol{\Phi})$ of a mixed tensor is a scalar: trace Φ (i.e., formed by one contraction). The right covariant unit tensor $\boldsymbol{\delta}$ is a mixed "numerical tensor" represented by the same unit matrix I in every B. The zero tensor $\mathbf{0}$ (of any kind) is a numerical tensor represented by the 3×3 zero matrix 0 in every B. The outer product of any two vectors at P is a second rank tensor at P constructed from the appropriate matrix products: e.g., the outer product $\boldsymbol{\theta\psi}$ is a right covariant mixed tensor represented in B by $\theta\tilde{\psi}$.

Neighboring particles P, P_1 have infinitesimal coordinate differences $d\xi^i$, $d\bar{\xi}^i, \ldots$ in B, $\bar{B}, \ldots$ respectively (it being assumed that the coordinate transformation functions $\bar{\xi}^i(\xi)$ are "sufficiently differentiable") and thus determine a contravariant body vector $\mathbf{d\xi}$ at P, where $\mathbf{d\xi}B = d\xi$, which is independent of time. We assume that the body possesses a time-dependent (Riemannian) metric, i.e., a correspondence between neighboring pairs of particles and non-negative real numbers ds such that, in any $B : P \rightarrow \xi$,

$$(ds)^2 = d\xi^i \gamma_{ij}(\xi,t) d\xi^j = \widetilde{d\xi}\gamma(\xi,t)d\xi \ . \tag{6.6}$$

ds is called the separation of P and P_1.

γ_{ij} are thus the coefficients of a positive-definite quadratic form which depends on time because $d\xi$ does not. $d\xi$ and γ denote matrices $[d\xi^r]$, $[\gamma_{rc}]$, respectively. Without loss of generality, γ will be taken to be symmetric. Because there is a similar equation for every B with the same value for ds, it follows that the corresponding matrices $\gamma, \bar{\gamma}, \ldots$ represent a covariant body tensor at P – the body metric tensor $\boldsymbol{\gamma}(P,t)$, and that we may write

$$(ds)^2 = d\boldsymbol{\xi}\cdot\boldsymbol{\gamma}(P,t)\cdot d\boldsymbol{\xi} \ , \tag{6.7}$$

a coordinate-free equation.

A <u>strain</u> $t' \rightarrow t$ is a change of metric $\boldsymbol{\gamma}(P,t') \rightarrow \boldsymbol{\gamma}(P,t)$ at P. <u>Principal axes of strain</u> at P are contravariant vectors $\boldsymbol{\theta}_i$ which are non-zero roots in $\boldsymbol{\theta}$ of the equation

$$\{\boldsymbol{\gamma}(P,t) - \lambda^2\boldsymbol{\gamma}(P,t')\}\cdot\boldsymbol{\theta} = \mathbf{0} \tag{6.8}$$

This equation for λ^2 has three positive roots, λ_1^2, λ_2^2, λ_3^2, say; the positive roots λ_1, λ_2, λ_3 are called <u>principal elongation ratios</u> for the strain $t' \rightarrow t$. To each corresponds a principal axis $\boldsymbol{\theta}_i$, a solution of (6.8) defined to within an arbitrary scalar multiplier. Other principal axes and values may be defined similarly; for example, the <u>principal strain rates</u> at P,t are the roots in κ of the equation

$$|\partial\boldsymbol{\gamma}(P,t)/\partial t - \kappa\boldsymbol{\gamma}(P,t)| = 0 \ . \tag{6.10}$$

Tensors at P are said to be <u>coaxial</u> if their principal axes coincide.

A material surface (i.e., a surface always composed of the same particles) moves <u>isometrically</u> if, for every pair of neighboring particles in the surface, their separation is independent of time. A constant volume flow is said to be a <u>shear flow</u> if there exists a one-parameter family of isometric surfaces; these are called <u>shear surfaces</u>. Neighboring shear surfaces move at constant separation; <u>shear lines at t</u> are lines in a shear surface that are instantaneously everywhere tangential to the local direction of relative motion of neighboring shear surfaces. If the shear lines are material lines throughout a time interval, the shear flow is said to be <u>unidirectional</u>. For any shear flow, the principal strain rates are of the form $0, \dot{s}, -\dot{s}$; $\dot{s}$ is called the <u>shear rate</u>. In unidirectional shear flow, there is at any given P,t a set of three <u>shear flow base vectors</u> $\boldsymbol{\alpha}^1$, $\boldsymbol{\alpha}^2$, $\boldsymbol{\alpha}^3$ with the properties

$$\boldsymbol{\gamma}(P,t') - \boldsymbol{\gamma}(P,t) = s(\boldsymbol{\alpha}^2\boldsymbol{\alpha}^1 + \boldsymbol{\alpha}^1\boldsymbol{\alpha}^2) + s^2\boldsymbol{\alpha}^2\boldsymbol{\alpha}^2 \ ; \tag{6.11}$$

$$\boldsymbol{\gamma}(P,t) = \boldsymbol{\alpha}^1\boldsymbol{\alpha}^1 + \boldsymbol{\alpha}^2\boldsymbol{\alpha}^2 + \boldsymbol{\alpha}^3\boldsymbol{\alpha}^3 \; ; \tag{6.12}$$

$$\boldsymbol{\alpha}^i \cdot \boldsymbol{\gamma}^{-1}(P,t) \cdot \boldsymbol{\alpha}^j = \delta^{ij} \; . \tag{6.13}$$

$\boldsymbol{\alpha}^2$ is normal to a shear surface (at P,t);
$\boldsymbol{\alpha}_1 = \boldsymbol{\gamma}^{-1}(P,t)\cdot\boldsymbol{\alpha}^1$ is tangential to a shear line at P,t. (6.13) is the condition that the vectors $\boldsymbol{\alpha}^i$ are <u>t-orthonormal</u>; (6.12) follows from (6.13). $s = s(P,t',t)$ is a scalar called the <u>magnitude of shear strain</u> for the shear $t' \to t$. If $\dot{s}(P,t'')$ exists and is integrable throughout the interval (t',t), it can be shown that

$$s(P,t',t) = \int_t^{t'} \dot{s}(P,t'')dt'' \; . \tag{6.14}$$

A flow is <u>shear free</u> throughout any time interval for which there exists a body coordinate system, $B : P \to \xi$, say, which is <u>orthogonal</u>, i.e., for which

$$\gamma_{ij}(\xi,t) = 0 \qquad (i = j) \quad (\text{all} \;\; t) \tag{6.15}$$

It can be shown that the only shear-free shear flow is a <u>rigid motion</u>, i.e., one for which $\boldsymbol{\gamma}(P,t)$ is independent of t, and that, for any shear-free flow, the coordinate lines of the always-orthogonal B are always tangential to principal strain rate directions and to principal strain directions for every pair of states t',t.

A material surface can be described by equations of the form

$$\phi(\xi) = \text{const.}, \quad \bar{\phi}(\bar{\xi}) = \text{const.}, \quad \text{in} \;\; B, \bar{B}, \ldots \; . \tag{6.16}$$

$\partial\phi/\partial\xi^i$, $\partial\bar{\phi}/\partial\xi^i, \ldots$ are components in B, $\bar{B}, \ldots$ of a covariant vector at P which is a normal to the surface at P in the sense that

$$\frac{\partial\phi}{\partial\xi^i} d\xi^i = 0, \quad \frac{\partial\bar{\phi}}{\partial\bar{\xi}^i} d\bar{\xi}^i = 0, \; \ldots \qquad \text{etc.} \tag{6.17}$$

where P, P_1 are any two neighboring particles in the surface and ξ^i and $\xi^i + d\xi^i$ are their coordinates in B. This normal is independent of time; its <u>magnitude</u> (defined by contraction with $\boldsymbol{\gamma}^{-1}(P,t)$) varies with time, and so also does the unit normal

$$\boldsymbol{\nu}(P,t), \text{ where } \nu B = \left[\frac{\partial\phi}{\partial\xi^r}\right]\left\{\frac{\partial\phi}{\partial\xi^i}\,\gamma^{ij}(\xi,t)\,\frac{\partial\phi}{\partial\xi^j}\right\}^{-1/2}. \quad (6.18)$$

It is assumed that, for any material surface, material on one side exerts a distribution of "contact" force on material on the other side; that for small surface elements the resultant of the contact force distribution tends to a single resultant force proportional to the area as the area tends to zero; and that there are no external couples proportional to the volume of a material element. The limiting value of contact force per unit area is called the traction.

The area here cannot in fact be permitted to tend to zero if one is to stay within the macroscopic continuum description for real materials. In attempting to derive such continuum properties from molecular structure, use has been made of the idea of a physically infinitesimal material element [45]: i.e., one which is large enough (on a molecular scale) to contain enough molecules for application of suitable averaging methods to give valid and typical results, and small enough (on a macroscopic scale) so that the variation over the element of relevant macroscopic field variables shall be negligible.

It can be shown that the traction at P,t for an element of unit normal $\boldsymbol{\nu}$ is expressible in the form of a contravariant body vector $\boldsymbol{\pi}(P,t)\cdot\boldsymbol{\nu}(P,t)$ where $\boldsymbol{\pi}$ is a symmetric, contravariant body tensor - a stress tensor - of second rank which is independent of $\boldsymbol{\nu}$. It follows that

$$\text{the normal component of traction} = \boldsymbol{\nu}\cdot\boldsymbol{\pi}\cdot\boldsymbol{\nu}\ , \quad (6.19)$$

(a quadratic form similar to (6.7)) and hence also that $\boldsymbol{\pi}$ is determined if the values of this quadratic form are given for six suitably chosen normals $\boldsymbol{\nu}$. One can thus define stress at P,t as a function whose values are scalars (the normal components of traction) and whose arguments are material surface elements. According to this definition, stress is a one-state quantity independent of the orientation of the material relative to space-fixed directions.

The principal values and principal axes of stress are the roots $\sigma_1, \sigma_2, \sigma_3$ in ε and the corresponding solutions $\boldsymbol{\theta}_1, \boldsymbol{\theta}_2, \boldsymbol{\theta}_3$ in $\boldsymbol{\theta}$ of the equation

$$(\boldsymbol{\pi}\cdot\boldsymbol{\gamma} - \varepsilon\boldsymbol{\delta})\cdot\boldsymbol{\theta} = \mathbf{0}\ , \tag{6.20}$$

where the same arguments P,t are used throughout. There are always three real principal values whose corresponding principal axes are mutually orthogonal.

For unidirectional shear flow, the first and second normal stress differences N_1, N_2 and the shear stress σ may be regarded as scalars formed by contracting $\boldsymbol{\pi}$ with the appropriate shear flow base vectors:

$$N_1 = (\boldsymbol{\alpha}^1\boldsymbol{\alpha}^1 - \boldsymbol{\alpha}^2\boldsymbol{\alpha}^2) : \boldsymbol{\pi}\ ;$$
$$N_2 = (\boldsymbol{\alpha}^2\boldsymbol{\alpha}^2 - \boldsymbol{\alpha}^3\boldsymbol{\alpha}^3) : \boldsymbol{\pi};\quad \sigma = (\boldsymbol{\alpha}^1\boldsymbol{\alpha}^2) : \boldsymbol{\pi}\ . \tag{6.21}$$

This definition is applicable whether the shear flow is rectilinear or curvilinear. A shear flow is steady if it is unidirectional with the shear rate s independent of time; if, as time increases in steady shear flow, the shear stress becomes independent of time, the ratio of shear stress to shear rate is then called the viscosity.

An oscillatory shear flow is unidirectional with a shear rate of the form $a\omega \cos \omega t$; if the shear stress is of the form

$$\sigma(P,t) = a\{\omega\eta'(\omega)\cos \omega t + G'(\omega)\sin \omega t\} + O(a^3) \tag{6.22}$$

then η' is called the dynamic viscosity and G' is called the dynamic rigidity. Polymers having high enough molecular weights and narrow enough molecular weight distributions exhibit a remarkable form for the function $\log G'$v. $\log \omega$: over an appreciable intermediate frequency range, $\log G'$ varies comparatively slowly with $\log \omega$ and with molecular weight; the value of G' in this zone, called the plateau modulus, is typically of the order 0.1 MPa [9] and lies within the range of values of shear modulus typically found for lightly cross-linked polymers (at temperatures above the "glass transition temperature" T_g below which the polymer becomes hard). As ω increases beyond the plateau zone, G' increases to large "glass-like" values.

For a one-step shear strain of magnitude s at $t = 0$ in a material whose metric tensor is otherwise constant, the ratio $\sigma(s,t)/s = G(s,t)$ is called the <u>relaxation modulus</u>; the <u>relaxation spectrum</u> H is defined by the equation

$$G(s,t,T) = \int_{\tau=0}^{\infty} H(\tau,s,T)e^{-t/\tau}d\tau/\tau \ , \qquad (6.23)$$

where the temperature T has now been included in the arguments. Many polymeric liquids satisfy the <u>time-temperature superposition principle</u> [46] which may be expressed in the form

$$H(\tau,s,T) = T\rho(T)f(s,\tau/a(T)) \qquad (6.24)$$

for some functions f, $a(T)$ determined empirically for each material. a is called the <u>shift factor</u>.

A shear-free flow is an <u>elongational flow</u> if two principal strain rates, κ_2 and κ_3, say, are equal and the third, κ_1, is positive. If these rates and the differences of the corresponding principal stresses σ_1, σ_2, σ_3 are independent of time (at long times, if necessary), then the ratio $2(\sigma_1 - \sigma_2)/\kappa_1$ is called the <u>elongational viscosity</u> or <u>Trouton viscosity</u>.

<u>Cartesian space vectors and tensors</u> may be defined (in a manner similar to that used above for body vector and tensors) by making the following changes: the space manifold $\{Q\}$ is used instead of the body manifold $\{P\}$ and the space coordinate systems used in the definitions are restricted to be rectangular Cartesian, i.e. are such that $(ds)^2 = \delta_{ij}dx^i dx^j$ for every pair of neighboring places (dx^i being their coordinate differences). It follows, in particular, that the space coordinate transformation matrix L is orthogonal and, in consequence, there is no distinction to be drawn between covariant and contravariant Cartesian space vectors.

A <u>convected family of space coordinate systems</u> is generated by a single body coordinate system B and a motion of the body in space, i.e., a one-parameter family of states. At each instant t, B is congruent to one space coordinate system of the convected family. Convected components of a space tensor are components referred to each space coordinate system of the convected family. In the literature, a convected family of space coordinate systems is miscalled "a convected coordinate system".

A correspondence between body and space fields, generated by any state t, may be defined as follows, using the stress tensor $\boldsymbol{\pi}$ as an illustration. Let $S : Q \to x$ be any rectangular Cartesian space coordinate system. Let B be that body coordinate system that is congruent to S at time t. Let $\mathbf{p}$ be that second rank Cartesian space tensor whose components p^{ij} in S are numerically equal to π^{ij}, the components of $\boldsymbol{\pi}$ in B. We may write this correspondence in the form $\boldsymbol{\pi} \xrightarrow{t} \mathbf{p}$. It is many-one. $\mathbf{p}$ is the Cauchy stress tensor. It is easy to show that $\boldsymbol{\gamma}(P,t) \xrightarrow{t} \mathbf{I}$, $\boldsymbol{\gamma}^{-1}(P,t) \xrightarrow{t} \mathbf{I}$, where $\mathbf{I}$ is the unit Cartesian space tensor, whose representative matrix is the 3×3 unit matrix I in every rectangular cartesian space coordinate system.

REFERENCES

1. Oldroyd, J. G., Proc. Roy. Soc. (Lond.), A200 (1950), 523.
2. Lodge, A. S., Body Tensor Fields in Continuum Mechanics, Academic Press, 1974.
3. Brioullin, L., Ann. Phys., Paris (10e Serie), 3 (1925), 251.
4. James, H. M. and E. Guth, J. Chem. Phys. 15 (1947), 669.
5. Lodge, A. S., Proc. 7th Int. Cong. Rheol., ed. C. Klason and J. Kubat, Chalmers University of Technology, Gothenburg, Sweden (1976), 79.
6. Treloar, L. R. G., The Physics of Rubber Elasticity, 3rd ed., Clarendon Press, Oxford, 1975.
7. Price, C., G. Allen, F. de Candia, M. C. Kirkham and A. Subramaniam, Polymer 11 (1970), 486.
8. Janeschitz-Kriegl, H., Polymer Melt Rheology and Flow Birefringence, Springer-Verlag, Berlin, 1983.

9. Onogi, S., T. Masuda, and K. Kitagawa, Macromol. 3 (1970), 109.
10. Meissner, J., Trans. Soc. Rheol. 16 (3) (1972), 405.
11. Treloar, L. R. G., Trans. Faraday Soc. 36 (1940), 538.
12. Busse, W., J. Phys. Chem., 36 (1932), 2862.
13. Green, M. S. and A. V. Tobolsky, J. Chem. Phys. 14 (1946), 80.
14. Lodge, A. S., Trans. Faraday Soc. 52 (1956), 120.
15. Yamamoto, M., J. Phys. Soc. Japan 11 (1956), 413.
16. Armstrong, R. C., A. S. Lodge, M. H. Wagner, and H. H. Winter, Pure and Applied Chem. 54 (7) (1982), 1349.
17. Doi, M. and S. F. Edwards, J. Chem. Soc., Faraday Trans. II, 74 (1978), 1789.
18. Curtiss, C. F. and R. B. Bird, J. Chem. Phys. 74 (3) (1981), 2016.
19. Doi, M., J. Poly. Sci., Polymer Phys. ed. 18 (1980), 1905.
20. _______, 18 (1980), 1005.
21. Marrucci, G. and J. J. Hermans, Macromol. 13 (1980), 380.
22. Lodge, A. S. and J. Meissner, Rheologica Acta 11 (1972), 351.
23. Lodge, A. S., J. Non-Newtonian Fluid Mech., 14 (1984), 67.
24. Laun, H. M., Rheologica Acta 17 (1978), 1.
25. _______, personal comm.
26. Booij, H. C., (Proc. 9th Int. Congr. Rheol.), Adv. in Rheology, ed. B. Mena, A. Garcia-Rejon and C. Rangel, National University of Mexico, 1984.
27. Kimura, S., K. Osaki, and M. Kurata, J. Polymer Sci., Polymer Phys. ed., 19 (1981), 151.
28. Osaka, K., N. Bessho, T. Kojimoto, and M. Kurata, J. Rheol. 23 (1979), 617.
29. Vrentas, C. and W. W. Graessley, J. Non-Newtonian Fluid Mech., 9 (1981), 239.
30. Booij, H. C., personal comm. (this Symposium), 1984.
31. Chang, Hui and A. S. Lodge, Rheologica Acta 10 (1971), 448.

32. Huang, T. A., Ph.D. Thesis, Univ. of Wisconsin-Madison, 1976.

33. Wales, J. L. S., Ph.D. Diss., Delft Univ. Press, 1976.

34. Kaye, A., College of Aeronautics Note No. 134, (1962), Cranfield, Bletchley, Bucks., U.K.

35. Bernstein, B., E. A. Kearsley, and L. Zapas, Trans. Soc. Rheol. 7 (1963), 391.

36. Wagner, M. H., Rheologica Acta 15 (1976), 133, 136.

37. _______ and H. M. Laun, ibid. 17 (1978), 138.

38. _______, J. Non-Newtonian Fluid Mech., 4 (1978), 39.

39. _______, Rheologica Acta 18 (1979), 33.

40. Fukuda, M., K. Osaki, and M. Kurata, J. Soc. Rheol. Japan 2 (1974), 110.

41. Wagner, M. H. and S. E. Stephenson, J. Rheol., 23 (4) (1979), 489.

42. _______, T. Raible and S. E. Stephenson, ibid. 25 (1), 1.

43. Raible, T., A. Demarmels, and J. Meissner, Polymer Bull. 1 (1979), 397.

44. Meissner, J., S. E. Stephenson, A. Demarmels and P. Portmann, J. Non-Newtonian Fluid Mech., 11 (1982), 221.

45. Rosenfeld, L., Theory of Electrons, 2nd ed., Dover, New York (1965), p. 14.

46. Ferry, J. D., Viscoelastic Properties of Polymers, 3rd ed., Wiley, New York (1980).

47. Bruker, I. and A. S. Lodge, J. Rheol., in press.

48. Hutton, J. F., Nature, London, 200 (1963), 646.

49. Weissenberg, K., Nature, London, 159 (1947), 310.

The author was partially supported by the Army Research Office Contract No. DAAG29-84-K-0046.

ON VISCOPLASTIC MODELS

W. O. Williams

Department of Mathematics
Carnegie-Mellon University
Pittsburgh, Pennsylvania

1. General Considerations.

Constitutive relations falling under the general term 'plasticity' model two significant behaviors long observed in materials. The first is the transition at elevated stress levels from (essentially) elastic and reversible response to (relatively) free and irreversible flow under additional loading: plastic yield. The second is the persistence after release of stress of a residual deformation: permanent deformation. The traditional models of plastic behavior have included these effects in a rate-independent differential (hypoelastic) relationship between stress, strain, and some measure of the past damage done the material. For a simple example let us consider a purely compressive one-dimensional model with linear elastic behavior and idealized Bauschinger effect, which can be described by the equation

$$\frac{\dot{\sigma}-E\dot{\varepsilon}}{\sigma-y(\varepsilon)} = B(\sigma-y(\varepsilon)) \tag{1.1}$$

where y is a function describing the elastic limit and B is the barrier function, zero for negative arguments and infinite for positive arguments. Clearly (1.1) is rate-independent and the part of the quadrant outside the closure of the elastic range

$$\mathcal{E} = \{(\sigma,\varepsilon)\,|\,\sigma < y(\varepsilon)\}$$

is inaccessible. It is assumed that the elastic modulus E is less than $y'(\varepsilon)$ for all ε.[1]

Notice that a single simple experimental test serves to suggest this type of behavior. If the stress is gradually increased to a high level and then gradually reduced again to zero the $\sigma - \varepsilon$ trajectory will ascend a straight line, intersect and then traverse the curve $\sigma = y(\varepsilon)$ and then descend another straight line, leaving a residual strain. In principle a test consisting of a repetition of this proceedure for a sequence of higher maxima of stress will completely determine the model.

However, any material, including the pollycrystalline metals usually modeled as elastic-plastic in this way, displays some measure of rate-dependence in testing. If the test described above is run at several different constant loading rates typically these materials show a slight difference in behavior in the elastic zone but more significant deviations at the yield limit. These include excursions above the yield curve with relaxation back to the yield curve if the process is slowed as well as creep and relaxation in the general vicinity of that curve. Moreover, it is found that for a stress-strain state on the yield curve a superposed sound wave propagates at a velocity consistent with the modulus E rather than with the tangent modulus y' as would be predicted by (1.1).

For other types of materials these rate effects are much more pronounced and indeed may dominate the elastic-plastic behavior: Figure 1 shows experimental results for a highly-filled polymer studied in [1]. Similar behavior has been observed in such diverse materials as silk thread and phosphor bronze ([2], pp. 97 and 98).

Moreover, even if a rate-independent model such as (1.1) is thought to be satisfactory for a given material it actually may turn out that the corresponding field equations

[1]For convenience in graphing, we write the stress and strain as positive although they are compressive.

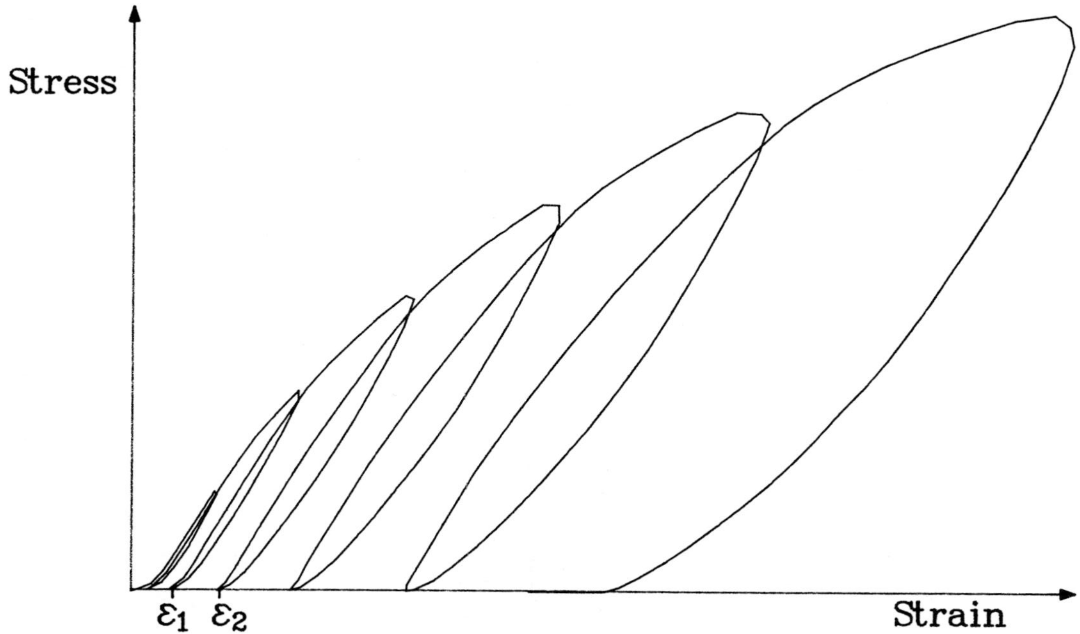

FIGURE 1. AN EXPERIMENTAL STRESS–STRAIN DIAGRAM FOR A HIGHLY FILLED POLYMER.

can most easily be analyzed by first introducing an artificial rate-dependence via a viscosity μ and later returning to the original model by means of the limit $\mu \to \infty$. (*Cf.* [3], [4], [5].) One may speculate that numerical analysis also may prove more stable for such lightly damped systems. Thus (1.1) might be replaced by

$$\frac{\dot{\sigma}-E\dot{\varepsilon}}{\sigma-y(\varepsilon)} = \mu H(\sigma-y(\varepsilon)) \tag{1.3}$$

where H is the Heaviside function, zero for negative arguments and unity for positive ones. This model features damping only outside the closure of the elastic zone and thus might be regarded as appropriate for polycrystalline metals.[1] As $\mu \to \infty$ clearly (1.3) formally approaches (1.1) and in fact it can be shown that solutions to the field equations corresponding to (1.3) approach the solutions responding to (1.1). In [3] it is shown, for a three-dimensional generalization of (1.3) (with $y(\varepsilon)$ a constant), that solutions to a certain class of boundary value problems approach, in the weak-* sense, solutions to the corresponding rate-independent theory. In [5] we have shown in the zero-dimensional case, *i.e.*, when the field equations reduce to ordinary differential equations, that solutions of (1.3) converge uniformly to solutions of (1.1). It does not appear difficult to extend these results to the case in which the viscosity also is active in the elastic zone (*cf.* [5]).

2. A specific model: general description.

A model introduced by Browning, Gurtin and Williams in [1] offers another method of introducing rate dependence. It was devised in [1] to describe behavior such as that illustrated in Figure 1 with viscous behavior throughout and with zero initial yield ($y(0) = 0$).[2] The basic premise

---

[1] See [6] for a discussion: Bell in [2] argues against the utility of the viscosity method for introducing rate effects, but on the basis of the behavior in the elastic zone.

[2] It is not difficult to see how to redefine the model to reflect essentially undamped behavior in the elastic zone, but the model then loses some of the virtues arising from its simplicity. Of course the $y(0) \neq 0$ case requires only a trivial modification.

is that the rate-independent model is modified not adding viscous terms but by functional composition, specifically by convolution in time with a relaxation function. As illustrated below this method offers some advantage in terms of characterization, since a single test characterizes the entire rate dependence. It likewise offers the advantage in analysis that it can borrow both from classical plasticity and classical linear viscoelasity. For example there is a correspondence principle exactly analogous to that one used to derive solutions to problems in linear viscoelasticity from the corresponding solutions in elasticity (cf. [7]). It is also not difficult to implement numerically: at Los Alamos National Laboratory the standard ADINA elastic-plastic code has been modified by superposing a convolution and this is now proving effective.[1]

The materials considered in [1] and discussed below were high explosives, a dense packing of explosive crystals in a urethane rubber matrix. These materials are very weak in tension and so only the compressive regime was considered. Safety rules dictate that casting, machining and testing of these materials be done by remote control in isolated locations and hence experimental data is quite expensive to obtain. Thus the relative simplicity of the model was in part dictated by paucity of data. Nonetheless, as shown below, the model proved quite successful, suggesting that it may be useful in less exotic settings.

3. Construction of the model.

We first introduce a pseudo-stress π which is presumed related to the strain ε by a classical rate-independent elastic-plastic law. We assume the form

$$\pi = F(\varepsilon, \varepsilon_m) \tag{3.1}$$

where $\varepsilon_m(t)$ is the maximum of ε on $[0,t]$. This is separated into convenient parts by setting

---

[1] If the relaxation function is expressed as a Prony series (series of exponentials) then the modified code can be reduced to be very little more time-consuming than the original one.

$$y(\varepsilon) = F(\varepsilon,\varepsilon) \tag{3.2}$$

and

$$\pi = y(\varepsilon)f(\varepsilon,\varepsilon_m) \tag{3.3}$$

so that y is the yield function ('virgin response' in [1]) and $y(\varepsilon_m)f(\cdot,\varepsilon_m)$ describes the response within the elastic zone "below" ε_m. Its value at 0 is the permanent strain induced by ε_m.

The rate effect is now introduced by postulating a <u>relaxation function</u> g such that the true stress σ and π are related by

$$\sigma(t) = \int_0^t g(t-s)d\pi(s), \tag{3.4}$$

presuming the history is zero prior to time zero. We also write $\sigma = g*d\pi$.

4. Characterization.

Three simple experimental tests now suffice to determine g, y, and f. First a stress relaxation test ($\varepsilon = \varepsilon_m$ constant) yields g to within an irrelevant multiplicative constant.[1] A computation yields its convolution-inverse the creep function j which is used in the next two cases to compute π from the experimental values of σ. Doing this, a single loading test suffices then to determine y. Finally y is used with j to allow computation of f from a test which consists of a series of loadings, unloadings, and reloadings to sucessively higher values. Such a characterization was carried out for three materials, and then the models tested by comparison to further experimental tests.

Figure 2 shows the π-ε behavior of the model fit to one of these. As can be seen a linear elastic relation was

---

[1] In the characterization in [1] a creep test was used to determine the creep function j, but g and j are easily computed one from the other.

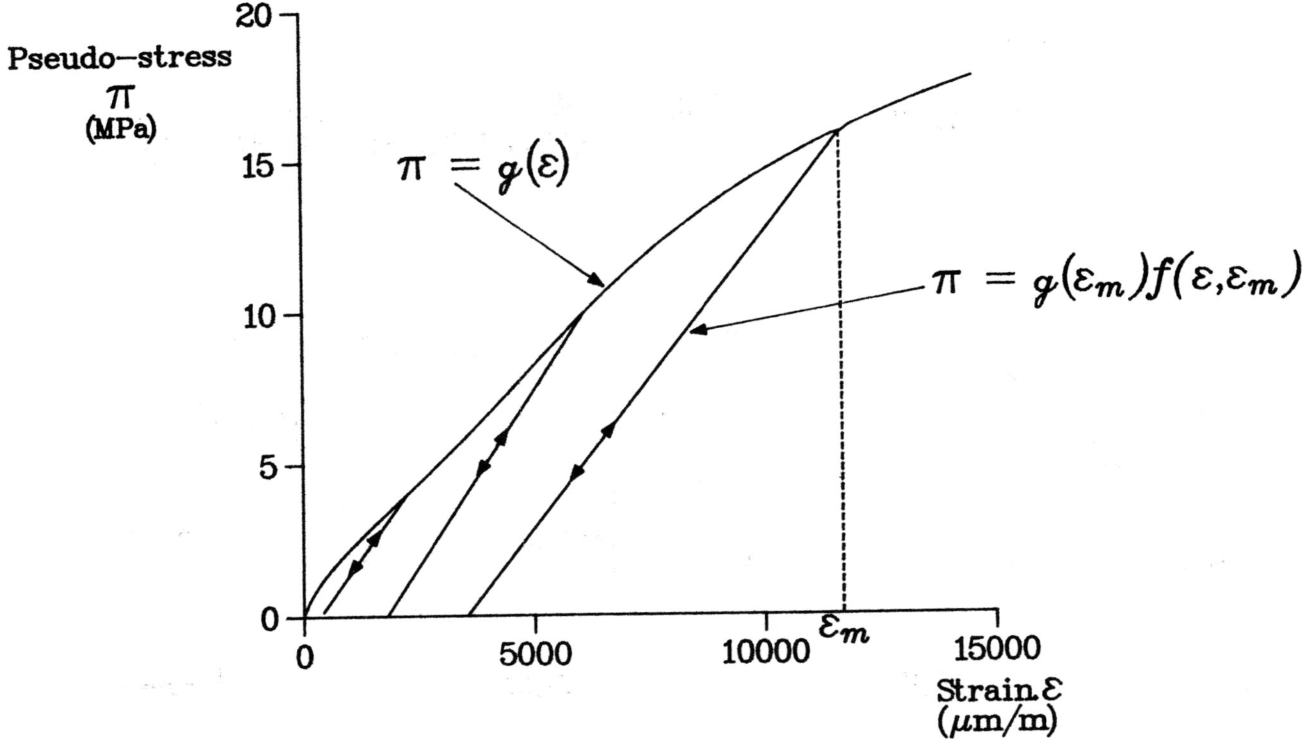

FIGURE 2. THE UNDERLYING RATE–INDEPENDENT RELATION.

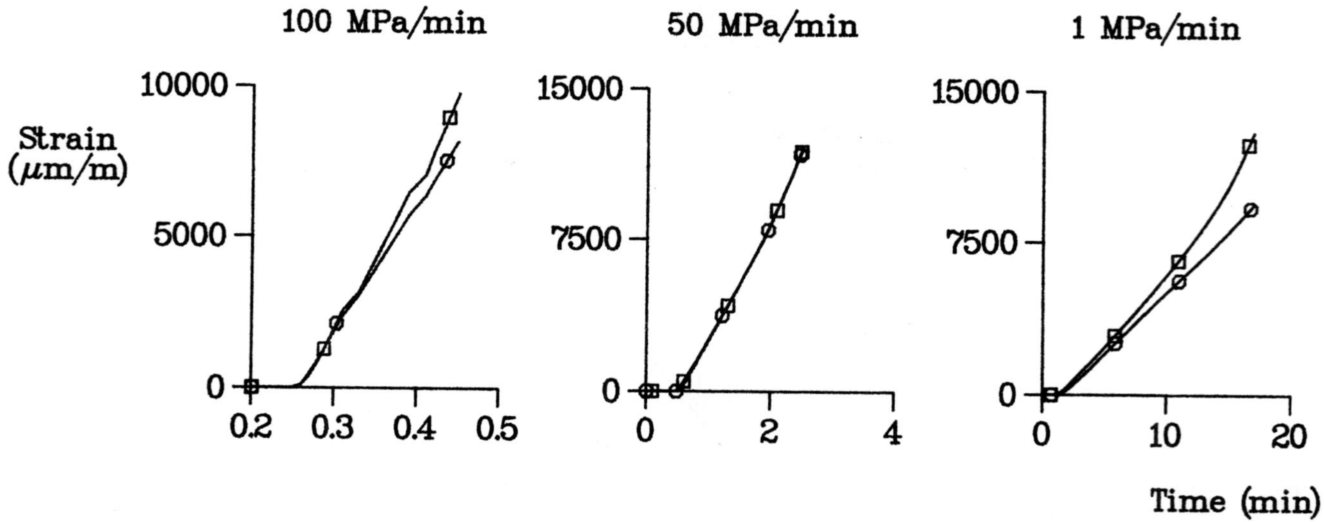

FIGURE 3. EXPERIMENTAL (□) AND CALCULATED (○) RESPONSE TO UNIFORM LOADINGS AT VARIOUS RATES.

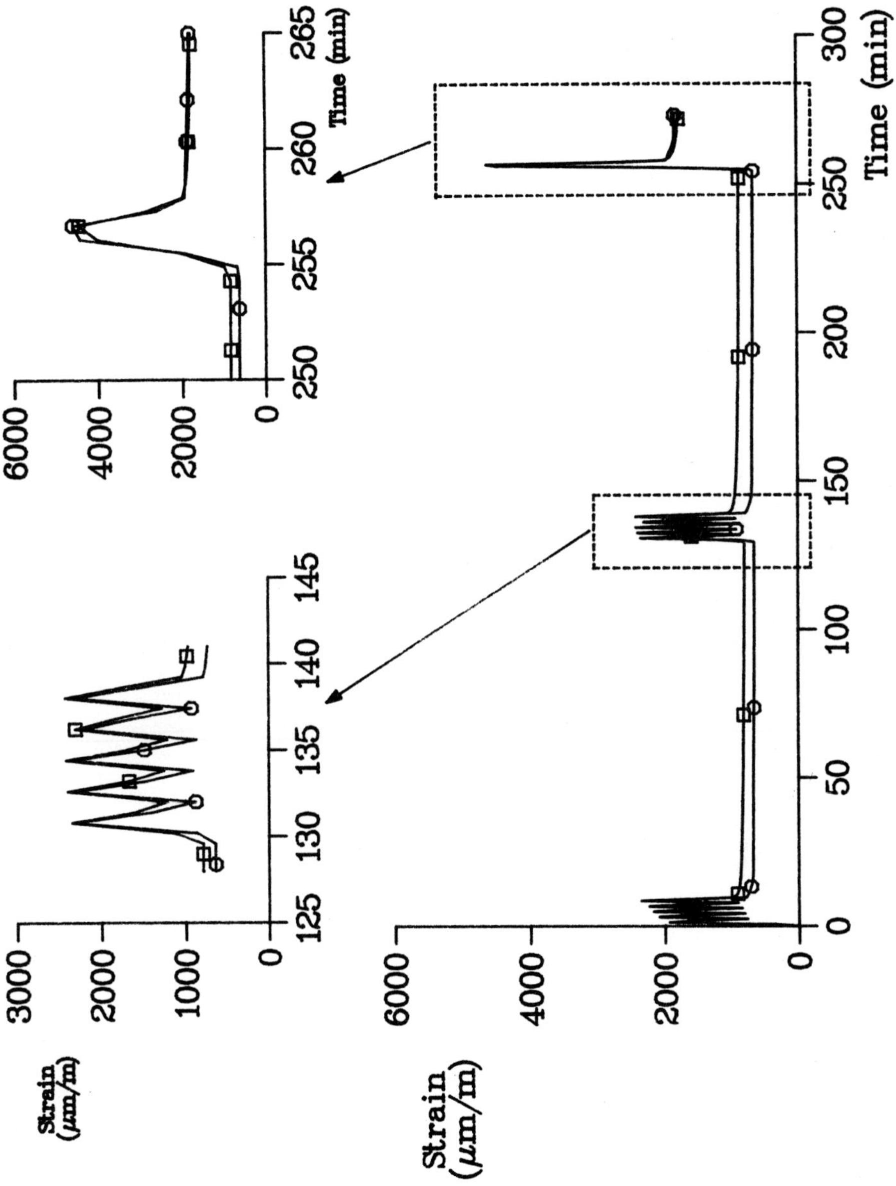

FIGURE 4. EXPERIMENTAL (□) AND CALCULATED (○) RESPONSE TO A COMPLEX LOADING.

used in the elastic zone; a small amount of residual hysterisis rendered this as accurate as any other fit in all three cases. Figure 3 illustrates the degree to which the rate-dependence was adequately portrayed, while Figure 4 shows a representative test of a model against an experiment not used in its characterization.

5. A three-dimensional extension.

True three-dimensional tests of the three materials were not available, so the following extension proceedure was used: the object was to produce the most general three-dimensional law derivable from one-dimensional results.[1] The assumptions made:

i) The bulk response was assumed to be elastic. Using hoop strain as well as longitudinal strain measurements it is possible to compute the bulk behavior exactly. These materials, like other viscoplastic materials, show rather irregular behavior in bulk,[2] but because this effect did not seem significant for applications a best fit $(\pi-\varepsilon)$-elastic law was assumed.

ii) The rate dependence was assumed isotropic, so that the one dimensional relaxation function g serves to relate the tensors $\underset{\sim}{\sigma}$ and $\underset{\sim}{\pi}$;

$$\underset{\sim}{\sigma} = g * d\underset{\sim}{\pi} \tag{5.1}$$

iii) The yield function was assumed to be of the Von Mises form

$$\varphi(\underset{\sim}{\pi}_0) = \underset{\sim}{\pi}_0 \cdot \underset{\sim}{\pi}_0 \tag{5.2}$$

where $\underset{\sim}{\pi}_0$ denotes the deviator

$$\underset{\sim}{\pi}_0 = \underset{\sim}{\pi} - \frac{1}{3}(\mathrm{tr}\underset{\sim}{\pi})\underset{\sim}{1}. \tag{5.3}$$

Note that work-hardening (the Bauschinger effect) is present, as φ_m may increase; φ_m replaces ε_m as the determinant of yield.

---

[1] Cf. [8] for a different approach to a three-dimensional theory from the same one-dimensional model.

[2] This phenomena is described by Cristescu [9] and appears to be of significance in predicting failure in rock mechanics.

iv) The $\underset{\sim}{\pi}_0 - \underset{\sim}{\varepsilon}_0$ relation was assumed hypoelastic both within and on the yield surface so that

$$\dot{\underset{\sim}{\varepsilon}}_0 = A(\underset{\sim}{\pi}_0)[\dot{\underset{\sim}{\pi}}_0] \quad \text{if} \quad \varphi < \varphi_m \tag{5.4}$$

$$\dot{\varepsilon}_0 = B(\underset{\sim}{\pi}_0)[\dot{\underset{\sim}{\pi}}_0] \quad \text{if} \quad \varphi = \varphi_m \tag{5.5}$$

with $A(\underset{\sim}{\pi})[\cdot]$, $B(\underset{\sim}{\pi})[\cdot]$ linear. Moreover if these are each symmetric and the relations assumed continuous at $\varphi = \varphi_m$ it follows (cf. [9]) that

$$B(\underset{\sim}{\pi}_0)[\dot{\underset{\sim}{\pi}}_0] = A(\underset{\sim}{\pi}_0)[\dot{\underset{\sim}{\pi}}_0] + \beta(\varphi)\dot{\varphi}\underset{\sim}{\pi}_0 \tag{5.6}$$

where β is a scalar function. (The continuity condition is equivalent to Drucker's normality condition.)

v) This relation was further assumed isotropic, so that $A(\underset{\sim}{\pi}_0)$ too reduces to a scalar multiplication $\alpha(\underset{\sim}{\pi}_0)$ and both α and β reduce to functions of the two invariants of $\underset{\sim}{\pi}_0$. Finally the quadratic invariant, φ, was selected to be the only one present. This last assumption is completely unjustifiable on physical or mathematical grounds and was made just for simplicity.

As a final form then we have for all values of φ

$$\dot{\underset{\sim}{\varepsilon}}_0 = \alpha(\varphi_m)\dot{\underset{\sim}{\pi}}_0 + \beta(\varphi_m)\dot{\varphi}_m\underset{\sim}{\pi}_0 \tag{5.7}$$

and

$$\text{tr}\dot{\underset{\sim}{\varepsilon}} = K(\varphi_m)(\text{tr}\dot{\underset{\sim}{\pi}}) \tag{5.8}$$

where K is a bulk modulus as noted before. The important property of (5.7) and (5.8), gained at the cost of the unjustified assumption on invariants, is that α and β are obtained algebraically, from the functions y, f, and K obtained from the one-dimensional tests.

Remark: All of the above discussion presumes that strains are small. In a large strain version all of the strain time-derivatives must be replaced by properly invariant deformation rates.

REFERENCES

1. Browning, R. V., M. E. Gurtin, and W. O. Williams, A viscoplastic constitutive theory for filled polymers, Int. J. Solid. Struct., (to appear).
2. Bell, F., The Experimental Foundations of Solid Mechanics, Handbuch der Physik, V. Vla/1, Berlin, Springer, (1973).
3. Duvant, G. and J.-L. Lions, Inequalities in Mechanics and Physics, New York, Springer, (1976).
4. Anzellotti, G. and S. Luckhaus, Dynamical evolution of elasto-perfectly plastic bodies, preprint.
5. Mihaelescu-Suliciu, M., D. R. Owen, I. Suliciu, and W. Williams, On solutions of the dynamic visco-plastic oscillator, (to appear).
6. Cristescu, N. and I. Suliciu Viscoplasticity, The Hague, Martinus Nijhoff/Dr. W. Junk, (1983).
7. Gurtin, M. and E. Sternberg, On the linear theory of viscoelasticity, Arch. Rational Mech. Anal., 11, (1962), pp. 291-356.
8. Simo, J. C. and R. L. Taylor, A simple 3-dimensional viscoelastic model accounting for damage effects, Report No. UCB/SESM-83/10, Dept. Civ. Eng'g., UCB, (1983)
9. Cristescu, I., Rock dilatancy in uniaxial tests, Rock Mech., 15, (1982), pp. 133-144.
10. Gurtin, M., On the hypoelastic formulation of plasticity using the past maximum of stress, J. Appl. Mech. (to appear).

Acknowledgement: The work described was supported in part by the National Science Foundation.

DISSIPATION IN MATERIALS WITH MEMORY

Constantine Dafermos

Lefschetz Center for Dynamical Systems
Division of Applied Mathematics
Brown University
Providence, Rhode Island

Dedicated to John Nohel on his sixtieth birthday.

1. INTRODUCTION.

The equations of motion of elastic materials are of hyperbolic type and so signals propagate with characteristic speed. When the elastic response of the material is nonlinear, characteristic speeds depend on deformation gradient. Consequently, compressive profiles get steeper and eventually break forming discontinuities in the velocity and the deformation gradient (shock waves).

It is of interest to compare and contrast the behavior of elastic materials with the behavior of materials with fading memory. Even though, generally speaking, memory dependence exerts a soothing influence, it is not clear in advance whether in materials with fading memory, in which the instantaneous response is elastic, the dissipative effect of viscosity is sufficiently powerful to overcome the destabilizing action of nonlinearity.

The first illustration of the competition between memory effects and nonlinear elastic response in materials with fading memory was provided by the theory of propagation of acceleration waves [1]. It was found that the amplitude $V(t)$ of an acceleration wave that propagates into a rest state evolves according to the Bernoulli equation

$$\dot{V} + \alpha V - \beta V^2 = 0 \tag{1.1}$$

where α and β are material constants. When the material is elastic, $\alpha = 0$ and so compressive waves of arbitrary initial amplitude explode in finite time. By contrast, in the presence of fading memory it is $\alpha > 0$ and so compressive waves do not explode unless their initial amplitude exceeds the threshold value α/β.

In view of the above result, it is tempting to conjecture that the dissipative effect of viscosity dominates in motions near equilibrium, preserving the smoothness of such motions, while in motions far from equilibrium the destabilizing action of nonlinear instantaneous elastic response may overpower the dissipative effect of viscosity causing the explosion of the acceleration and the velocity gradient.

To test the above conjecture one should consider the equations of motion

$$\rho(\underset{\sim}{x})\partial_t^2 \underset{\sim}{\chi}(\underset{\sim}{x},t) = \mathrm{Div}\, S(\underset{\sim}{x},t) + \rho(\underset{\sim}{x})\underset{\sim}{b}(\underset{\sim}{x},t) \tag{1.2}$$

for material bodies with fading memory:

$$S(\underset{\sim}{x},t) = \underset{\tau=0}{\overset{\infty}{\mathscr{S}}}(\nabla \underset{\sim}{\chi}(\underset{\sim}{x},t-\tau)), \tag{1.3}$$

and investigate existence, uniqueness and regularity of solutions under assigned initial history data and boundary conditions. A number of recent analytical studies [2,3,4,5,6,7] of (1.2) for various types of constitutive relations of the form (1.3), mainly in one-dimensional bodies, seem to support the aforementioned conjecture on the outcome of the competition between dissipation and nonlinearity. The article by Renardy in this volume describes some of the relevant results and gives a comprehensive list of references.

The analysis has revealed that the theory of the complicated system (1.2), (1.3) can be illuminated with the help of the simple model integrodifferential equation

$$\partial_t u(x,t) + \partial_x \phi(u(x,t)) - \int_0^t K(t-\tau)\partial_x \psi(u(x,\tau))d\tau = 0 \tag{1.4}$$

which has been studied by several authors [8,9,10]. Our intent here is to look at (1.4) from a somewhat different angle so as to bring out the relevance of the Bernoulli equation (1.1), which, as we have seen, governs the evolution of the amplitude of acceleration waves, while avoiding the use of characteristics.

We will consider (1.4) on the upper half-plane $\{-\infty < x < \infty, 0 \leq t < \infty\}$ under initial conditions

$$u(x,0) = u_0(x), \quad -\infty < x < \infty. \tag{1.5}$$

In order to prevent "resonance" between the instantaneous "elastic" response and the fading memory response, we will assume $\varphi'(u) > 0$. We show that when the instantaneous "elastic" response is nonlinear, say $\varphi''(u) > 0$, then solutions of (1.4), (1.5) with $\sup|u_0(\cdot)|$ small but $-\inf u_0'(\cdot)$ large explode in a finite time, no matter what properties the memory term has. On the other hand, we prove that when the memory term is appropriately dissipative, $\sup|u_0(\cdot)|$ is small and $-\inf u_0'(\cdot)$ does not exceed a threshold positive value, then a smooth solution of (1.4), (1.5) exists on the entire upper half x-t plane. The precise form of these results will be stated in Section 2 and the proofs will be presented in Section 3.

2. STATEMENT OF RESULTS.

We consider the initial value problem (1.4), (1.5) with $\varphi \in C^2(-\infty,\infty)$, $\psi \in C^1(-\infty,\infty)$, $K \in C^2[0,\infty)$, $u_0 \in C^1(-\infty,\infty)$ and u_0, u_0' bounded on $(-\infty,\infty)$. Our results will depend on various combinations of assumptions out of the list

$$\varphi'(u) > 0, \tag{2.1}$$

$$\varphi''(u) > 0, \tag{2.2}$$

$$\psi'(u) > 0, \tag{2.3}$$

$$K(t) > 0, \quad t \in [0,\infty), \tag{2.4}$$

$$K \in W^{2,1}(0,\infty), \tag{2.5}$$

$$\varphi'(u)K'(t) + \psi'(u)K(0)K(t) \leq 0, \quad t \in [0,\infty), \tag{2.6}$$

which are to hold for u in sets to be specified below.

Assumption (2.1) would rule out "resonance" between the instantaneous "elastic" response and the fading memory response. Assumption (2.2) would indicate that the instantaneous "elastic" response is genuinely nonlinear and thus potentially destabilizing. Assumptions (2.3) through (2.6) would express that the memory response is dissipative and thus stabilizing. In the special case $\varphi \equiv \psi$, (2.6) may be replaced with the weaker assumption

$$N'(t) \geq 0,\ t \in [0,\infty), \tag{2.7}$$

where N is the resolvent kernel of K (cf. (3.13) below).

The first proposition states that, no matter what the memory response is, when φ is genuinely nonlinear then solutions of (1.4), (1.5) with $\sup|u_0(\cdot)|$ small and $-\inf u'(\cdot)$ large must blow up in a finite time.

Theorem 2.1. *Assume (2.1), (2.2) hold at $u = 0$. Given any $T > 0$ there are positive constants ε and μ, depending on T, such that when*

$$\sup|u_0(\cdot)| < \varepsilon, \tag{2.8}$$

$$-\inf u_0'(\cdot) > \mu, \tag{2.9}$$

then the length of the maximal interval of existence of C^1-smooth solutions to (1.4), (1.5) cannot exceed T.

The next proposition establishes a *maximum principle* for solutions of (1.4), (1.5) under the assumption that the memory response is appropriately dissipative (compare with [8]).

Theorem 2.2. *Assume (2.1), (2.3), (2.4) and (2.6) (or (2.7) when $\varphi \equiv \psi$) hold for u in a neighborhood of the interval $[\inf u_0(\cdot), \sup u_0(\cdot)]$. If for some $T \in (0,\infty]$ (1.4), (1.5) admits a bounded solution $u \in C^1((-\infty,\infty) \times [0,T))$, then*

$$\inf u_0(\cdot) \leq u(x,t) \leq \sup u_0(\cdot),\quad -\infty < x < \infty,\quad 0 \leq t < T. \tag{2.10}$$

The following proposition establishes the existence of a global solution to (1.4), (1.5), provided the memory response be dissipative and $\sup|u_0(\cdot)|$, $-\inf u_0'(\cdot)$ be sufficiently small.

Theorem 2.3. *Assume (2.1), (2.2), (2.3), (2.4), (2.5) and (2.6) (or (2.7) when $\varphi \equiv \psi$) hold for u in a neighborhood of 0. There are positive constants ε and δ such that if*

$$\sup|u_0(\cdot)| < \varepsilon, \tag{2.11}$$

$$-\inf u_0'(\cdot) < \delta, \tag{2.12}$$

<u>then there exists a unique solution</u> $u \in C^1((-\infty,\infty) \times [0,\infty))$ <u>of</u> (1.4), (1.5) <u>which is bounded and has bounded first derivatives on</u> $(-\infty,\infty) \times [0,\infty)$.

3. PROOF OF THEOREMS.

Let us assume (1.4), (1.5) admits a solution $u \in C^1((-\infty,\infty) \times [0,T))$ for some $T \in (0,\infty]$. We set

$$Y(x,t) := \int_0^t K(t-\tau)\partial_x \psi(u(x,\tau))d\tau, \quad -\infty < x < \infty, \quad 0 \leq t < T. \tag{3.1}$$

Our first goal is to show that when (2.1) holds for every u in the range of the solution then, for any $x \in (-\infty,\infty)$, $t \in (0,T)$, $Y(x,t)$ is determined solely by the "history" function $u(x,\cdot): [0,t] \to (-\infty,\infty)$. (A result of this type was first obtained by MacCamy [8] in the special case $\varphi \equiv \psi$). To this end we use (1.4) to get

$$\partial_x u = -\frac{1}{\varphi'(u)}\partial_t u + \frac{1}{\varphi'(u)} Y. \tag{3.2}$$

Substituting $\partial_x u$ from (3.2) into (3.1) we deduce that Y satisfies the linear integral equation

$$Y(x,t) - \int_0^t K(t-\tau)g(u(x,\tau))Y(x,\tau)d\tau \tag{3.3}$$

$$= -\int_0^t K(t-\tau)\partial_\tau H(u(x,\tau))d\tau$$

where we have set

$$g(u) := \frac{\psi'(u)}{\varphi'(u)}, \tag{3.4}$$

$$H(u) := \int_0^u g(v)dv. \tag{3.5}$$

For fixed $x \in (-\infty,\infty)$, $t \in [0,T)$, we introduce the resolvent kernel $L(x,t,\cdot)$ through

$$L(x,t,s) + K(t-s) = \int_s^t K(\tau-s)g(u(x,\tau))L(x,t,\tau)d\tau, \quad 0 \leq s \leq t. \tag{3.6}$$

A simple calculation then shows that (3.3) admits the solution

$$Y(x,t) = -\int_0^t K(t-\tau)\partial_\tau H(u(x,\tau))d\tau + \int_0^t L(x,t,\tau)g(u(x,\tau))\int_0^\tau K(\tau-s)\partial_s H(u(x,s))dsd\tau. \tag{3.7}$$

Interchanging the order of integrations in (3.7) and integrating by parts we deduce

$$\begin{aligned} Y(x,t) &= -K(0)H(u(x,t)) \\ &+ \{K(t) - \int_0^t L(x,t,\tau)g(u(x,\tau))K(\tau)d\tau\}H(u_0(x)) \\ &+ \int_0^t \{-K'(t-s) + K(0)L(x,t,s)g(u(x,s)) \\ &+ \int_s^t L(x,t,\tau)g(u(x,\tau))K'(\tau-s)d\tau\}H(u(x,s))ds. \end{aligned} \tag{3.8}$$

Combining (3.8) with (3.6) we conclude

$$Y(x,t) = -K(0)H(u(x,t)) - L(x,t,0)H(u_0(x)) + \int_0^t M(x,t,s)H(u(x,s))ds \tag{3.9}$$

where

$$M(x,t,s) := -K'(t-s) + K(0)L(x,t,s)g(u(x,s)) + \int_s^t L(x,t,\tau)g(u(x,\tau))K'(\tau-s)d\tau,\ 0 \le s \le t. \tag{3.10}$$

Equations (3.9), (3.10), (3.6) verify that, as claimed above, $Y(x,t)$ is indeed determined solely by the "history" $\tau \mapsto u(x,\tau)$, $\tau \in [0,t]$.

For future reference we note that, by virtue of (3.6), (3.10) yields

$$\begin{aligned} M(x,t,s) &= -K'(t-s) - g(u(x,s))K(0)K(t-s) \\ &+ \int_s^t L(x,t,\tau)g(u(x,\tau))\{K'(\tau-s) \\ &\qquad + g(u(x,s))K(0)K(\tau-s)\}d\tau. \end{aligned} \tag{3.11}$$

Also for future use, we integrate (3.10), with respect to s, over $[0,t]$. Interchanging the order of integrations and using (3.6) we get

$$\begin{aligned} \int_0^t M(x,t,s)ds &= K(0) - K(t) + \int_0^t L(x,t,\tau)g(u(x,\tau))K(\tau)d\tau \\ &= K(0) + L(x,t,0). \end{aligned} \tag{3.12}$$

In the special case $\varphi \equiv \psi$, whence $g \equiv 1$, it is $L(x,t,s) = N(t-s)$, $-\infty < x < \infty$, $0 \leq t < T$, $0 \leq s \leq t$, where

$$N(t) + K(t) = \int_0^t N(t-\tau)K(\tau)d\tau. \tag{3.13}$$

For that case a simple calculation shows that (3.10) reduces to

$$M(x,t,s) = N'(t-s), \quad 0 \leq s \leq t. \tag{3.14}$$

Proof of Theorem 2.2. We fix constants $a \in (0,\infty)$, $b \in (0,T)$ and consider the trapezoid

$$\mathscr{T}: = \{(x,t): 0 \leq t \leq b, -a-k(b-t) \leq x \leq a+k(b-t)\}, \tag{3.15}$$

where k is an upper bound of $|\varphi'|$ on the range of the solution. We claim that

$$\underline{u} \leq u(x,t) \leq \overline{u}, \quad (x,t) \in \mathscr{T}, \tag{3.16}$$

where $\underline{u}: = \min\{u_0(x): -a-kb \leq x \leq a+kb\}$, $\overline{u}: = \max\{u_0(x): -a-kb \leq x \leq a+kb\}$. Note that (3.16) would sharpen the assertion of the theorem by establishing also a local domain of dependence property.

We determine $\eta > 0$ sufficiently small that (2.1), (2.3) and (2.6) hold for $u \in [\underline{u}-\eta,\overline{u}+\eta]$. Arguing by reductio ad absurdum, suppose the range of the restriction of the solution on $\mathscr{T}$ contains the point $\hat{u} \in (\overline{u},\overline{u}+\eta)$. Let t be the smallest number in $(0,b]$ with the property that there is $x \in [-a-k(b-t),a+k(b-t)]$ such that $u(x,t) = \hat{u}$. Directional derivatives of u at (x,t) in any direction inside $\mathscr{T}$ must be nonnegative. In particular, using (1.4) and (3.1), we get

$$Y(x,t) = \partial_t u(x,t) + \varphi'(u(x,t))\partial_x u(x,t) \geq 0. \tag{3.17}$$

By account of (3.6) and (2.4),

$$L(x,t,s) < 0, \quad 0 \leq s \leq t. \tag{3.18}$$

Then, using (3.11), (2.6), (3.4), (2.3), (2.1) (or (3.14), (2.7) in the special case $\varphi \equiv \psi$) we deduce

$$M(x,t,s) \geq 0, \ 0 \leq s \leq t. \tag{3.19}$$

From (3.9), (3.18), (3.19), (3.4), (3.5), (2.1), (2.3) and (3.12) we get

$$Y(x,t) < \{-K(0) - L(x,t,0) + \int_0^t M(x,t,s)ds\}H(\hat{u}) = 0 \quad (3.20)$$

and this is a contradiction to (3.17).

The same argument shows that the range of the restriction of the solution on $\mathscr{T}$ cannot contain any point $\hat{u} \in (\underline{u}-\eta, \underline{u})$. This completes the proof of (3.16) and thereby of the theorem. □

In preparation for the the proof of Theorems 2.1 and 2.3, let us assume, temporarily, that $u_0 \in C^2(-\infty,\infty)$. In this case, if $u \in C^1((-\infty,\infty) \times [0,T))$ is a solution of (1.4),(1.5), and since $Y \in C^1((-\infty,\infty) \times [0,T))$ by virtue of (3.9), (3.10), (3.6), it follows by standard theory that $u \in C^2((-\infty,\infty) \times [0,T))$. Differentiating (1.4) with respect to t we obtain

$$\partial_t\partial_t u + \varphi'(u)\partial_t\partial_x u + \varphi''(u)\partial_t u\partial_x u - K(0)\psi'(u)\partial_x u$$
$$- \int_0^t K'(t-\tau)\psi'(u)\partial_x u \, d\tau = 0. \quad (3.21)$$

We substitute $\partial_x u$ from (3.2) into (3.21) and use (3.4), (3.5). After an integration by parts with respect to τ, we deduce

$$\partial_t\partial_t u + \varphi'(u)\partial_x\partial_t u + A\partial_t u - B(\partial_t u)^2 = \Gamma, \quad (3.22)$$

where

$$A(x,t) := K(0)g(u(x,t)) + \frac{\varphi''(u(x,t))}{\varphi'(u(x,t))} Y(x,t), \quad (3.23)$$

$$B(x,t) := \frac{\varphi''(u(x,t))}{\varphi'(u(x,t))} \quad (3.24)$$

$$\Gamma(x,t) := -K'(0)H(u(x,t)) + K'(t)H(u_0(x)) + K(0)g(u(x,t))Y(x,t)$$
$$- \int_0^t K''(t-\tau)H(u(x,\tau))d\tau + \int_0^t K'(t-\tau)g(u(x,\tau))Y(x,\tau)d\tau. \quad (3.25)$$

<u>Proof of Theorem 2.3</u>. Let, temporarily, $u_0 \in C^2(-\infty,\infty)$. Assuming there exists a solution $u \in C^2((-\infty,\infty) \times [0,T))$, we proceed to establish a priori bounds on $|u|$ and $|\partial_t u|$, $|\partial_x u|$.

Fix $\varepsilon > 0$ so small that (2.1), (2.2), (2.3) and (2.6) hold on $[-\varepsilon,\varepsilon]$. When (2.11) holds, Theorem 2.2 yields

$$|u(x,t)| < \varepsilon, \quad -\infty < x < \infty, \quad 0 \le t < T. \tag{3.26}$$

Using (3.9), (3.18), (3.19), (3.12), (3.4), (3.5), (2.1), (2.3), (2.4), and (3.26), we get

$$|Y(x,t)| < K(0)[H(\varepsilon) - H(-\varepsilon)], \quad -\infty < x < \infty, \quad 0 \le t < T. \tag{3.27}$$

Combining (3.23), (3.24), (3.25) with (3.26), (3.27), (2.1), (2.2), (2.3), (2.4), (2.5), (3.4) and (3.5) we deduce that when ε is sufficiently small then

$$A(x,t) \ge \alpha > 0, \; 0 < B(x,t) \le \beta, \; |\Gamma(x,t)| \le \gamma, \quad -\infty < x < \infty, \; 0 \le t < T, \tag{3.28}$$

where α, β, γ are positive constants, independent of T, such that

$$\gamma < \frac{\alpha^2}{4\beta}. \tag{3.29}$$

We now fix constants $a \in (0,\infty)$, $b \in (0,T)$ and consider the trapezoid $\mathscr{T}$ defined by (3.15). We proceed to derive upper and lower bounds for $\partial_t u$ on $\mathscr{T}$. For $t \in [0,b]$, we set

$$V(t) := \max\{0,\max\{\partial_t u(x,t) : -a-k(b-t) \le x \le a+k(b-t)\}\}. \tag{3.30}$$

V is a nonnegative Lipschitz function on $[0,b]$. If at some $t \in (0,b]$, $V(t) > 0$, we identify the point $x \in [-a-k(b-t), a+k(b-t)]$ such that

$$V(t) = \partial_t u(x,t). \tag{3.31}$$

Then, for $\eta \in (0,t)$,

$$V(t) - V(t-\eta) \le \partial_t u(x,t) - \partial_t u(x-\eta\varphi'(u(x,t)), t-\eta) \tag{3.32}$$

whence

$$D_t^- V(t) \le \partial_t \partial_t u(x,t) + \varphi'(u(x,t))\partial_x \partial_t u(x,t). \tag{3.33}$$

Therefore, combining (3.33) with (3.22), (3.31) and (3.28), we deduce that at every $t \in (0,b]$ with $V(t) > 0$ it is

$$D_t^- V + \alpha V - \beta V^2 \le \gamma. \tag{3.34}$$

It is clear that (3.34) holds, a fortiori, at point $t \in (0,b]$ with $V(t) = 0$. We set

$$V_0 : = \frac{\alpha + (\alpha^2 - 4\beta\gamma)^{1/2}}{2\beta} \tag{3.35}$$

and note that when (2.12) holds with $\delta = V_0/\varphi'(\varepsilon)$ then $V(0) = \max\{0, \max \partial_t u(\cdot,0)\} = \max\{0, -\min\{\varphi'(u_0(\cdot))u_0'(\cdot)\}\} < V_0$ in which case (3.34) yields the upper bound

$$\partial_t u(x,t) < V_0, \quad (x,t) \in \mathcal{T}. \tag{3.36}$$

In order to get lower bounds for $\partial_t u$ on $\mathcal{T}$, we define

$$W(t) : = \min\{0, \min\{\partial_t u(x,t) : -a-k(b-t) \leq x \leq a+k(b-t)\}\}, \tag{3.37}$$

for each $t \in [0,b]$. As above, we can show that if $W(t) < 0$ at a point $t \in (0,b]$ then

$$D_t^- W(t) \geq \partial_t \partial_t u(x,t) + \varphi'(u(x,t)) \partial_x \partial_t u(x,t), \tag{3.38}$$

where x is a point at which $\partial_t u(\cdot,t)$ is minimized over the interval $[-a-k(b-t), a+k(b-t)]$. Hence, using (3.38), (3.22) and (3.28),

$$D_t^- W + \alpha W \geq -\gamma. \tag{3.39}$$

It is clear that (3.39) holds, a fortiori, at points $t \in (0,b]$ with $W(t) = 0$. The differential inequality (3.39) yields a lower bound for $W(t)$ even when $-W(0)$ is large, i.e., even when $\max u_0'(\cdot)$ is large. Specifically, (3.39) implies

$$\partial_t u(x,t) > -\frac{\gamma}{\alpha} - \varphi'(\varepsilon) \sup u_0'(\cdot), \quad (x,t) \in \mathcal{T}. \tag{3.40}$$

Since the upper and lower bounds on $\partial_t u$, given by (3.36) and (3.40), are independent of a and b, the validity of these estimates is immediately extended onto the entire domain $(-\infty,\infty) \times [0,T)$ of the solution.

By virtue of (3.2), (3.27), the above derived estimates for $|\partial_t u|$ yield immediately an L^∞ estimate for $|\partial_x u|$ on $(-\infty,\infty) \times [0,T)$. Once L^∞ estimates for $|u|$, $|\partial_t u|$ and $|\partial_x u|$, independent of T, have been derived, the existence of a global solution $u \in C^1((-\infty,\infty) \times [0,\infty))$ can be established by means of standard continuation arguments that need not be elaborated here. Furthermore, since the estimates for $|u|$, $|\partial_t u|$ and $|\partial_x u|$ obtained above depend solely upon $\sup|u_0(\cdot)|$ and $\sup|u_0'(\cdot)|$, i.e., they are independent of the size of $u_0''(\cdot)$, we may remove the extraneous assumption $u_0 \in C^2(-\infty,\infty)$ by means of a standard completion argument. □

In proving the remaining Theorem 2.1, the first step is to show that the modulus of the solution, on its domain of existence, can be estimated solely in terms of $\sup|u_0(\cdot)|$, i.e., independently of the size of $u_0'(\cdot)$.

Lemma 3.1. *Assume that* (2.1) *holds for all* u *and that* $u \in C^1((-\infty,\infty) \times [0,T))$ *is a solution of* (1.4), (1.5) *for some* $T \in (0,\infty)$. *Given any* $\delta > 0$, *there is* $\varepsilon > 0$, *depending on* δ *and* T, *such that if* (2.8) *holds then*

$$|u(x,t)| < \delta, \quad -\infty < x < \infty, \quad 0 \le t < T. \tag{3.41}$$

Proof. We fix constants $a \in (0,\infty)$, $b \in (0,T)$ and consider the trapezoid $\mathscr{T}$ defined by (3.15). We define

$$U(t) := \max\{|u(x,\tau)| : 0 \le \tau \le t, -a-k(b-\tau) \le x \le a+k(b-\tau)\}, \quad 0 \le t \le b. \tag{3.42}$$

U is a nonnegative, nondecreasing Lipschitz function on $[0,b]$. We pick any $t \in (0,b]$ and identify the point (x,τ), $\tau \in [0,t]$, $x \in [-a-k(b-\tau), a+k(b-\tau)]$, such that $U(t) = |u(x,\tau)|$. Leaving aside the trivial case $\tau = 0$, we note that for any $\eta \in (0,\tau)$,

$$\begin{aligned} U(t) - U(t-\eta) &\le |u(x,\tau)| - |u(x-\eta\varphi'(u(x,\tau)), \tau-\eta)| \\ &\le |u(x,\tau) - u(x-\eta\varphi'(u(x,\tau)),\tau-\eta)| \end{aligned} \tag{3.43}$$

whence, upon using (1.4) and (3.1),

$$D_t^- U(t) \le |\partial_\tau u(x,\tau) + \varphi'(u(x,\tau))\partial_x u(x,\tau)| = |Y(x,\tau)|. \tag{3.44}$$

On the other hand, by account of (3.9), (3.10), (3.6), (3.4) and (3.5), we deduce

$$\max\{|Y(x,\tau)| : 0 \le \tau \le t, -a-k(b-\tau) \le x \le a+k(b-\tau)\} \le G(U(t),t), \tag{3.45}$$

where $G: [0,\infty) \times [0,T] \to [0,\infty)$ is a Lipschitz function, independent of a, b, and such that

$$G(0,t) = 0, \quad 0 \leq t \leq T. \tag{3.46}$$

It follows from (3.44), (3.45) that U satisfies the differential inequality

$$D_t^- U \leq G(U,t), \quad 0 < t < T. \tag{3.47}$$

Therefore, given $\delta > 0$ there is ε, depending on δ and T, such that any solution of (3.47) with $U(0) < \varepsilon$ satisfies $U(t) < \delta$, $0 \leq t < T$. Since G is independent of a, b, this establishes (3.41). □

<u>Proof of Theorem 2.1</u>. To simplify the analysis, let us assume $u_0 \in C^2(-\infty,\infty)$ in which case, as noted above, any C^1 solution of (1.4), (1.5) is automatically C^2-smooth. Proceeding by reductio ad absurdum, suppose a solution $u \in C^2((-\infty,\infty) \times [0,T))$ exists, for the given $T > 0$, for any initial data $u_0 \in C^2(-\infty,\infty)$.

We fix $\delta > 0$ so small that (2.1), (2.2) hold on $[-\delta,\delta]$ and then, applying Lemma 3.1, we determine $\varepsilon > 0$ such that (2.8) implies (3.41). Combining (3.23), (3.24), (3.25), (2.1), (2.2), (3.4), (3.5), (3.9), (3.10) and (3.6), we conclude that, so long as (3.41) holds,

$$|A(x,t)| \leq \alpha, \; B(x,t) \geq \beta > 0, \; |\Gamma(x,t)| \leq \gamma, \tag{3.48}$$
$$-\infty < x < \infty, \; 0 \leq t < T,$$

where α, β, γ are constants depending on δ and T.

We now proceed to show that when (2.9) holds with μ sufficiently large then $\partial_t u$ must blow up to ∞ at some $t_0 < T$. For that purpose we fix any $x_0 \in (-\infty,\infty)$ and define

$$V(t) := \max\{\partial_t u(x,t) : x_0 \leq x \leq x_0 + t\varphi'(\delta)\} \tag{3.49}$$
$$0 \leq t < T.$$

For any fixed $t \in [0,T)$ we identify the point x at which $\partial_t u(\cdot,t)$ is maximized over the interval $[x_0, x_0+t\varphi'(\delta)]$. Then

$$V(t) = \partial_t u(x,t). \tag{3.50}$$

Therefore, for any $\eta \in (0,T-t)$,

$$V(t+\eta)-V(t) \geq \partial_t u(x+\eta\varphi'(u(x,t)),t+\eta)-\partial_t u(x,t) \tag{3.51}$$

whence

$$D_t^+ V(t) \geq \partial_t \partial_t u(x,t) + \varphi'(u(x,t))\partial_x \partial_t u(x,t). \tag{3.52}$$

Combining (3.52) with (3.22) and using (3.48), (3.50), we get the differential inequality

$$D_t^+ V + \alpha|V| - \beta V^2 \geq -\gamma, \quad 0 \leq t < T. \tag{3.53}$$

There is a threshold number $V_0 > 0$ such that any solution of (3.53) with initial value $V(0) > V_0$ must blow up to ∞ at some $t_0 < T$. Consequently, when (2.9) holds with $\mu := V_0/\varphi'(-\varepsilon)$ we select $x_0 \in (-\infty,\infty)$ such that $-u_0'(x_0) > \mu$ and observe that $V(0) = \partial_t u(x_0,0) = -\varphi'(u_0(x_0))u_0'(x_0) > V_0$ in which case $\partial_t u$ must blow up at some $t_0 < T$. The proof is complete. □

REFERENCES

1. Coleman, B. D. and M. E. Gurtin, Waves in materials with memory II. On the growth and decay of one-dimensional acceleration waves, Arch. Rational Mech. Anal. 19 (1965), 239-265.
2. Dafermos, C. M. and J. A. Nohel, A nonlinear hyperbolic Volterra equation in viscoelasticity, Am. J. Math. Supplement dedicated to P. Hartman (1981), 87-116.
3. Hattori, H., Breakdown of smooth solutions in dissipative nonlinear hyperbolic equations, Q. Appl. Math. 40 (1982), 113-127.
4. Hrusa, W. J., A nonlinear functional differential equation in Banach space with applications to materials with fading memory, Arch. Rational Mech. Anal. 84 (1983), 99-137.
5. Hrusa, W. J. and J. A. Nohel, The Cauchy problem in one-dimensional nonlinear viscoelasticity, J. Diff. Equations, (to appear).
6. Hrusa, W. J. and M. Renardy, On a class of quasilinear partial integrodifferential equations with singular kernels, J. Diff. Equations, (submitted).

7. MacCamy, R. C., A model for one-dimensional nonlinear viscoelasticity, Q. Appl. Math. 35 (1977), 21-33.
8. MacCamy, R. C., A model Riemann problem for Volterra equations, Arch. Rational Mech. Anal. 82 (1983), 71-86.
9. Malek-Madani, R. and J. A. Nohel, Formation of singularities for a conservation law with memory, SIAM J. Math. Anal., (to appear).
10. Nohel, J. A., A nonlinear conservation law with memory, Volterra and Functional Differential Equations, Lecture Notes in Pure and Applied Mathematics, Volume 81, Dekker, New York, 1982, 91-123.

The author was partially supported by the National Science Foundation under Grant No. DMS 8205355, the U. S. Army Research Office under Contract No. DAAG 29-83-K-0029, and the Office of Naval Research under Contract No. N00014-83-K-0542.

HYPERBOLIC PHENOMENA IN THE FLOW OF VISCOELASTIC FLUIDS

Daniel D. Joseph

University of Minnesota
Department of Aerospace Engineering and Mechanics
Minneapolis, Minnesota

ABSTRACT

This paper treats the problem of hyperbolicity, change of type and nonlinear wave propagation in the flow of viscoelastic fluids. Rate equations for fluids with and without instantaneous elasticity are derived and discussed. The equations of fluids with instantaneous elasticity are hyperbolic in unsteady flow and can change type in steady flow. The wave speeds depend on velocities and stresses. Some estimates of wave speeds into states of rest are given. For many of the popular models of fluids the vorticity is the field variable which changes type. The vorticity of all fluids with instantaneous elasticity can change type in motions which perturb rigid ones. Experiments and analysis exhibiting vorticity of changing type are exhibited. The linearized viscoelastic problem is governed by equations having the properties of a telegraph equation. The damping is small when the fluid is very elastic. Elastic fluids have a long memory, a large time (Weissenberg number) for relaxation. The damping is rapid when the relaxation time is small even when the flow is very supercritical. It is shown that steady flow around a body is of "transonic" type. The linearized problem for flow over a flat plate is reduced to an integral equation for the vorticity distribution on the plate. The problem of nonlinear wave propagation is discussed and the problems of nonlinear smoothing and shocking are considered. It is shown (by M. Slemrod) that the shocks of vorticity can arise from smooth data in some models and shocks of velocity in other models.

1. INTRODUCTION

The equations of gas dynamics have a hyperbolic structure which support waves of compression and rarefaction. The linearized theory of gas dynamics leads to the wave equation in which compression and rarefaction waves are on an equal footing in that both types of waves may propagate without change of form. The nonlinear theory of gas dynamics leads to striking new qualitative understandings of wave propagation. Initially smooth waves of compression must shock up. Initial shocks of rarefaction must smooth out. The equations of steady gas dynamics change type from subsonic to supersonic when the speed of the fluid at some point exceeds the speed of sound. The field of flow can be partitioned into subsonic parts, governed by an elliptic equation, and a supersonic part, governed by a hyperbolic equation. The problem of change of type in gas dynamics is called transonic flow.

Quasilinear systems of equations governing the flow of incompressible viscoelastic fluids are also of mixed type when the constitutive equation admits an instantaneous elastic response. As in gas dynamics, steady flows of such fluids can change type.

I am interested in the possibility that many effects in the flow of viscoelastic liquids, as well as difficulties in numerical simulation are associated with wave propagation, the nonlinear smoothing of shocks, the shock up of smooth solutions and with the appearance of real characteristics and a change of type analogous to the sonic transition in gas dynamics.

The notes for this lecture are a compendium for hyperbolic things arising in the theory of flow of viscoelastic fluids with instantaneous elasticity.

Fluids with instantaneous elasticity have no instantaneous viscosity in the same way that Maxwell models have no instantaneous viscosity and Jeffrey's models do. For example

$$\lambda \frac{\mathbf{D}\underline{\tau}}{\mathbf{D}t} + \underline{\tau} = 2\eta \underline{D}[\underline{u}] \tag{1.1}$$

is an Oldroyd rate equation for the determinate part of stress $\underline{\tau}$, $\mathbf{D}/\mathbf{D}t$ is an invariant time derivative, $\underline{D}[\underline{u}] = \frac{1}{2}(\nabla\underline{u} + \nabla\underline{u}^T)$, where $\underline{u}(\underline{x},t)$ is the velocity, λ is a relaxation time and η is the "viscosity". Equation (1.1) is a Maxwell model. It has a purely elastic instantaneous response. On the other hand

$$\lambda \frac{D\underline{\tau}}{Dt} + \underline{\tau} = 2\eta \underline{D}[\underline{u}] + \Lambda \frac{D\underline{D}[\underline{u}]}{Dt} \tag{1.2}$$

is an Oldroyd model with a retardation time Λ. This equation is of the Jeffrey's type. The retardation term produces an instantaneous viscous response superimposed on the elastic response. Λ, rather than η, is the measure of the viscous response to impulsive motion. The most general class of simple fluids with instantaneous elasticity are those whose histories are convergent in the weighted L_2 spaces of Coleman and Noll. Joseph, Renardy and Saut (1984), hereafter called JRS, derived the general form of the rate equation which is implied by the fading memory theory of Coleman and Noll. More general theories of fading memory based on different topologies containing, for example, models like (1.2), have been presented by Saut and Joseph (1983).

It is probable that most polymer solutions have some instantaneous viscous response following, say, a model like (1.2) with $\Lambda \neq 0$, but perhaps small. If the instantaneous viscous response is small, say Λ/λ is small, the smoothing effects of viscosity will also be small leading to shock structure of the type which is well known in gas dynamics. A theory for such viscous smoothing of shocks in viscoelastic fluids has been discussed by Saut and Joseph (1983) and Narain and Joseph (1983). It follows that the study of fluids with instantaneous elasticity ought to apply to many real fluids which have a small viscous response.

The visco part of viscoelasticity does not mean that the fluid has a viscous response. It means that the amplitude of impulsive, shock solutions will decay but that the shocks will not smooth. This property of flow is not shared by gas dynamics where there is no damping. When the damping is very large (e.g; when λ is small), the elastic response is short. When the damping is small (e.g.; when λ is large), the elastic response is long. In most problems with damping, even the simple ones which arise in the one dimensional theory of kinematic waves (Whitham, 1974, p. 62) it is not possible to form a shock without first satisfying a criterion for a critical amplitude. Such critical amplitude results are now well known in the theory of nonlinear wave propagation in viscoelastic fluids (see Section 13).

There are possibly many works on wave propagation in viscoelastic fluids. Of the ones known to me, the works of Coleman and Gurtin (1968) and Slemrod (1978) seem to me to be the most important. Problems which perturb uniform flows of Oldroyd fluids have been studied by Giesekus (1970), Ultmann and Denn (1970) and Luskin (1984). In these problems the entire field of flow is either subcritical or supercritical. There is no problem of transonic type. The possibility of transonic type in the flow of Oldroyd fluids was first mentioned by Rutkevich (1970), Joseph, Renardy and Saut (1984) treated the problem of hyperbolicity and change generally. They introduced the notion that in many models and in all models on motions perturbing uniform ones, the vorticity is the hyperbolic variable which changes type. Yoo, Ahrens and Joseph (1984) tried to explain the striking results of Metzner, Uebler & Fong (1969) with an analysis of the vorticity perturbing irrotational sink flow. This type of hyperbolic flow with zero vorticity in a cone like region and nonzero vorticity outside was observed in the flow into a hole.

The first detailed solution of flow exhibiting a change of type has recently been given by Yoo and Joseph (1984). They consider the linearized problem which arises when Poiseuille flow of an upper convected Maxwell model is perturbed with wavy walls. The vorticity of this flow will change type when the velocity in the center of the pipe is larger than a critical value defined by the propagation of shear waves. There is then a region around the pipe axis in which the vorticity equation is hyperbolic and a low speed region near the walls where the vorticity equation is elliptic. They linearize the problem for small amplitude waviness and the linearized problem is solved in detail. The characteristic nets depend on the viscoelastic "Mach" number, which is the ratio ($M = U/C$) of the unperturbed maximum velocity U to the speed of shear waves C into the fluid at rest and the elasticity number E. There is a supercritical (hyperbolic) region around the axis of the pipe when $M > 1$. When $M \gg 1$, the diameter of this hyperbolic region is small when E is large, and large when E is small. Regions of positive and negative vorticity are swept out along forward facing characteristics in the hyperbolic region. There is rapid damping of vorticity through a narrow layer to small vorticity in the core when $M \gg 1$ and the Weissenberg number $W = M\sqrt{E}$ tends to zero. (The Weissenberg number is proportional to the relaxation time of the fluid.)

The rate of damping of vorticity decreases as W is increased. Flows with high M appear to be more "elastic" when W is large in the sense that the damping is suppressed as the relaxation time of the fluid is increased.

I am going to discuss some selected topics from the aforementioned papers. I have introduced some new thoughts in these notes. The problem of flow around bodies treated in Section 11 and the flat plate problem of Section 12 seem not to have been considered as transonic type problems before. In Section 12 I have reduced the linearized supercritical problem for flow over a flat plate to a single integral equation. I would like to know if this problem can be solved. Various parts of the discussion of nonlinear wave propagation in Section 13 are new.

I wish to express my gratitude to my colleagues and collaborators in different aspects of this work; these are Michael Renardy, Jean Claude Saut and, more recently, John Yoo. I also want to acknowledge the help which I have received from my students Mark Ahrens, Edmond O'Donovan and Oliver Riccius. Marshal Slemrod has got a fine result, right in this ballpark, which appears as Appendix A to this work. Mark Ahrens prepared Appendix B.

2. RATE EQUATIONS FOR FLUIDS WITH INSTANTANEOUS ELASTICITY

It is perhaps useful to try to understand hyperbolicity in fluids with instantaneous elasticity without choosing special models. We want to know which phenomena are intrinsic and which are accidental arising only for one or the other of the special models.

A general theory of rate equations for fluids with instantaneous elasticity was given by JRS. Here we give a slightly different derivation leading to the general form of quasilinear problems associated with motions of fluids with instantaneous elasticity.

Let $\underline{C}_t(\underline{x},t) = \nabla\underline{\chi}_t^T \, \nabla\underline{\chi}_t$ be the relative Cauchy strain where $\underline{\chi}_t(\underline{x},t)$ is the position of $\underline{x} = \underline{\chi}_t(\underline{x},t)$ at $\tau=t$. The determinate stress $\underline{\tau}$ may be expressed by a functional of the history of the Cauchy strain,

$$\underline{\tau} = \underset{\tau=-\infty}{\overset{t}{\tilde{\underline{\mathbf{F}}}}} \, [t, \underline{C}_t(\underline{x},\tau)]. \tag{2.1}$$

For the moment I state only that the domain of $\tilde{\underline{\mathbf{F}}}$ is equipped with the topology of a Hilbert space. Saut and Joseph (1983) show how different choices of the topological domain of $\tilde{\underline{\mathbf{F}}}$ lead to different types of constitutive equations. I want to make this point again, under (2.4), in the context of this discussion of the general form of a rate equation. $\tilde{\underline{\mathbf{F}}}$ depends on the history of $\underline{C}_t(\underline{x},\tau)$ and explicitly on the present time, perhaps through a kernel function and in other ways.

It follows from differentiating (2.1) that

$$\frac{d\underline{\tau}}{dt} = \frac{\partial}{\partial t}\tilde{\underline{\mathbf{F}}} + \tilde{\underline{\mathbf{F}}}_1[t,\underline{C}_t \,|\, \frac{d\underline{C}_t}{dt}] \tag{2.2}$$

where $\tilde{\underline{\mathbf{F}}}_1[t,\underline{C}_t|\cdot]$ is a Frechet derivative, evaluated on $\underline{C}_t(\underline{x},t)$. The representation theorem for a linear functional on a Hilbert space leads us to

$$\tilde{\underline{\mathbf{F}}}_1\left[t,\underline{C}_t \,\middle|\, \frac{d\underline{C}_t}{dt}\right] = \int_{-\infty}^{t} \underline{K}(t-\tau,\underline{C}_t(\underline{x},\tau))\, \frac{d\underline{C}_t}{dt}(\underline{x},\tau)dt \tag{2.3}$$

where $\underline{K}(t-\tau,\underline{C}_t(\underline{x},\tau))$ is a fourth order, tensor valued kernel function, not too singular, satisfying some invariance conditions arising from isotropy (see JRS). The term $\frac{\partial\tilde{\mathbf{F}}}{\partial t}$ is a derivative holding the history of $\underline{C}_t$ fixed. This term is again a functional of the form (2.1); in the case of single integral models with a kernel proportional to $e^{-(t-\tau)/\lambda}$ we have $\partial\tilde{\underline{\mathbf{F}}}/\partial t = -\underline{\tau}/\lambda$ (cf. (2.19)). Motivated by this last observation we introduce the suggestive notation

$$\tilde{\underline{\tau}} \overset{\text{def}}{=} -\lambda \frac{\partial\tilde{\underline{\mathbf{F}}}}{\partial t} \tag{2.4}$$

for a constant λ. $\tilde{\underline{\tau}}$ is stress-like and λ is a "relaxation time".

Saut and Joseph (1983) have shown that the smoothness of the kernel $\underline{K}$ in (2.3) depends on the choice of the topology of convergence for the history $\underline{C}_t(\underline{x},\tau)$. If $\frac{d\underline{C}_t}{dt}(\underline{x},\cdot)$ lies in the weighted $L_h^2(0,\infty)$ (fading memory) spaces of Coleman and Noll, then the kernel $\underline{K}$ must have at least the smoothness which is required for the validity of the application of the Schwarz inequality to the integral in (2.3). If $\frac{d\underline{C}_t}{dt}$ lies in the (fading memory) Sobolev spaces used by Saut and Joseph (1983), then various derivatives of Dirac measures can appear in the kernel $\underline{K}$. These give rise to "viscosity" terms like the retardation term in (1.2). The fading memory spaces of Coleman and Noll allow only those constitutive equations which exhibit instantaneous elasticity. The $L_h^2(0,\infty)$ fading memory spaces of Coleman and Noll therefore do not form a general basis for discussion of problems of rheology. They do form a general basis for discussion of problems of instantaneous elasticity. In the work below dom $\tilde{\underline{\mathbf{F}}} = L_h^2(0,\infty)$.

Since the position at $\tau < t$ of all particles $\underline{x}$ on the same path line is the same

$$\frac{d\chi_t}{dt} = \frac{\partial \chi_t}{\partial t} + \underline{u}(x,t) \cdot \nabla \chi_t = 0 \tag{2.5}$$

we have

$$\frac{d\underline{F}_t}{dt} = \nabla \frac{\partial \chi_t}{\partial t} + (\underline{u}\cdot\nabla)\nabla\chi_t = -\nabla(\underline{u}\cdot\nabla\chi_t) + (\underline{u}\cdot\nabla)\nabla\chi_t = -\underline{F}_t\underline{L}$$

where $\underline{L} = \underline{L}(\underline{x},t) = \nabla\underline{u}(\underline{x},t)$ and $\underline{F}_t = \nabla\chi_t(\underline{x},t)$ Hence,

$$\frac{d\underline{C}_t(x,\tau)}{dt} = \frac{d\underline{F}_t^T}{dt}\underline{F}_t + \underline{F}_t^T\frac{d\underline{F}_t}{dt} = -\underline{L}^T(\underline{x},t)\underline{C}_t(\underline{x},\tau) - \underline{C}_t(\underline{x},\tau)\underline{L}(\underline{x},t) \tag{2.6}$$

and

$$\left(\int_{\infty}^{t} \underline{K}(t-\tau,\underline{C}_t)\frac{d\underline{C}_t}{dt}\,d\tau\right)_{ij} = \int_{-\infty}^{t} K_{ijk\ell}\frac{dC_{k\ell}}{dt}\,dt = M_{ijkp}(\underline{x},t)L_{pk}(\underline{x},t) \tag{2.7}$$

where

$$M_{ijkp}(\underline{x},t) = -\int_{-\infty}^{t}(K_{ijk\ell} + K_{ij\ell k})\,\underline{C}_{p\ell}\,(\underline{x},\tau)d\tau \tag{2.8}$$

is symmetric in the first pair of indices and $L_{pk} = \dfrac{\partial u_p}{\partial x_k}$. We may write (2.8) as

$$\frac{d\underline{\tau}}{dt} = \underline{M}(\underline{x},t)\ \underline{L}(\underline{x},t) - \frac{1}{\lambda}\tilde{\underline{\tau}} \tag{2.9}$$

where $\underline{M}$ is the fourth order tensor valued functional defined by (2.8). Together with the equations of motion

$$\rho \frac{d\underline{u}}{dt} = -\nabla p + \text{div } \underline{\tau} \text{ , div } \underline{u}=0, \tag{2.10}$$

(2.9) and (2.10) are a quasilinear system in derivatives of $\underline{u}$ and $\underline{\tau}$. The stress-like term $\tilde{\underline{\tau}}(\underline{x},t)$ and the coefficients $\underline{M}(\underline{x},t)$ are of lower order (see JRS) since they do not involve derivatives of $\underline{u}$ or $\underline{\tau}$, and have other required properties. Hence, the system

$$\frac{\partial \tau_{ij}}{\partial t} + u_e \frac{\partial \tau_{ij}}{\partial x_e} - M_{ijke} \frac{\partial u_e}{\partial x_k} + \frac{1}{\lambda} \tilde{\tau}_{ij} = 0,$$

$$\rho \left[\frac{\partial u_i}{\partial t} + u_e \frac{\partial u_i}{\partial x_e} \right] + \frac{\partial p}{\partial x_i} - \frac{\partial}{\partial x_j} \tau_{ij} = 0,$$

$$\frac{\partial u_e}{\partial x_e} = 0 \tag{2.11}$$

is linear in the derivatives of u_i, τ_{ij} and p.

We may decompose the fourth order tensor $\underline{M}$ into symmetric and skew symmetric parts,

$$M_{ijke} = S_{ijke} + A_{ijke}$$

where $\underline{S}$ is symmetric and $\underline{A}$ is axisymmetric in k and e. We note that

$$S_{ijke} \frac{\partial u_e}{\partial x_k} = S_{ijke} D_{ek},$$

$$A_{ijke} \frac{\partial u_e}{\partial x_k} = A_{ijke} \Omega_{ek}$$

where

$$D_{ek} = \frac{1}{2} \left[\frac{\partial u_e}{\partial x_k} + \frac{\partial u_k}{\partial x_e} \right]$$

and

$$\Omega_{ek} = \frac{1}{2}\left[\frac{\partial u_e}{\partial x_k} - \frac{\partial u_k}{\partial x_e}\right] = \varepsilon_{ekm}\zeta_m$$

can be expressed in terms of the vorticity $\underline{\zeta}$ = curl $\underline{u}$. Bernard Coleman has shown that his thermodynamics imply that the fourth order tensor $\underline{S}$ is derivable from an energy and, as a consequence

$$S_{ijke} = S_{keij}$$

3. WAVE SPEEDS I, THEORETICAL

The quasilinear system is called evolutionary if the Cauchy initial value problem for perturbation of arbitrary motion is well posed (see Gelfand, 1963). We look at periodic initial data whose period is so small that $\underline{\tau},\underline{u},p$ are essentially constant over the period $2\pi/k$, $k=|\underline{k}|$ over which

$$(\delta\underline{\tau},\delta\underline{u},\delta p) = (\sigma,v,\pi)\exp i\underline{k}\cdot\underline{x}$$

are rapidly varying, $k\to\infty$. We seek propagating waves of the form

$$(\delta\underline{\tau},\delta\underline{u},\delta p) = (\underline{\sigma},\underline{v},\underline{\pi})\exp i(\underline{k}\cdot\underline{x}+\omega t) \tag{3.1}$$

and $(\underline{\sigma},\underline{v},\pi)$ depend only on the components of $\underline{u}$ and $\underline{M}$ which appear as coefficients of derivatives in the quasilinear system (2.11). The resulting equations are then divided by k and the right side of (2.11) vanishes as $k\to\infty$. This leads to the homogeneous equations for the components of $(\underline{\sigma},\underline{v},\pi)$ with $\underline{k}/k \overset{\text{def}}{=} \underline{n}$ fixed

$$\rho\left[\frac{\omega}{k} + \underline{u}\cdot\underline{n}\right]v_i + n_i\pi - n_j\sigma_{ij} = O(\frac{1}{k})$$

$$\left[\frac{\omega}{k} + \underline{u}\cdot\underline{n}\right]\sigma_{ij} - M_{ijke}v_k n_e = O(\frac{1}{k}) \tag{3.2}$$

$$v_e n_e = 0$$

The homogeneous system corresponding to (3.2) can be solved if and only if the determinant of the coefficient vanishes. The system is evolutionary if there are real wave speeds

$$C = \lim_{k\to\infty} \frac{\omega}{k}\,.$$

There may of course be more than one wave speed. If C is not real there will be rapidly growing solutions. The reader may see more details of this type of analysis in the studies by Rutkevich (1969, 1972) of Oldroyd models.

4. WAVE SPEEDS II, PHYSICAL

The subject of gas dynamics would be in deep trouble if it were not possible to know the speed of sound. This type of deep trouble should be a concern in the study of hyperbolicity in the flow of viscoelastic fluids. In fact the way to know if a fluid has instantaneous elasticity is to show it has a wave speed. Now showing that a certain fluid has a wave speed is not easy. First of all, there are many fluids for which there are no wave speeds. These common cases are for fluids like Newtonian ones, or even viscoelastic ones with a big viscous response (say, Λ/λ is not small), which do not have an instantaneous elastic response, or a predominantly elastic response to impulsive data. Secondly, the fluids which admit a wave speed admit infinitely many wave speeds. Just as the speed of sound in a gas at a point depends on the thermodynamic condition there, a wave in a viscoelastic fluid at a point depends on the velocity and the state of stress there. So each problem gives rise to its own field of wave speeds.

The simplest type of wave propagation is into a region at rest. The wave speed for this is

$$C = \sqrt{\frac{G(0)}{\rho}} \tag{4.1}$$

where G(0) is the instantaneous value of the shear relaxation modulus G(s). In fact all kinds of small amplitude discontinuities will propagate with this speed, jumps in acceleration (Coleman and Gurtin, 1968), jumps in velocity and even in displacement (Narain and Joseph, 1982). The same propagation speed (4.1) holds for waves propagating into regions undergoing rigid motions. We shall show that flows which perturb rigid motions go supercritical when $U/C > 1$. This wave speed is a fundamental material parameter. There are at present no rheometers for measuring G(0) (or C) and there are no tables of values. E. H. Lieb, (1975), in his Ph.D. thesis, measured $C = 8.0$ cm/sec in one fluid (carbomethylcellulose in 50% glycerin and 49% water solution) using a rotating cylinder apparatus. For Maxwell models with a single relaxation time we have $G(0) = \eta/\lambda$, where η is the viscosity and λ is the relaxation time. There are some papers which report viscosities and relaxation times which are obtained by fitting data with some impusle experiments (for example, see Papanastasiou, Scriven and Macosko (1983), Dodson, Townsend and Walters (1971), Laun (1978). This type of rheometry suffers from two defects. Usually a special form for the kernel in the linear theory is assumed. The kernel is represented by an exponential or by a sum of exponentials. A more serious defect is that it is usually assumed, wrongly, that the early time response to impulse which gives G(0) satisfies an overly simplified asymptotic (for large time) theory (see Narain and Joseph (1984) for a full discussion and remedy for this defect). From the impulse data reported in many different papers we have estimated that wave speeds for water-based polymer solutions are of the order of 10 cm/sec. The wave speeds of polymer melts are of the order of 100 cm/sec. At the same time, the large viscosities of melts reduce the possibility of achieving large velocities. Hence for the flow of melts we expect that the flow is usually subcritical $M < 1$.

We have developed a wave speed meter which appears to give direct, reproducible values for C. A patent has been applied for.

5. VORTICITY

Some important special models give rise to a hyperbolic vorticity equation. It is of interest to derive the equation for the vorticity without making assumptions other than that the fluid has instantaneous elasticity, i.e., the quasilinear system (2.11) governs.

The vorticity $\underline{\zeta}$ satisfies

$$\rho(\frac{\partial}{\partial t} \underline{\zeta} + \underline{u}\cdot\nabla\underline{\zeta}) = \rho\underline{\zeta}\cdot\nabla\underline{u} + \text{curl div } \underline{\tau}. \tag{5.1}$$

Equation (5.1) is satisfied by $\underline{\zeta} = 0$ if and only if curl div $\underline{\tau} = 0$. This condition is always satisfied by Newtonian fluids for which $\underline{\tau} = \eta(\nabla\underline{u} + \nabla\underline{u}^T)$ for a constant viscosity η giving rise to curl div$\underline{\tau} = \eta\nabla^2\underline{\zeta}$. In this case potential flow is always a solution of (5.1). More generally we may verify that potential flow $\underline{u} = \nabla\Phi$, $\forall\ \Phi\ \ C^1(\underline{x})$ is a solution of (5.1) if and only if div $\underline{\tau}$ is the gradient of a potential. Given the potential Φ we may obtain the velocity as a gradient $\underline{u} = \nabla\Phi$. Then $\underline{\zeta} = \text{curl } \underline{u} = 0$. The velocity $\underline{u}$ determines the path lines χ_t $(\underline{x}, \tau)$ and the past values of the relative Cauchy strain $\underline{C}_t(\underline{x}, \tau)$. This gives rise, through (2.11), to the values $\underline{\tau}(\underline{x},t)$ of the extra stress at the present time t. It is not guaranteed that this computation will lead to a $\underline{\tau}(\underline{x},t)$ such that curl div $\underline{\tau} = 0$ at each and every point at which $\underline{\zeta} = 0$.

Whenever $\underline{u} = \nabla\Phi$ there are two more equations than unknowns. Evidently only certain special potential flows are compatible with (5.1). Uniform motions, say $\underline{u} = \underline{e}_x U$, for constant U, have $\underline{\tau} = 0$ and are always potential flows.

Another form of the vorticity equation is fundamental in the study of hyperbolicity and change of type. We first apply the substantial derivative $\frac{d}{dt}$ to (5.1). Then we apply $\frac{d}{dt}$ curl to (2.10), eliminate curl div $\frac{d\underline{\tau}}{dt}$ and find that

$$\rho\frac{d^2\underline{\zeta}}{dt^2} - \text{curl div } (\underline{ML_T}) = \underline{\Theta}_1 + \underline{\Theta}_2 + \underline{\Theta}_3 \tag{5.2}$$

where

$$\underline{L} = \nabla\underline{u}, \ (\nabla u)_{ij} = \partial_j u_i$$

$$\underline{\zeta} = \text{curl } \underline{u},$$

$$\underline{\Theta}_1 = -\frac{1}{\lambda} \text{ curl div } \underline{\tilde{\tau}},$$

$$\underline{\Theta}_2 = \frac{d}{dt} (\text{curl div } \underline{\tau}) - \text{curl div } \frac{d\underline{\tau}}{dt}, \tag{5.3}$$

$$\underline{\Theta}_3 = -\rho\frac{d}{dt} (\text{curl}\frac{d\underline{u}}{dt} - \frac{d}{dt} \text{ curl } \underline{u})$$

Third derivatives of $\underline{u}$ are all on the left side of (5.2). The right side contains second derivatives of $\underline{u}$ and $\underline{\tau}$, at most. ($\underline{\Theta}_2$ is essentially a second derivative of $\underline{\tau}$.) Equation (2.9) shows that $\underline{\tau}$ and $\underline{u}$ are of the same differential order, so that the left side has third order derivatives of $\underline{u}$ and the left side has second order derivatives of $\underline{u}$, at most.

$$
\begin{aligned}
&\rho\Big[\partial_{tt}\zeta_k + 2\underline{u}\cdot\nabla\partial_t\zeta_k + u_e u_j \zeta_{k,ej}\Big] - \varepsilon_{k\ell j} M_{jmpq} u_{q,pm\ell} \\
&= \rho\Big[(\partial_t + \underline{u}\cdot\nabla)(\underline{\zeta}\cdot\nabla)u_k - (\partial_t \underline{u}\cdot\nabla)\zeta_k - u_e u_{j,e}\zeta_{k,j}\Big] \\
&\quad + \varepsilon_{kej}[M_{jmpq,e} u_{q,pm} + (M_{kjmpq,m} u_{q,p})_{,e} \\
&\quad - u_{p,e}\tau_{jq,pq} + (u_{p,q}\tau_{jq,p})_{,e} + \frac{1}{\lambda}\tilde{\tau}_{jm,me}]
\end{aligned}
\tag{5.4}
$$

where $(\cdot)_{,m} \overset{\text{def}}{=} \frac{\partial}{\partial x_m}(\cdot)$. Equation (5.4) reduces to a second order equation for the vorticity plus lower order terms whenever the last term on the left of (5.4) can be expressed in terms of second order derivatives of $\underline{\zeta}$ (see (6.3)). Potential flow is possible only when (5.4) is satisfied with $\zeta=0$, $\underline{u}=\nabla\Phi$.

6. SPECIAL MODELS

Special constitutive models of fluids with instantaneous elasticity arise from (2.9) by choosing special forms for the fourth order tensor $\underline{M}$ and the stress-like term $\tilde{\underline{\tau}}$. In fact $\tilde{\underline{\tau}} = \underline{\tau}$ for all single integral models with separable kernels in exponential form with time constant λ.

We will consider four special families of constitutive models.

(1) JRS models (introduced by JRS (1984)).

(2) Three parameter family of "Maxwell" models introduced by Oldroyd.

(3) Models of the Giesekus type.

(4) BKZ models.

The JRS models arise when the fourth-order tensor $\underline{M}$ is expressed by the most general expression which involves second order tensors. This leads to

$$
\frac{d\underline{\tau}}{dt} = \underline{PD} + \underline{D}\underline{P}^T + \frac{1}{2}(\underline{A\Omega} + (\underline{A}\Omega)^T) + \frac{1}{\lambda}\tilde{\tau}
\tag{6.1}
$$

for an arbitrary second order tensor $\underline{P}$; the fourth order tensor $\underline{A}$ has already been introduced. Coleman's thermodynamics implies that $\underline{P} = \underline{P}^T$.

The Oldroyd models with three constants arises from (6.1) when
The Oldroyd models with three constants arises from (6.1) when
$\underline{P} = a\underline{\tau} + \frac{\eta}{\lambda}\underline{1}$, $-1 \leq a \leq 1$, $\tilde{\underline{\tau}} = \underline{\tau}$ and $\underline{A}$ chosen so that

$$\lambda\left[\frac{d\underline{\tau}}{dt} - a(\underline{D}\;\underline{\tau} + \underline{\tau}\;\underline{D}) + \underline{\tau}\;\underline{\Omega} - \underline{\Omega}\;\underline{\tau}\right] + \underline{\tau} = 2\eta\underline{D} \tag{6.2}$$

The upper, lower and corotational Maxwell model correspond to a=1,-1,0, respectively.

We may say that the JRS models are of the Giesekus type when

$$\underline{\tilde{\tau}} = \underline{\tau} + \alpha\underline{\tau}^2$$

with mobility constant α. The model actually proposed by Giesekus (1982) has retardation time on the right side which we put to zero for instantaneous elasticity, $\alpha\underline{\tau}^2$ is added to the left side and a = 1

$$\lambda\left[\frac{d\underline{\tau}}{dt} - \nabla\underline{u}\;\underline{\tau} - \underline{\tau}\;\nabla\underline{u}^T\right] + \underline{\tau} + \alpha\underline{\tau}^2 = 2\eta\underline{D} \tag{6.3}$$

Equations of the form (6.1), (6.2) and (6.3) permit hyperbolic waves of vorticity. The last term on the left of (5.4) may be expressed as a second order derivative of the vorticity

$$\varepsilon_{kej}M_{jmpq}u_{q,pme} = -\frac{1}{2}\varepsilon_{kej}P_{jq}\;(\text{curl}\;\underline{\zeta})_{q,e}$$

$$+\;\frac{1}{2}P_{mp}\zeta_{k,mp} + \frac{1}{2}\varepsilon_{kej}\varepsilon_{pqr}A_{jmpq}\zeta_{r,me}$$

Then (5.4) may be regarded as a second order equation for the vorticity.

The vorticity equation for the three parameter family of Oldroyd models can be written as follows.

$$\rho\left[\partial_{tt}\zeta_k + 2\underline{u}\cdot\nabla\partial_t\zeta_k + u_e u_j \zeta_{k,ej}\right] + \frac{1}{2}(a-1)\varepsilon_{kej}\tau_{jq}[\text{curl }\underline{\zeta}]_{q,e}$$

$$-\frac{1}{2}(a+1)\tau_{mp}\zeta_{k,mp} - \frac{\eta}{\lambda}\nabla^2\zeta_k = -\frac{1}{\lambda}\varepsilon_{kej}\tau_{jm,me}$$

$$+ \rho\left[\left(\frac{\partial}{\partial t} + \underline{u}\cdot\nabla\right)(\underline{\zeta}\cdot\nabla u_k) - (\partial_t\underline{u}\cdot\nabla)\zeta_k - u_e u_{j,e}\zeta_{k,j}\right]$$

$$+ \varepsilon_{kej}\left[\frac{a+1}{2}\tau_{mp,m}\zeta_{k,p} + \frac{1-a}{2}\tau_{jq,e}(\text{curl }\underline{\zeta})_q\right]$$

$$+ \varepsilon_{kej}\Big[\frac{a+1}{2}(\tau_{mp,e}u_{j,mp} + \tau_{mp,me}u_{j,p} - \tau_{jm,mp}u_{p,e})$$

$$+ \frac{a-1}{2}(\tau_{mp,e}u_{p,jm} + \tau_{mp,me}u_{p,j} + \tau_{jp,me}u_{p,m}$$

$$+ \tau_{jp,m}u_{p,me} + \tau_{jp,me}u_{m,p} + \tau_{jp,m}u_{m,pe})\Big] \tag{6.4}$$

Moreover

$$\varepsilon_{kej}\tau_{jm,me} = (\text{curl div }\underline{\tau})_k = \rho\left[\text{curl }\frac{d\underline{u}}{dt}\right]_k = \rho\left[\frac{d\zeta_k}{dt} + (\zeta\cdot\nabla)u_k\right] \tag{6.5}$$

Potential flow with $\zeta=0$ can be a solution of (6.4) only if the last bracket in (6.4) vanishes $\underline{u}=\nabla\Phi$.

The BKZ model is a single integral model which is motivated by the nonlinear theory of elasticity. Let $\underline{C}(\tau)$ and $\underline{C}^{-1}(\tau)$ be the right relative Cauchy strain tensor with invariants

$$(I,II) = (\text{tr}\underline{C}(\tau),\text{tr}\underline{C}^{-1}(\tau))$$

and "strain energy"

$$W(t-\tau,I,II).$$

The determinate stress is given by

$$\underline{\tau} = \int_{-\infty}^{t} \{-W_{,II}(t-\tau,I,II)\ \underline{C}^{-1}(\tau) + W_{,I}(t-\tau,I,II)\underline{C}(\tau)\}d\tau \tag{6.6}$$

where $W_{,I} = \partial W/\partial I$, etc. The determinate stress vanishes on the rest state ($\underline{C}=\underline{C}^{-1}=\underline{1}$); hence

$$W_{,II}(t-\tau,3,3) = W_{,I}(t-\tau,3,3). \tag{6.7}$$

The rate equation satisfied by $\underline{\tau}$ may be obtained from (6.4) by differentiating with respect to t, using

$$\frac{d\underline{C}}{dt} = -\underline{L}^{T}(t)\underline{C}(\tau) - \underline{C}(\tau)\underline{L}(t),$$

$$\frac{d\underline{C}^{-1}}{dt} = \underline{L}(t)\underline{C}^{-1}(\tau) + \underline{C}^{-1}(\tau)\underline{L}^{T}(t)$$

$$\frac{dI}{dt} = \text{tr}\ \frac{d\underline{C}}{dt} = -2\ \text{tr}[\underline{C}(\tau)\underline{L}(t)]$$

$$\frac{dII}{dt} = \text{tr}\frac{d\underline{C}^{-1}}{dt} = 2\ \text{tr}[\underline{C}^{-1}(\tau)\underline{L}^{T}(t)] \tag{6.8}$$

We find that

$$\frac{d\underline{\tau}}{dt} + \frac{1}{\lambda}\ \tilde{\underline{\tau}} = \underline{L}\ \underline{P}_1 + \underline{P}_1\underline{L}^{T} + \underline{P}_2\underline{L}$$

$$+ \underline{L}^{T}\underline{P}_2 + \int_{-\infty}^{t} \{\underline{F}_1(\tau)\text{tr}[\underline{C}(\tau)\underline{L}(t)]$$

$$+ \underline{F}_2(\tau)\text{tr}[\underline{C}^{-1}(\tau)\underline{L}^{T}(t)]\}d\tau \tag{6.9}$$

where

$$\tilde{\underline{\tau}} = \int_{-\infty}^{t}\left[\frac{\partial W_{,II}}{\partial t}\ \underline{C}^{-1} - \frac{\partial W_{,I}}{\partial t}\underline{C}\right]d\tau$$

$$\underline{P}_1 = -\int_{-\infty}^{T} W_{,II}(t-\tau,I,II)\underline{C}^{-1}(\tau)d\tau$$

$$\underline{P}_2 = -\int W_{,I}(t-\tau,I,II)\underline{C}(\tau)d\tau$$

$$\underline{F}_1(\tau) = 2\left[W_{,I\ II}\underline{C}^{-1}(\tau) - W_{,II}\underline{C}(\tau)\right]$$

$$\underline{F}_2(\tau) = 2\left[-W_{,II\ II}\underline{C}^{-1}(\tau) + W_{,I\ II}\underline{C}(\tau)\right]$$

The last terms of (6.9) (under the integral) are of leading order in hyperbolic analysis because they contain first derivatives $L_{ij} = \partial_j u_i(\underline{x},t)$ of $\underline{u}_i(\underline{x},t)$ at the present time. These terms are spoilers because they lead to third derivatives of $\underline{u}$, as in (5.4), which do not reduce to second derivatives of $\underline{\zeta}$. This shows that not every popular fluid model has a vorticity of changing type.

It may be helpful to draw a distinction between two types of rate equations. The first type among our special examples depends only on the velocity and the stress. We may regard the stress components as unknown, to-be-determined, field variables, Equations (6.2) and (6.3). The other type, like (6.1) and (6.9), has coefficients which are functionals on the history of the Cauchy tensor. One could say that these equations are not "true" rate equations because the coefficients do not depend directly on the field variables and problems based on these equations must be posed as initial history rather than initial value problems. Rate equations of the first type are clearly convenient because the balance equations of mechanics are also relations among the velocity and stresses and the whole system of balance and constitutive euqations is then closed. Points of convenience are not necessarily points of principle.

7. CLASSIFICATION OF TYPE IN STEADY PLANE FLOW

In plane flows there are six equations for six unknowns; u_1, u_2, τ_{22}, τ_{11}, $\tau_{21}=\tau_{12}$, p. The quasilinear equations which governs the flow of fluids with instantaneous elasticity may be written as

$$\rho\frac{\partial u_1}{\partial t} + \rho u_1\frac{\partial u_1}{\partial x} + \rho u_2\frac{\partial u_1}{\partial y} + \frac{\partial p}{\partial x} - \frac{\partial \tau_{11}}{\partial x} - \frac{\partial \tau_{12}}{\partial y} = 0$$

$$\rho\frac{\partial u_2}{\partial t} + \rho u_1\frac{\partial u_2}{\partial x} + \rho u_2\frac{\partial u_2}{\partial y} + \frac{\partial p}{\partial y} - \frac{\partial \tau_{12}}{\partial x} - \frac{\partial \tau_{22}}{\partial y} = 0$$

$$\frac{\partial \tau_{11}}{\partial t} + u_1\frac{\partial \tau_{11}}{\partial x} + u_2\frac{\partial \tau_{11}}{\partial y} - \left[M_{1111}\frac{\partial u_1}{\partial x} + M_{1112}\frac{\partial u_2}{\partial x} + M_{1121}\frac{\partial u_1}{\partial y} + M_{1122}\frac{\partial u_2}{\partial y}\right] = \tilde{\tau}_{11}/\lambda$$

$$\frac{\partial \tau_{12}}{\partial t} + u_1\frac{\partial \tau_{12}}{\partial x} + u_2\frac{\partial \tau_{12}}{\partial y} - \left[M_{1211}\frac{\partial u_1}{\partial x} + M_{1212}\frac{\partial u_2}{\partial y} + M_{1221}\frac{\partial u_1}{\partial y} + M_{1222}\frac{\partial u_2}{\partial y}\right] = \tilde{\tau}_{12}/\lambda$$

$$\frac{\partial \tau_{22}}{\partial t} + u_1\frac{\partial \tau_{22}}{\partial x} + u_2\frac{\partial \tau_{22}}{\partial y} - \left[M_{2211}\frac{\partial u_1}{\partial x} + M_{2212}\frac{\partial u_2}{\partial x} + M_{2221}\frac{\partial u_1}{\partial y} + M_{2222}\frac{\partial u_2}{\partial y}\right] = \tilde{\tau}_{22}/\lambda$$

$$\frac{\partial u_1}{\partial x} + \frac{\partial u_2}{\partial y} = 0 \tag{7.1}$$

where $(x_1,x_2) = (x,y)$. We may write this system as

$$\frac{\partial \tilde{q}}{\partial t} + \underline{A}\frac{\partial \tilde{q}}{\partial x} + \underline{B}\frac{\partial \tilde{q}}{\partial y} = \tilde{\underline{q}} \tag{7.2}$$

where $\tilde{\underline{q}}$ stands for the column vector whose components are $[u_1,u_2,p,\tau_{11},\tau_{21},\tau_{22}]$. The characteristic directions $\alpha = dy/dx$ for steady flows are determined as the roots of

$$\det |\underline{A}\alpha - \underline{B}| = 0.$$

This may be written as

$$\begin{bmatrix} \rho(\alpha u_1 - u_2) & 0 & \alpha & -\alpha & 1 & 0 \\ 0 & \rho(\alpha u_1 - u_2) & -1 & 0 & -\alpha & 1 \\ -\alpha M_{1111} + M_{1121} & -\alpha M_{1112} + M_{1122} & 0 & \alpha u_1 - \alpha u_2 & 0 & 0 \\ -\alpha M_{1211} + M_{1221} & -\alpha M_{1212} + M_{1222} & 0 & 0 & \alpha u_1 - u_2 & 0 \\ -\alpha M_{2211} + M_{2221} & -\alpha M_{2212} + M_{2222} & 0 & 0 & 0 & \alpha u_1 - u_2 \\ \alpha & -1 & 0 & 0 & 0 & 0 \end{bmatrix} \tag{7.3}$$

$$= (\alpha u_1 - u_2)^2 [a_4 \alpha^4 + a_3 \alpha^3 + a_2 \alpha^2 + a_1 \alpha + a_0] = 0$$

where

$$a_4 = M_{1212} - \rho u_1^2,$$

$$a_3 = -M_{2212} + M_{1211} - M_{1222} + M_{1112} + 2\rho u_1 u_2,$$

$$a_2 = -M_{2211} + M_{2222} - M_{1212} - M_{1221} + M_{1111} - M_{1122} - \rho(u_1^2 + u_2^2)$$

$$a_1 = M_{2221} - M_{1211} + M_{1222} - M_{1121} + 2\rho u_1 u_2,$$

$$a_0 = M_{1221} - \rho u_2 .$$

In general the roots of the 6th order polynomial (7.3) are neither all real nor all complex. The real roots correspond to real characteristic lines across which discontinuities in certain of the variables may propagate.

It is necessary to identify the real characteristic directions and the variables to which they associate. It is clear from (7.3) that streamlines are always doubly characteristic,

$$\left[\frac{dy}{dx} - \frac{u_2}{u_1}\right]^2 = 0. \tag{7.4}$$

The streamlines do not change type, from hyperbolic to elliptic because dy/dx is always real and never complex.

For problems of changing type we have to analyze the roots of the quartic

$$\alpha^4 a_4 + \alpha^3 a_3 + \alpha^2 a_2 + \alpha a_1 + a_0 = 0 \tag{7.5}$$

The roots of a quartic may be factored into the roots of two quadratics. The roots of the quadratics are real or complex. Real roots correspond to hyperbolicity with real characteristic directions.

For JRS fluids one of the quadratics has imaginary roots and the other can have complex or real roots depending on the values of parameter, velocities and stresses at a point. In this situation we encounter a change of type with real characteristic directions only in certain regions of the flow. The roots which govern a change of type are associated with the vorticity. The imaginary roots are associated with the relation $\nabla^2\psi = -\zeta$, where ζ is the vorticity and ψ is the streamfunction. The streamfunction is associated with an elliptic operator. Discontinuous initial data for the streamfunction will be smoothed.

The analysis of hyperbolicity in plane steady flows is easiest when formed in terms of the vorticity $\zeta = \zeta_3$, where ζ is governed by (6.4) in the plane 1,2. After a little calculation we find that

$$\rho u_j u_k \frac{\partial^2\zeta}{\partial x_j \partial x_k} - \frac{1}{2} P_{jj} \frac{\partial^2\zeta}{\partial x_k \partial x_k} - \frac{1}{2} A_{kjke} \frac{\partial^2\zeta}{\partial x_j \partial x_e} + \frac{1}{2} A_{ejke} \frac{\partial^2\zeta}{\partial x_j \partial x_k}$$

$$= \text{L.O.T.}, \qquad j,k,e = 1,2 \tag{7.6}$$

where L.O.T. stands for lower order terms and $A_{ijk\ell}$ is symmetric in ij and skew symmetric in $k\ell$. Eq (7.6) may be written as

$$\left[\rho u_1^2 - A_{1221} - \frac{1}{2}(P_{11} + P_{22})\right] \frac{\partial^2 \zeta}{\partial x_1^2} + [2\rho u_1 u_2 - (A_{1112} + A_{2221})] \frac{\partial^2 \zeta}{\partial x_1 \partial x_2}$$

$$+ \left[\rho u_2^2 + A_{1221} - \frac{1}{2}(P_{11} + P_{22})\right] \frac{\partial^2 \zeta}{\partial x_2^2} = \text{L.O.T.} \tag{7.7}$$

We may apply the theory of 2nd order quasilinear equations to (7.7). We look for real characteristic surfaces $\phi(x_1,x_2) = \text{const.}$ Let $\nabla\phi = (\phi_1,\phi_2,0)$, then (7.7) generates real characteristic surfaces given by

$$\left[\rho u_1^2 - A_{1221} - \frac{1}{2}(P_{11} + P_{22})\right]\phi_1^2 + [2\rho u_1 u_2 - (A_{1112} + A_{2221})]\phi_1\phi_2$$

$$+ \left[\rho u_2^2 + A_{1221} - \frac{1}{2}(P_{11} + P_{22})\right]\phi_2^2 = 0 \tag{7.8}$$

We note that if $\phi(x_1,x_2) = \text{const.}$ gives a plane characteristic curve $x_2(x_1)$, then $\phi_1 + \phi_2 \frac{dx_2}{dx_1} = 0$, $\frac{dx_2}{dx_1} = -\frac{\phi_1}{\phi_2}$. It follows from (7.8) that we have real characteristics given by

$$\frac{dx_2}{dx_1} = \frac{B \pm \sqrt{B^2 - AC}}{A} \tag{7.9}$$

provided that

$$B^2 - AC \geqq 0 \tag{7.10}$$

where

$$A = \rho u_1^2 - A_{1221} - \frac{1}{2}(P_{11} + P_{22}),$$

$$2B = 2\rho u_1 u_2 - (A_{1112} + A_{2221}),$$

$$C = \rho u_2^2 + A_{1221} - \frac{1}{2}(P_{11} + P_{2\,2})$$

The condition (7.10) for real characteristics may be written as

$$\rho u_1^2\left[\frac{P_{11}+P_{22}}{2}-A_{1221}\right]+\rho u_2^2\left[\frac{P_{11}+P_{22}}{2}+A_{1221}\right]-2\rho u_1u_2[A_{1112}+A_{2221}]$$

$$+(A_{1112}+A_{2221})^2+A_{1221}^2-\left[\frac{P_{11}+P_{22}}{2}\right]^2\geq 0 \qquad 7.11)$$

Equations governing change of type in Oldroyd models arise from those just given under the substitutions mentioned in (2.19). In particular the characteristic surface equation becomes

$$\left[\rho u_1^2-\left[\frac{a+1}{2}\tau_{11}+\frac{a-1}{2}\tau_{22}+\frac{\eta}{\lambda}\right]\right]\phi_1^2$$

$$+2[\rho u_1u_2-\tau_{12}]\phi_1\phi_2 \qquad (7.12)$$

$$+\left[\rho u_2^2-(\frac{a-1}{2}\tau_{11}+\frac{a+1}{2}\tau_{22}+\frac{\eta}{\lambda})\right]\phi_2^2=0$$

where we have put $[P_{11},P_{22},A_{1112},A_{2221},A_{1221}]=[a\tau_{11},a\tau_{22},\tau_{12},\tau_{12},\frac{1}{2}(\tau_{11}-\tau_{22})]$.

8. CONDITIONS FOR A CHANGE OF TYPE. PROBLEMS OF NUMERICAL SIMULATION.

In general the condition (7.11) for hyperbolicity appears to be easiest to satisfy when the speed is large and the normal stresses small, or large and positive. For the Oldroyd models (7.11) may be written as

$$\left[\rho u_1^2+\frac{\tau_{22}}{2}(1-a)-\frac{\tau_{11}}{2}(1+a)-\frac{\eta}{\lambda}\right]\left[(1+a)\frac{\tau_{22}}{2}+(a-1)\frac{\tau_{11}}{2}+\frac{\eta}{\lambda}-\rho u_2^2\right]$$

$$+(\rho u_1u_2-\tau_{12})^2>0 \qquad (8.1)$$

We may write (8.1) in dimensionless form by using a scale length and a scale velocity U. Then dimensionless variables related to dimensional variables by

$$[u,v,\tau,\sigma,\gamma,M]=\left[\frac{u_1}{U},\frac{u_2}{U},\frac{\tau_{12}}{\tau_0},\frac{\tau_{11}}{\tau_0},\frac{\tau_{22}}{\tau_0},\frac{U}{C}\right]$$

where $\tau_0=\eta/\lambda$ and

$$C = \sqrt{\frac{\eta}{\rho\lambda}} = \sqrt{\frac{\tau_0}{\rho}} \tag{8.2}$$

is the speed of a wave of vorticity into a fluid in uniform motion. The criterion (8.1) for hyperbolicity

$$\left[M^2u^2 + \frac{\Upsilon}{2}(1-a) - \frac{\sigma}{2}(1+a) - 1\right]\left[(1+a)\frac{\Upsilon}{2} + (a-1)\frac{\sigma}{2} + 1 - M^2v^2\right]$$

$$+ [M^2uv - \tau]^2 > 0 \tag{8.3}$$

depends only on the viscoelastic Mach number M (and the parameter $a\varepsilon[-1,1]$). When a = 1 (upper convected Maxwell model), the criterion (8.3) can be written as

$$M^2[v^2(1+\sigma) + u^2(1+\Upsilon) - 2\,\tau uv]$$

$$- (1+\Upsilon)(1+\sigma) + \tau^2 > 0 \tag{8.4}$$

This criterion is satisfied for high speed flow if the stresses σ, Υ, τ are not large and of the wrong sign.

The vorticity equation changes type at points at which the right side of (8.1) or (8.3) vanishes. Numerical simulations of problems of changing type can be very difficult. For example numerical simulations of transonic flows have only become satisfactory in recent years and they are still being improved. Problems of the transonic type can be expected in numerical simulations of the flow of viscoelastic fluid.

Steady flow of the Oldroyd fluids (6.2) is governed by the following system of dimensionless equations

$$\text{div } \underline{u} = 0,$$

$$M^2(\underline{u}\cdot\nabla)\,\underline{u} + \nabla p - \text{div } \underline{\tau} = 0,$$

$$\underline{u}\cdot\nabla\underline{\tau} - a(\underline{D\tau} + \underline{\tau D}) + \underline{\tau\Omega} - \underline{\Omega\tau} - \underline{D} = -\underline{\tau}/W \tag{8.5}$$

where

$$W = U\lambda/d$$

is the Weissenberg number, which may be regarded as a dimensionless relaxation time measured in units of d/U. Stresses relax slowly when W is large. The fluid has a long memory.

The high Weissenberg number problem is that people doing numerical analysis can't get answers when the stress levels are high. This problem is evidently independent of the "transonic" problem of change of type. Most of the computations in which one encounters this problem are for flows without inertia, $\rho = 0$. Yoo and Joseph (1984) also encountered this problem in high speed flows involving change of type. They found that regions of positive and negative vorticity decay rapidly when W is small (even if the "Mach" number is large). When W is large the supercritical regions are more "purely" hyperbolic in that the damping of the vorticity is suppressed. We have something like a telegraph equation with damping proportional to 1/W. (See (12.19).

Though the high Weissenberg number problem and problems of change of type are independent they may be related problems in certain special cases. Though flows without inertia will not ordinarily change type, the criterion for change of type wherein $B^2 - AC$ changes sign can be interesting even when $\rho = 0$.

For the upper convected Maxwell model (a = 1) we find from (8.1) that when $\rho = 0$

$$B^2 - AC = \tau_{12}^2 - \left[\tau_{11} + \frac{\eta}{\lambda}\right]\left[\tau_{22} + \frac{\eta}{\lambda}\right]. \tag{8.6}$$

The equations for steady flow are elliptic when this is negative. In fact, this quantity appears to approach zero very rapidly as λ (or W) is increased in numerical integrations of steady flows. In fact, $B^2 - AC$ is always negative for the upper convected Maxwell model

$$\underline{\tau} + \frac{\eta}{\lambda}\underline{1} = \frac{\eta}{\lambda^2}\int_{-\infty}^{t} e^{-(t-\tau)/\lambda}\, \underline{C}_t^{-1}(\tau)d\tau. \tag{8.7}$$

Since $\underline{C}_t^{-1}$ has positive eigenvalues, the principal values $\tau_{11} + \frac{\eta}{\lambda}$, $\tau_{22} + \frac{\eta}{\lambda}$ are positive and therefore

$$\det \begin{bmatrix} \tau_{11} + \frac{\eta}{\lambda} & \tau_{12} \\ \tau_{12} & \tau_{22} + \frac{\eta}{\lambda} \end{bmatrix} \geqq 0. \tag{8.8}$$

In numerical integrations using finite elements or finite differences, values of τ_{ij} and u_i are obtained from discrete steps in which the criterion (8.8) may be violated. In doing numerical integrations one should verify that the condition (8.8) is not violated.

Some constitutive models, like the corotational one with $a = 0$ may undergo real change of type even when $\rho = 0$.

People doing flow computations for viscoelastic fluids are able to go to higher Weissenberg numbers when they have constitutive equations with more Newtonian viscosity (non-zero retardation times). This observation suggests that viscosity methods for dealing with problem of change of type and shocks could also be useful for solving the high Weissenberg number problem. So far however the people doing numerical works have not used the viscosity method in the limits of small viscosity.

9. LINEARIZED PROBLEMS OF CHANGE OF TYPE

Up to now our study has been exact and fully nonlinear. We may advance our understanding of the problems of vorticity of changing type by considering simpler problems which arise under linearization.

We first identify a class of motions, say motions in the plane or axisymmetric motions. We find some special exact solution of all the equations which fall in the given class. Usually this special solution is featureless, like uniform flow, flow into a sink, flow in a channel, flow between cylinders, extensional flow and so on. The featureless solution does not exhibit the unusual features of change of type. We then perturb all the equations around the special one with perturbations in the given class, and we linearize. This leads us to linearized problems with variable coefficients depending alone on the special solution and not on the perturbation. We analyze the linear perturbed problem for hyperbolicity and change of type. The equations for the characteristics are given by (7.10) where A,B,C are evaluated on the unperturbed special solution.

The procedures of linearization are such that the characteristic directions are the same for each and every linear problem perturbing the special one. The characteristic surfaces for linearized problems are a gift, since nothing beyond the special solution is needed to compute them. We can know which are the elliptic regions and which the hyperbolic regions of flow cheaply, but to find other properties of the perturbed flow, say isovorticity and streamlines, we are obliged to solve linear PDE's.

JRS (1984) treated the linearized problem for a change of type in shear flow and extensional flow for the family of Oldroyd models characterized by the parameter a [1,1]. They also treated sink flow in the plane and circular Couette flow of an upper convected Maxwell model (a=1). They identified the regions of subcritical (elliptic) and supercritical (hyperbolic) flow, but they did not compute the characteristics or solve some boundary value problem. Shear flow $u = \kappa y$, κ is the rate of shear, is hyperbolic outside a strip centered on $y = 0$. Extensional flow $(u,v) = s(x,-y)$, s is the rate of extension, is either hyperbolic outside an ellipse (where u and v are large) or inside a region bounded by branches of a hyperbola $(x^2/A^2) - (y^2/B^2) < 1$ (where the velocity need not be large. Sink flow with radial velocity $u_r = -Q/r$ (potential flow) is hyperbolic when the radius r is small

$$r < \left[\lambda Q\left(\frac{\rho}{\eta}Q-2\right)\right]^{1/2}$$

provided that the source strength $Q > 2\eta/\rho$, where η is the viscosity. Couette flow outside a rotating cylinder of radius a is hyperbolic in an annulus whose inner radius is either a or is greater than a, depending on conditions. When the inner radius of the annulus of hyperbolicity is greater than a we have another example where the region of high speed flow is elliptic and regions of lower speed are hyperbolic. The variation of the stresses is important.

Yoo, Ahrens and Joseph (1984) have tried to explain some striking experimental results of Metzner, Uebler and Fong (1969) with an analysis of the vorticity perturbing irrotational sink flow. The main point at issue in the experiments of Metzner, et al., is that they observe a conical region of zero vorticity. Outside this region the flow is rotational. If we accept the experimental results at face value, we must conclude that there are surfaces across which some derivatives of the vorticity are discontinuous. This type of behavior says "look for hyperbolicity and change of type." Such a field could not be supported by an elliptic vorticity field.

In the experiments of Metzner, Uebler & Fong (1969) a fluid is sucked from a pipe of large diameter through a sudden contraction. If the hole into which the flow goes is small the problem may be thought to be a form of sink flow. Because there are boundary walls, the flow through a sudden contraction is not a sink flow in a strict sense. We shall imagine first that the flow into the hole is not strongly influenced by the walls of the large pipe. We then have a hole in the semi infinite region above a plane. This flow is then regarded as an axisymmetric perturbation of sink flow without boundaries. The characteristic surfaces for the vorticity of all axisymmetric linearized problems perturbing sink flow can then be obtained by integrating the differential equations for the characteristics. Yoo, Ahrens and Joseph then tried to determine if the characteristic surfaces computed in this way could be the locus for the discontinuity in the derivatives of the vorticity observed in the experiments. It is important to verify that the characteristics are cone like in the region where potential sink flow was observed and that the region of potential flow is in the spherical annulus of hyperbolicity. These issues are addressed below.

Metzner, Uebler and Fong (1969) consider high speed flow of viscoelastic fluids into a sudden contraction. They say that "A tentative analysis of the observed velocity field suggests the flow upstream of the small duct to be radially directed toward the origin of the spherical coordinate system. If this is so the continuity equation gives

$$ur^2 = f(\theta). \tag{1}$$"

They actually measure velocities in the cone and they report that their measurements were accurate and that $f(\theta)$ may be taken as constant when $0 \leq \theta \leq 10^\circ$. They also write that $u_\theta = u_\phi = 0$ in the cone. Outside of the cone there is secondary motion and nonzero vorticity.

The nature of the comparison of theory and experiment, explained in the captions of Figs. 9.1 and 9.2, is discussed in greater detail in the paper of Yoo, Ahrens and Joseph (1984). It turns out that the regions of hyperbolicity are model sensitive. Using the measured values of physical parameters in the equation one finds that the observed potential flow does lie in the hyperbolic region (some points are outside the hyperbolic region in Fig. 9.1(b)). The characteristic surfaces are cone like in the regions where potential flow was observed.

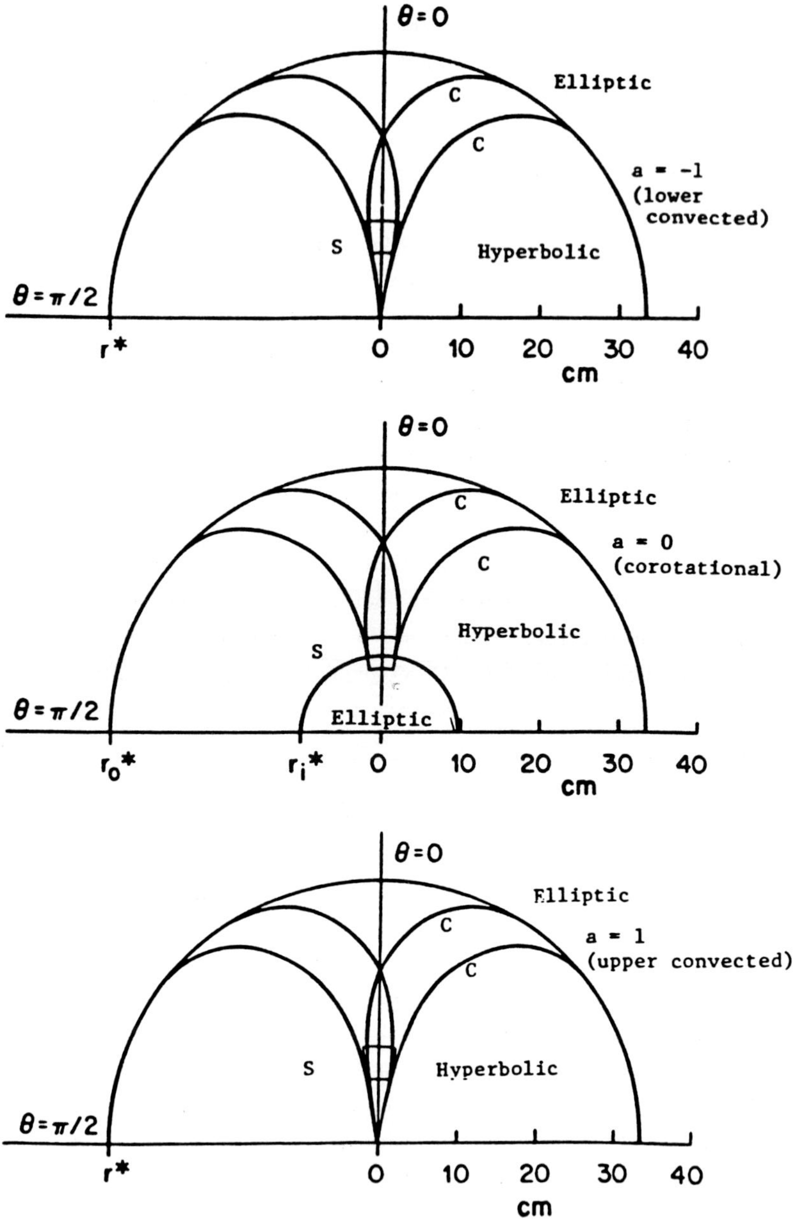

Fig. 9.1. Metzner, Uebler and Fong (1969) measured potential flow in the sectorial box designated S. The cross-sections of characteristic surfaces of revolution which are tangent to the cone of semi vertex angle 10° at the origin are called C. There are two such surfaces of revolution.

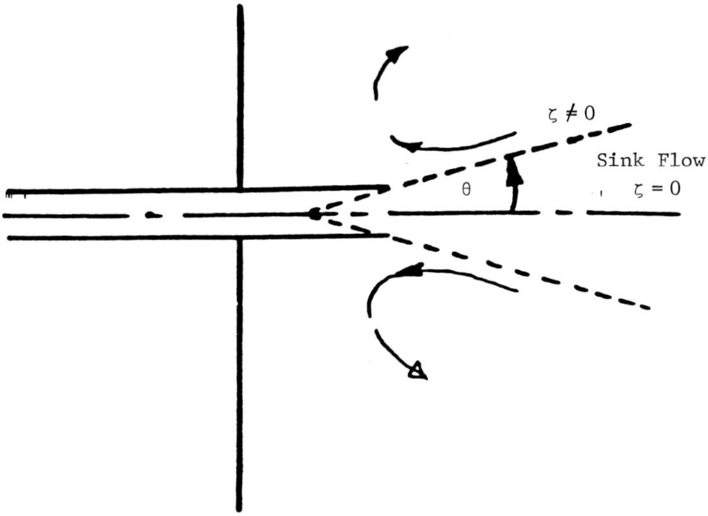

Fig. 9.2. Schematic diagram of flow into a sudden contraction (Metzner, Uebler and Fong (1969)). The vorticity appears to vanish in a cone with nonzero vorticity outside. Accepting this, a jump in some derivative of the vorticity on the cone is required. The measurements verifying potential flow were taken in a cone with $\theta = 10°$ and at a certain value in and out of the pipe. The regions where potential flow was measured are in the sectorial boxes shown in Fig. 9.2.

We regard the comparisons of theory and experiment in Fig. 9.2 as exploratory and not definitive. It is of course striking that the experiments of Metzner, Uebler and Fong (1969) do appear to involve a vorticity of changing type. It would be interesting to see if this striking type of experimental result could be repeated by other investigators using different fluids and experimental arrangements. We hasten to add that the Separan solution used in the experiment is not an Oldroyd model and surely cannot be characterized by a viscosity and relaxation time. In fact only special models give the vorticity precisely as the quantity which changes type. We have already remarked that models with true viscosity; e.g., retardation times, will smooth discontinuities, with only a little smoothing if the retardation "viscosity" parameter is small. Probably all the polymer solutions used in experiments have some small smoothing. In view of all these uncertainties in theory and experiments it would be premature to make strong claims.

10. CHANNEL FLOWS WITH WAVY WALLS

Yoo and Joseph (1984) solved the problem of flow of an upper convected Maxwell model through a channel with wavy walls linearized for small waviness. It is easy to find an exact solution of the equations when the walls are flat. The velocity profile is the same quadratic one that one finds in Newtonian fluids. There are some normal stresses which are absent in the Newtonian case. This Poiseuille flow is perturbed and linearized. The characteristics and regions of hyperbolicity can be computed without specifying the nature of the perturbation. Yoo and Joseph went further. They used a specific perturbation, the amplitude of the waviness of the walls. They defined the linearized problem for small waviness and solved it numerically. The Yoo-Joseph paper gives the first actual computation of a flow with change of type.

We are going to outline the analysis and some of the main results of Yoo and Joseph. We shall express the equations in terms of the "Mach" number $M = U/C$ and the elasticity number $E = \eta\lambda/\rho d^2$ (this is independent of U). The Weissenberg number $U_0\lambda/d$ is given by $W = M\sqrt{E}$.

$$u_x + v_y = 0, \tag{10.1a}$$

$$uu_x + vu_y + \frac{1}{M^2}(p_x - \sigma_x - \tau_y) = 0, \tag{10.1b}$$

$$uv_x + vv_y + \frac{1}{M^2}(p_y - \tau_x - \gamma_y) = 0, \tag{10.1c}$$

$$u\sigma_x + v\sigma_y - 2(\sigma+1)u_x - 2\tau u_y = -\frac{\sigma}{M\sqrt{E}}, \tag{10.1d}$$

$$u\tau_x + v\tau_y - (\gamma+1)u_y - (\sigma+1)v_x = -\frac{\tau}{M\sqrt{E}}, \tag{10.1e}$$

$$u\gamma_x + v\gamma_y - 2\tau v_x - 2(\gamma+1)v_y = -\frac{\gamma}{M\sqrt{E}}. \tag{10.1f}$$

We shall seek and find a solution of these equations satisfying no slip conditions at the walls

$$u = v = 0 \text{ at } y = \pm(1 + \varepsilon \sin nx) \tag{10.2}$$

with symmetric streamlines

$$\frac{\partial u}{\partial y} = v = 0 \text{ at } y = 0. \tag{10.3}$$

It is noteworthy that our solution is completely determined by data (10.2) and (10.3) on the velocity alone. It is not necessary, and it would be wrong to prescribe more about velocity or stresses. The vorticity $\underline{\omega}$ = curl $\underline{u}$ is related to the streamfunction

$$\underline{\omega} = \text{curl } \underline{u} = \omega \underline{e}_z = -\left[\frac{\partial^2\psi}{\partial x^2} + \frac{\partial^2\psi}{\partial y^2}\right]\underline{e}_z \tag{10.4}$$

where $\omega(x,y)$ satisfies

$$\begin{aligned}
&(M^2u^2-\sigma-1)\omega_{xx} + 2(M^2uv-\tau)\omega_{xy} + (M^2v^2-\gamma-1)\omega_{yy} \\
&\quad + \left[-P_x + \frac{M}{\sqrt{E}}u\right]\omega_x + \left[-P_y + \frac{M}{\sqrt{E}}v\right]\omega_y \\
&\quad + (u_x-v_y)(\sigma_{xy}+\gamma_{xy}+\tau_{xx}+\tau_{yy}) + (u_y+v_x)(\gamma_{yy}-\sigma_{xx}) \\
&\quad + \sigma_y u_{xx} + 2\tau_y u_{xy} + \gamma_y u_{yy} - \sigma_x v_{xx} - 2\tau_x v_{xy} - \gamma_x v_{yy} = 0.
\end{aligned} \tag{10.5}$$

Equation (10.5) may be written as

$$A\omega_{xx} + 2B\omega_{xy} + C\omega_{yy} + \text{L.O.T.} = 0 \tag{10.6}$$

where the terms L.O.T. are of lower order for hyperbolic analysis (see JRS, 1984). Characteristic directions

$$\frac{dy}{dx} = \frac{B}{A} \pm \frac{\sqrt{B^2-AC}}{A},$$

$$A = M^2u^2 - \sigma - 1,$$

$$B = M^2uv - \tau,$$

$$C = M^2v^2 - \gamma - 1 \tag{10.7}$$

for the vorticity exist whenever

$$\Sigma \overset{\text{def}}{=} \tau^2 - 2M^2\tau uv - (1+\gamma)(1+\sigma) + M^2v^2(1+\sigma) + M^2u^2(1+\gamma) > 0. \tag{10.8}$$

The expression (10.8) is expressed in terms of unknown velocity and stress fields. The criterion (10.8) for hyperbolicity can be satisfied in some regions of flow and not in others. The border $\Sigma = 0$ between the elliptic and hyperbolic regions of flow is like the sonic line in gas dynamics. Across this line the equations are said to change type.

Equation (10.1) and all the equations of this section are general in that they apply to every plane problem, not just the channel flow problem introduced in Section 1.

Now we shall solve the governing equations for flow in a channel with straight walls $\varepsilon = 0$.

$$(u_0, v_0) = (1-y^2, 0),$$

$$(p_0, \tau_0) = -2M(x,y)\sqrt{E},$$

$$(\sigma_0, \gamma_0) = (8M^2Ey^2, 0),$$

$$\omega_0 = 2y \tag{10.9}$$

The basic motion depends exclusively on the Weissenberg number $W = M\sqrt{E}$ measuring the size of stresses. The solution (10.9) is relatively featureless and, in particular, it gives no indication of hyperbolicity.

Now we consider any plane perturbation of (10.9). The problem with wavy walls is one such perturbation, but there are infinitely many others. We may linearize the formula (10.3) for the characteristics of any flow slightly perturbing the Poiseuille flow (10.9). The characteristics for all these perturbations

$$\frac{dy}{dx} = \frac{1}{M}\left[\frac{-2y\sqrt{E} \pm \sqrt{(y^2-1)^2 - 4Ey^2 - \frac{1}{M^2}}}{-(y^2-1)^2 + 8Ey^2 + \frac{1}{M^2}}\right] \tag{10.10}$$

are defined in terms of quantities defined for the basic flow (10.9) and are given once and for all, independent of the perturbation.

Equation (10.10) shows that flows perturbing plane Poiseuille flow can exhibit a change of type with a "sonic" line $\Sigma = 0$ given by

$$\Sigma(y^2) = (y^2-1)^2 - 2Ey^2 - \frac{1}{M^2} = 0 \qquad (-1 \leqq y \leqq 1).$$

Since $\Sigma(y^2)$ is monotonically decreasing, it has a maximum at $y^2 = 0$ and

$$\Sigma(0) = 1 - \frac{1}{M^2} > 0$$

if and only if the viscoelastic "Mach" number $M > 1$. The "sonic" line across which the flow changes type is $y = y^*$ where $\Sigma(y^{*2}) = 0$

$$y^* = \pm\left[1 + 2E - 2\left[E^2 + E + \frac{1}{4M^2}\right]^{1/2}\right]^{1/2}$$

The linearized problem for small ε is

$$u_x + v_y = 0, \tag{10.11a}$$

$$(1-y^2)u_x - 2yv + \frac{1}{M^2}(p_x - \sigma_x - \tau_y) = 0, \tag{10.11b}$$

$$(1-y^2)v_x + \frac{1}{M^2}(p_y - \tau_x - \gamma_y) = 0, \tag{10.11c}$$

$$(1-y^2)\sigma_x + 16M^2Eyv - 2(8M^2Ey^2+1)u_x + 4y\tau + 4Myu_y\sqrt{E} + \sigma/M\sqrt{E} = 0, \tag{10.11d}$$

$$(1-y^2)\gamma_x + 4Myv_x\sqrt{E} - 2v_y + \gamma/M\sqrt{E} = 0, \tag{10.11e}$$

$$(1-y^2)\tau_x - 2Mv\sqrt{E} + 2y\gamma - (8M^2Ey^2+1)v_x - u_y + \tau/M\sqrt{E} = 0 \tag{10.11f}$$

where

$$u(x,\pm 1) = 2 \sin nx$$

$$v(x,\pm 1) = 0,$$

$$v(x,0) = u_y(x,0) = 0. \tag{10.12}$$

Yoo and Joseph solved the equations (10.11) using only the velocity data (10.12) and the method of separation to reduce these equations to ordinary differential equations in y. The ordinary differential equations were solved numerically. The vorticity

$$\omega \overset{\text{def}}{=} v_x - u_y = -\nabla^2\psi \tag{10.13}$$

satisfies an equation

$$\left[(1-y^2)^2 - 8Ey^2 - \frac{1}{M^2}\right]\omega_{xx} + 4\frac{\sqrt{E}}{M}\, y\omega_{xy} - \frac{\omega_{yy}}{M^2} + \left[\frac{2\sqrt{E}}{M} + \frac{1}{M\sqrt{E}}(1-y^2)\right]\omega_x$$

$$= -\frac{v}{M\sqrt{E}} - 16Eyu_{xx} + \frac{4\sqrt{E}}{M}u_{xy} - 2(1-y^2)v_x$$

$$+ \frac{1}{M^2}[2\tau_x + 2y\gamma_{yy} - 2y\sigma_{xx} + 4\gamma_y] \tag{10.14}$$

of changing type whose characteristics are given by (10.7).

In Figs. 10.1, 10.2 and 10.3 we have graphed vorticity and streamlines for M = 10. In 10.1a and 10.2a we superimposed characteristics (light) on zero vorticity lines. We first observe that when W is small the decay of the vorticity is rapid. When W is large the vorticity decays only very slowly with oscillations all the way to the center.

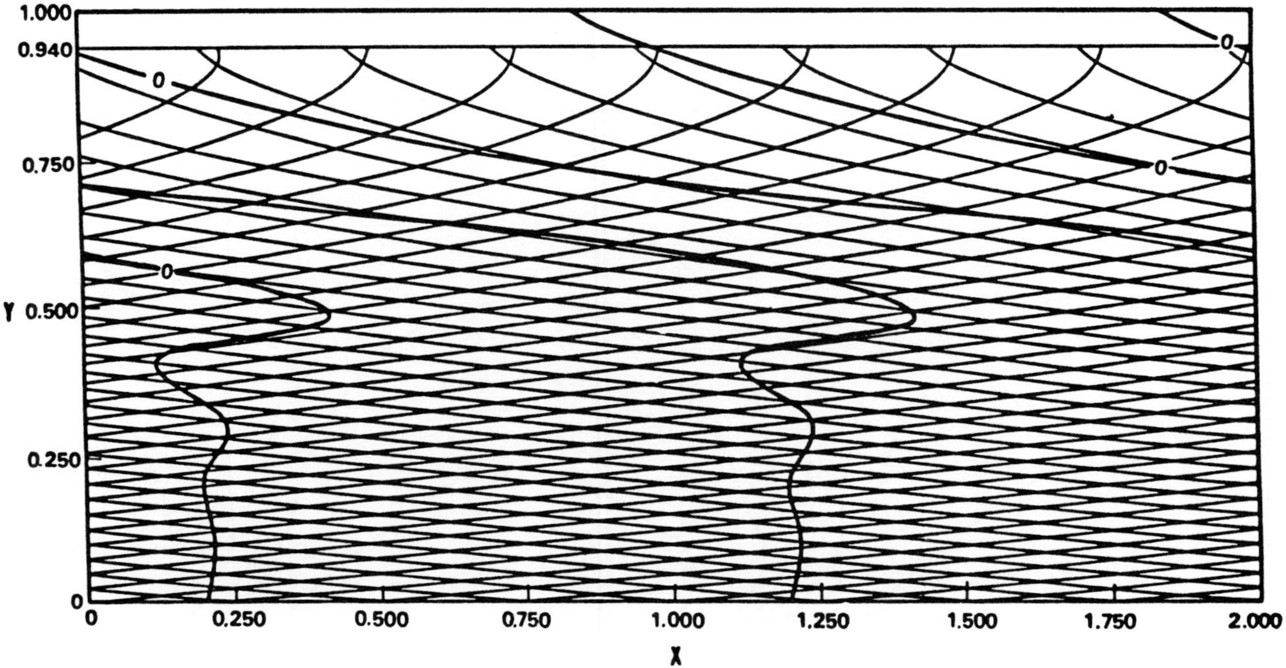

Fig. 10.1(a): Zero vorticity curves for (E,M,R,W) = (0.001,10,316,0.316)

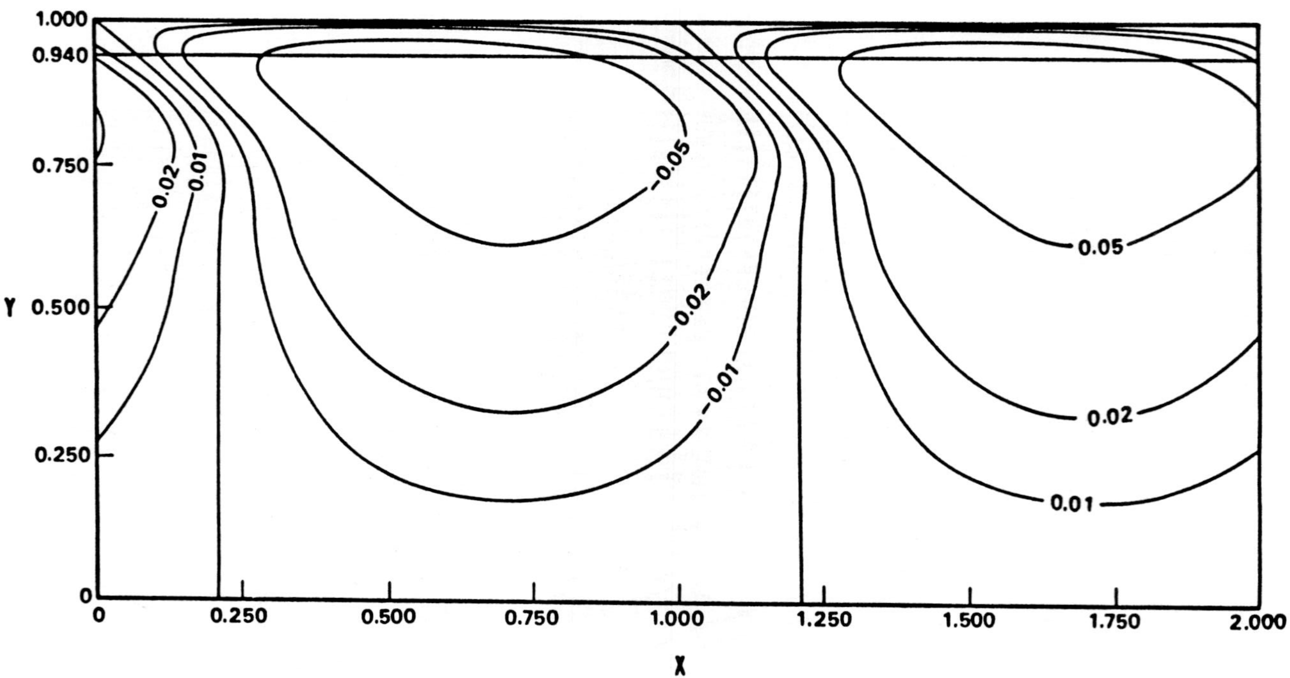

Fig. 10.1(b): Streamlines

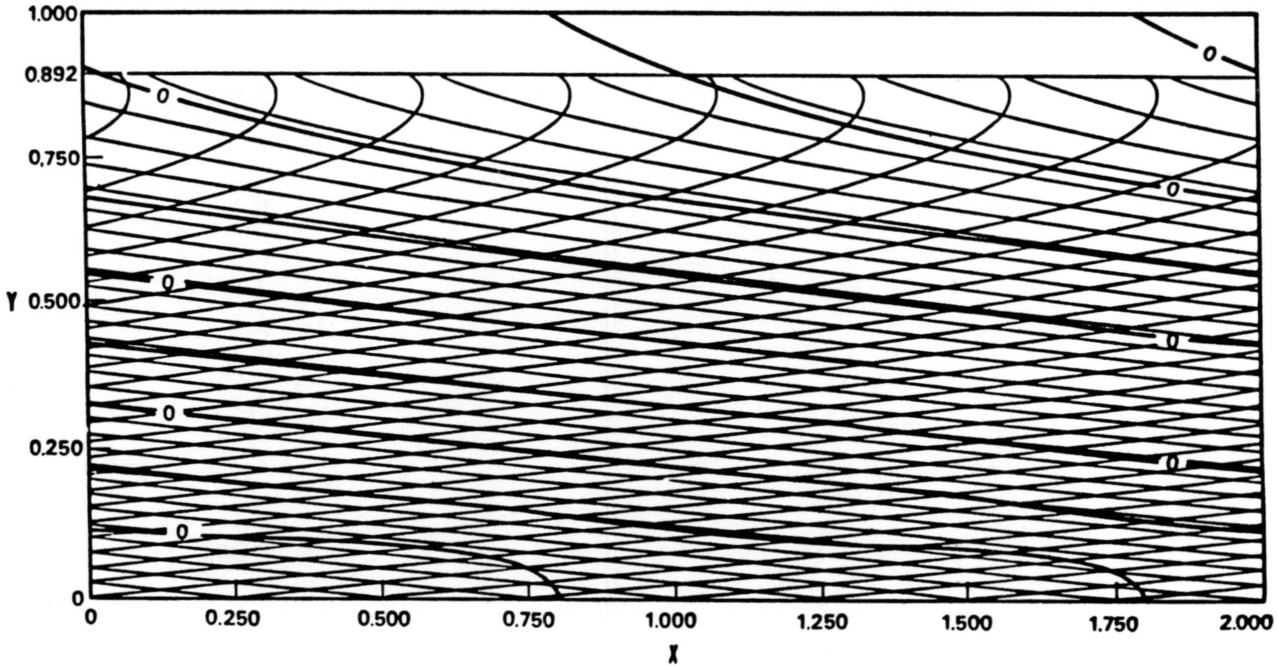

Fig. 10.2(a): Zero vorticity curves for (E,M,R,W) = (0.01,10,100,1)

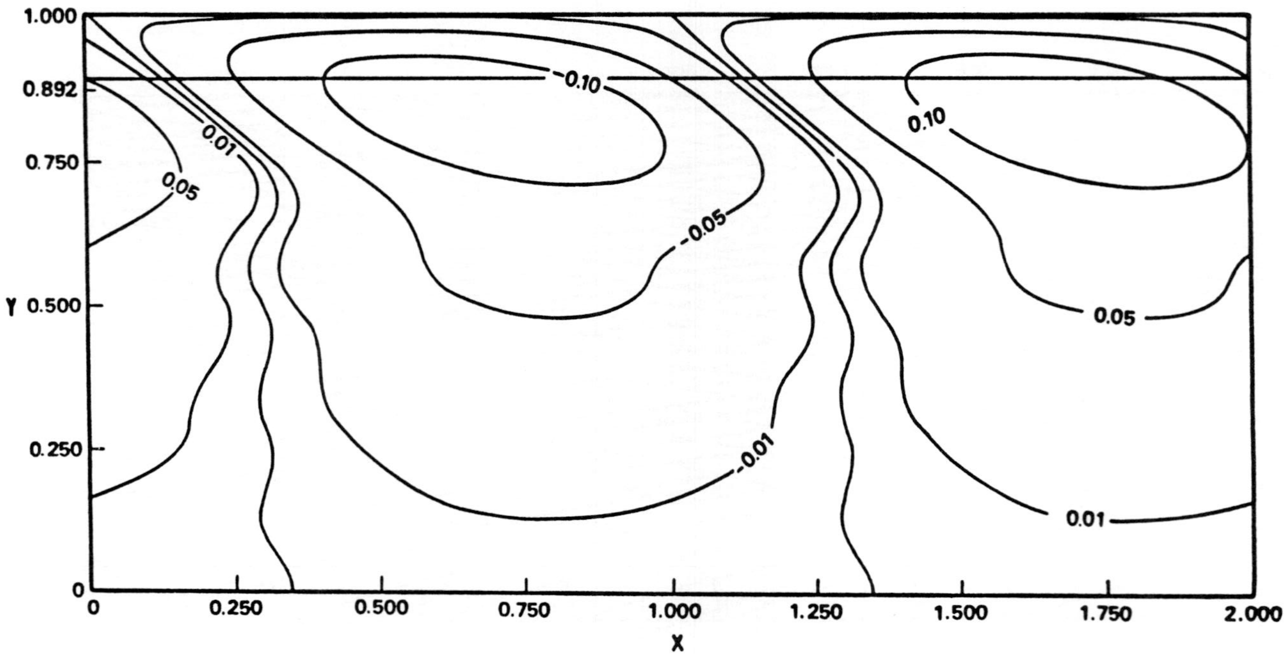

Fig. 10.2(b): Streamlines

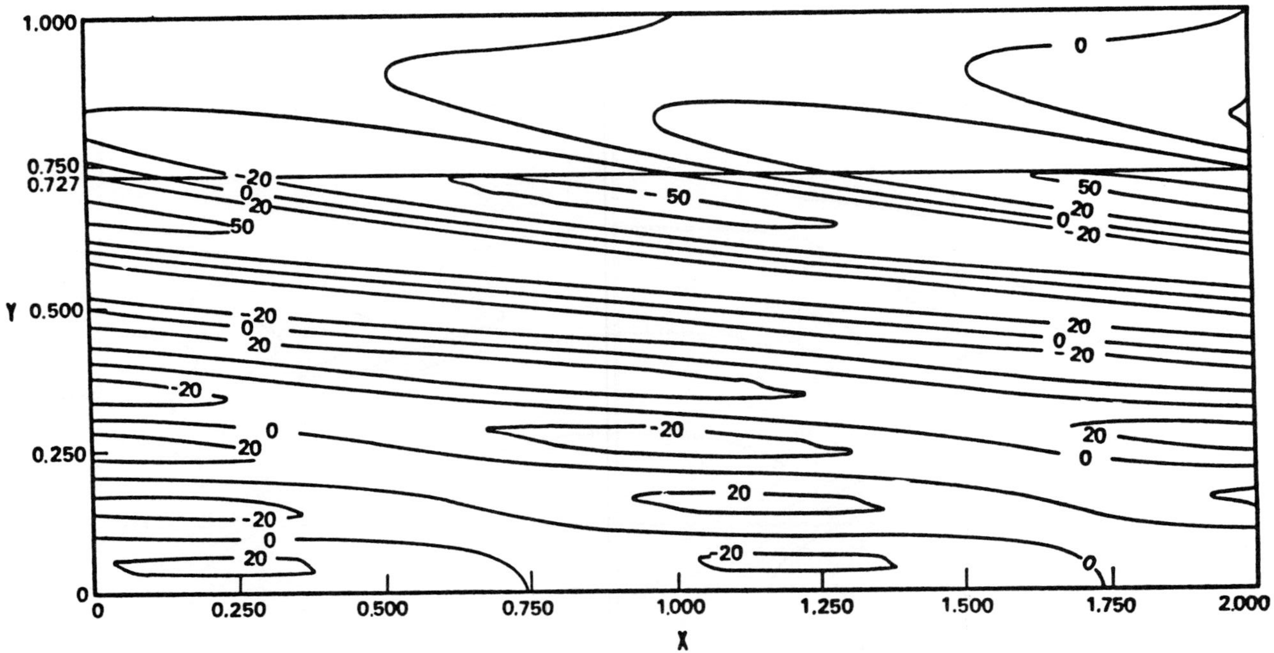

Fig. 10.3(a): Iso-vorticity curves for (E,M,W,R) = (0.1,10,31.6,3.16)

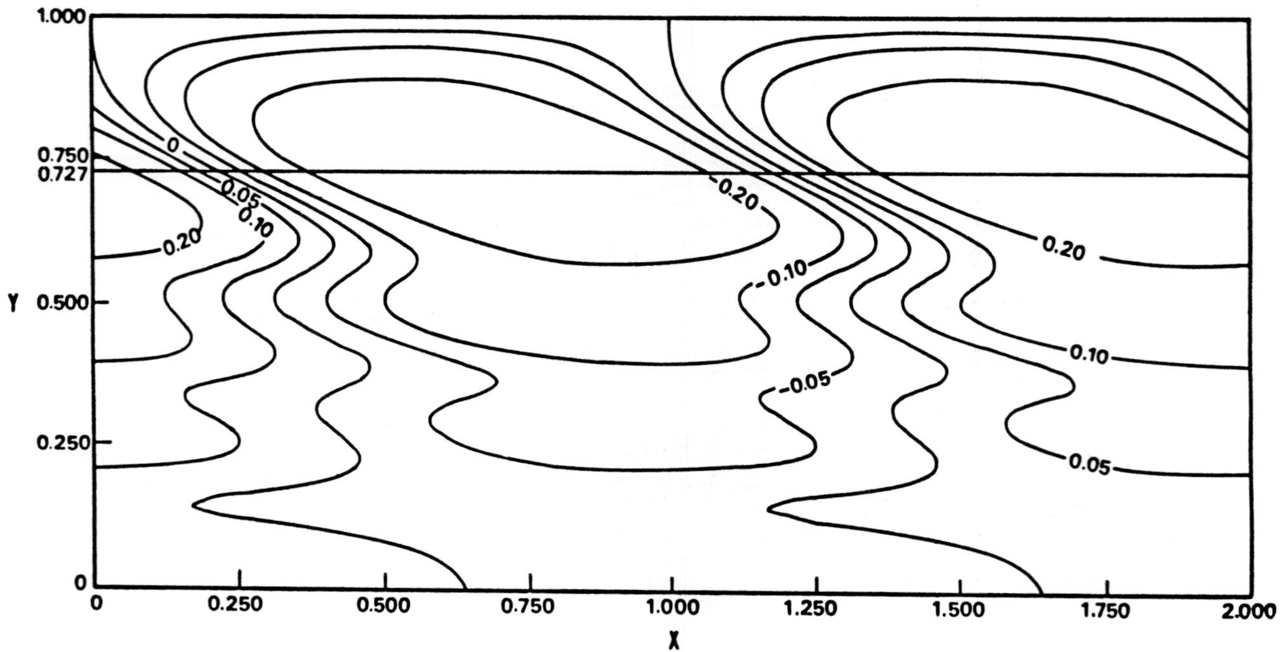

Fig. 10.3(b): Streamlines

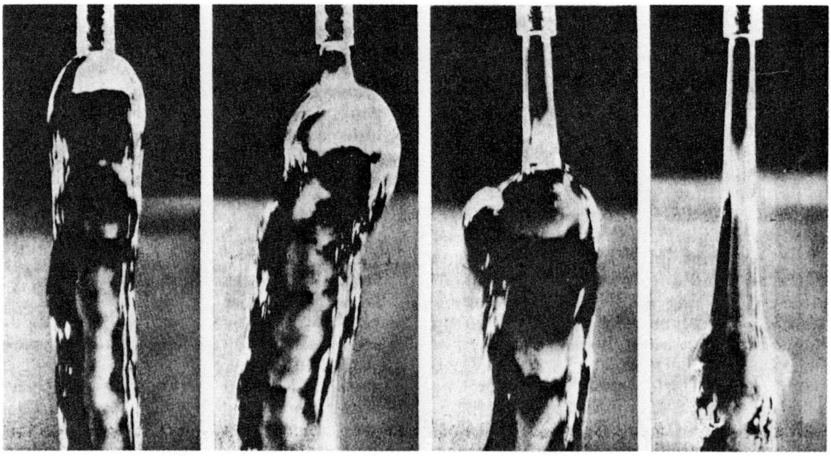

(a) (b) (c) (d)

Fig. 10.4. (After Brenschede and Klein, 1970). Delayed die swell in polyisobutilene solution is toluol. $D = 4\mathring{Q}/\pi R^3$ where $\mathring{Q}$ is the volume flow rate and R is the pipe radius. The critical value D is 14×10^4 reciprocal seconds.
(a) $D = 8 \times 10^4$ (b) $D = 15.2 \times 10^4$
(c) $D = 22.8 \times 10^4$ (d) $D = 29 \times 10^4$

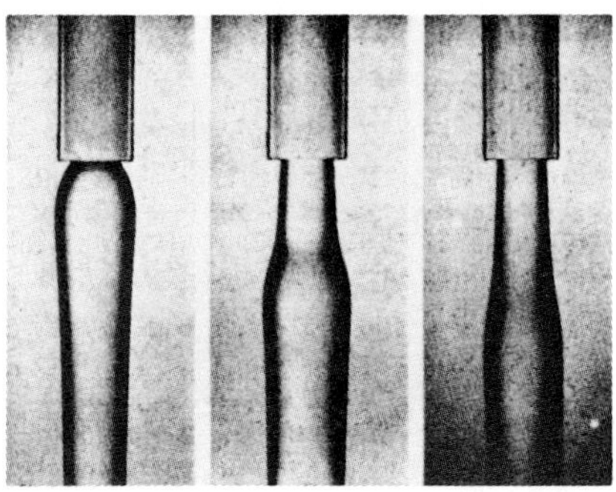

a b c

Fig. 10.5. (After H. Giesekus, 1968). Delayed die swell in a .5% polyacrylamide solution in water.
(a) slow speed, (b) post critical speed,
(c) high speed

Yoo and Joseph did an asymptotic analysis which give the frequency of the oscillations and other properties of the solution. The second property is that regions of positive and negative vorticity which are not damped are swept out along characteristics. It would be hard to understand these solutions from the streamline plots.

The same type of hyperbolic dynamics which Yoo and Joseph found in channels will occur in pipes. Mark Ahrens is working on this problem. The change of type which occurs in the center of a pipe may have applications in the problem of delayed die swell. The phenomenon of delayed die swell is not well known. At low speeds the jet will spread near the exit of the jet, as in Fig. 10.4(a) and 10.5(a). At yet higher speeds the jet does not spread near the exit, the swell is delayed, as in Fig. 10.4(b-d). Fig. 10.5 shows a form of delayed die swell which we have seen repeatedly in our own experiments. The delayed swell seems to occur at a critical speed, not so different than what one might expect from a change of type. Of course, the reason for the delayed swell is not understood. The form of the jet reminds one of a hydraulic jump, which is the shock phenomenon corresponding to shocks in gas dynamics.

11. PROBLEMS ASSOCIATED WITH THE FLOW OF VISCOELASTIC FLUIDS AROUND BODIES

It is perhaps not unreasonable to think that far from the body we have only a small perturbation of uniform motion with constant velocity $U\underline{e}_x$ in the direction x. The body is stationary.

The vorticity of all steady flows of viscoelastic fluids with instantaneous elasticity which perturb uniform flow can change type.

To be precise, the linearized equations for the vorticity of flows perturbing uniform flow vanish on states of constant vorticity and the type of this equation in steady flow changes when the ratio of the free stream velocity of the wave speed increases through unity. This criterion for a change in type may be expressed in terms of a viscoelastic Mach number.

Ultman and Denn (1970, 1971) consider the equations for two-dimensional steady flow of an upper convected Maxwell fluid. They linearize at a motion with uniform velocity and zero stress, and they show that these linearized equations change type when a viscoelastic "Mach" number

$$M = \frac{U}{C} = U/\sqrt{\eta/\lambda\rho}$$

exceeds one (see Eq. (2.4)). Here U is the velocity of the unperturbed uniform flow, and C is the wave velocity for propagation of shear waves in a Maxwell fluid. Ultman and Denn (1970) did not notice that it is precisely the vorticity which changes type. They refer to their linearization as an Oseen approximation. Oseen introduced his linearization around uniform flow for slow viscous flow, because Stokes equations have no solution for flow around bodies in two dimensions. Oseen's equations do not change type. Ultman and Denn (1971) use Oseens approximate method to compute subcritical flow and they say that their calculations agree with their experiments when the fluid parameters are properly chosen.

In a second paper, Ultman and Denn (1970) consider the supercritical flow but they do not give experimental results and they do not discuss or try to solve their equations. They attempt to correlate some experimental observations of D. F. James (1967) with the change of type. James observes a sudden change in the slope of the heat transfer curve as a function of velocity. This happens at a critical velocity which for the Polyox solution used by James, was about 1 cm/sec. It is not clear from the graphs how abrupt this change of slope is, but there is a change of slope. Ultman and Denn (1970) also suggest that the transition from subcritical to supercritical flow might explain abrupt changes in the drag coefficient they say was observed by A. Fabula (1966). Again, the idea is that the critical velocity at transition is the wave speed C. They make an estimate of C from a molecular theory and correlate this prediction with the data of James. Of course, any such estimate can at best be expected to give an order of magnitude, since the molecular theory is coarse and the fluids used in the experiment are not Maxwell fluids.

Recently Ambari, Deslouis and Tibollet (1984) have considered mass transfer in the flow of a viscoelastic fluid around a cylinder. They also find a critical value at which the mass transfer undergoes an anomolous transition.

Oseen's methods work because slow flow around bodies is in some sense a perturbation of uniform motion. The case of fast flow is different because the uniform stream is not small, but the velocity on the body must vanish. In supercritical flow the ratio of the stream velocity to the wave speed is greater than one, and the wave velocity which is a property of the fluid, cannot be controlled. It follows that even if the flow away from the body is supercritical the flow near the body will always be subcritical, and the underlying problem is really one of "transonic" type.

We are going to show that the equations which govern perturbations of uniform flow have a simple form, depending on the density ρ and relaxation modulus G(s) alone, and are model independent. The flow near the body depends on the constitutive model.

The stress in a simple incompressible fluid may be decomposed into an isotropic and a determinate part. The determinate part is the constitutive equation which relates stress and deformation. In linearization at uniform motion, the determinate stress in fluids with instantaneous elasticity is given by an integral (see JRS, 1984)

$$\underline{\tau} = \int_{-\infty}^{t} G(t-\tau)\underline{A}[\underline{u}(\underline{\chi},\tau)]\, d\tau \tag{11.1}$$

where

$$\underline{\chi} = \begin{bmatrix} x - U(t-\tau) \\ y \\ z \end{bmatrix},$$

$$\underline{A}[\underline{u}] = \nabla\underline{u} + \nabla\underline{u}^T, \quad \underline{A}_{ij} = \frac{\partial u_i}{\partial x_j} + \frac{\partial u_j}{\partial x_i}$$

and G is smooth, positive and monotone decreasing. The speed of various kinds of shear waves (Coleman and Gurtin, 1968, Narain and Joseph, 1982) into a fluid of density ρ at rest is

$$C = \sqrt{\frac{G(0)}{\rho}}. \tag{11.2}$$

The "Mach" number is U/C. By differentiating (2.1) with respect to t, holding $\underline{x}$ fixed, we find that

$$\frac{\partial \underline{\underline{\tau}}}{\partial t} + U \frac{\partial \underline{\underline{\tau}}}{\partial x} = G(0)\underline{A}[\underline{u}(\underline{x},t)] + \int_{-\infty}^{t} G'(t-\tau)\underline{A}[\underline{u}(\underline{\chi},\tau)]\, d\tau \tag{11.3}$$

where the last term is of lower differential order for hyperbolic analysis.

For all the Oldroyd models the perturbation of uniform motion leads to the linearized Maxell model with

$$G(s) = \frac{\eta}{\lambda} \exp(-s/\lambda) \tag{11.4}$$

The rate equation for a Maxwell model in a flow perturbing uniform flow is

$$\frac{\partial \underline{\underline{\tau}}}{\partial t} + U \frac{\partial \underline{\underline{\tau}}}{\partial x} = \frac{\eta}{\lambda} \underline{A}[\underline{u}(x,t)] - \frac{\underline{\underline{\tau}}}{\lambda}. \tag{11.5}$$

The equations of motion are

$$\rho\left[\frac{\partial \underline{u}}{\partial t} + U \frac{\partial \underline{u}}{\partial x}\right] = -\nabla p + \text{div}\ \underline{\tau}.$$

The vorticity $\underline{\zeta}$ = curl $\underline{u}$ satisfies

$$\rho\left[\frac{\partial \underline{\zeta}}{\partial t} + U \frac{\partial \underline{\zeta}}{\partial x}\right] = \text{curl div}\ \underline{\tau}. \tag{11.6}$$

Using (2.1) and curl div $\underline{A}[\underline{u}(\underline{x},\tau)] = \Delta\underline{\zeta}(\underline{\chi},\tau)$

where

$$\Delta = \frac{\partial^2}{\partial x^2} + \nabla_2^2, \qquad \nabla_2^2 = \frac{\partial^2}{\partial y^2} + \frac{\partial^2}{\partial z^2},$$

we get

$$\rho\left[\frac{\partial\underline{\zeta}}{\partial t} + U\frac{\partial\underline{\zeta}}{\partial x}\right] = \int_{-\infty}^{t} G(t-\tau)\Delta\underline{\zeta}(\underline{\chi},\tau)d\tau. \tag{11.7}$$

The x derivatives under the integral may be replaced with χ derivatives. Equation (11.7) shows that states of uniform vorticity are solutions of the equations of motion of fluids with instantaneous elasticity which perturb uniform flow. Potential flow is possible.

For steady flows it is convenient to change variables in the integrals. We write

$$\chi = x - U(t-\tau),\ d\chi = Ud\tau \tag{11.8}$$

Equation (11.3) may be written as

$$U\frac{\partial\underline{\tau}}{\partial x} = G(0)\underline{A}[\underline{u}(\underline{x})] + \frac{1}{U}\int_{-\infty}^{x} G'\left[\frac{x-\chi}{U}\right]\underline{A}[\underline{u}(\underline{\chi})]\,d\chi \tag{11.9}$$

Equation (11.6) is now

$$\rho U\frac{\partial\underline{\zeta}}{\partial x} = \text{curl div }\underline{\tau} \tag{11.10}$$

and (11.7) becomes

$$\rho U^2\frac{\partial\underline{\zeta}}{\partial x} = \int_{-\infty}^{x} G\left[\frac{x-\chi}{U}\right]\left[\frac{\partial^2}{\partial\chi^2} + \nabla_2^2\right]\underline{\zeta}(\underline{\chi})\,d\underline{\chi} \tag{11.11}$$

which, after integrating the first term under the integral by parts, becomes

$$\rho U^2 \frac{\partial \underline{\zeta}}{\partial x} = G(0) \frac{\partial \underline{\zeta}}{\partial x} + \frac{G'(0)}{U} \underline{\zeta} + \frac{1}{U^2} \int_{-\infty}^{x} G''\left[\frac{x-\chi}{U}\right] \underline{\zeta}(\chi)\, d\chi$$

$$+ \nabla_2^2 \int_{-\infty}^{x} G\left[\frac{x-\chi}{U}\right] \underline{\zeta}(\chi)\, d\chi \tag{11.12}$$

The analysis of supercritical flow is conveniently framed for equations of second order. We differentiate (2.12) with respect to x and find that

$$(M^2-1) \frac{\partial^2 \underline{\zeta}}{\partial x^2} - \frac{G''(0)}{UG(0)} \frac{\partial \underline{\zeta}}{\partial x} = \nabla_2^2 \underline{\zeta} + \frac{G''(0)}{U^2 G(0)} \underline{\zeta}$$

$$- \frac{1}{UG(0)} \int_{-\infty}^{x} \left[G'\left[\frac{x-\chi}{U}\right] \nabla_2^2 + \frac{1}{U^2} G'''\left[\frac{x-\chi}{U}\right]\right] \underline{\zeta}(\chi)\, d\chi. \tag{11.13}$$

This equation (11.13) changes type from elliptic to hyperbolic when

$$M^2 = \rho U^2/G(0) = U^2/C^2$$

increases through one.

For Maxwell models (11.13) reduces, using (11.4), to

$$\left[\rho U^2 - \frac{\eta}{\lambda}\right] \frac{\partial^2 \underline{\zeta}}{\partial x^2} - \nabla_2^2 \underline{\zeta} + \frac{\rho U}{\lambda} \frac{\partial \underline{\zeta}}{\partial x} = 0 \tag{11.15}$$

this equation changes type when $M^2 = U^2/C^2$, $C^2 = \eta/\rho\lambda$ increases through one.

The vorticity of steady flows with instantaneous elasticity perturbing uniform flow will change type from elliptic to hyperbolic when the ratio of the velocity of the free stream to the velocity of propagation of shear waves

into a fluid at rest exceeds unity. Characteristics for vorticity in plane flow are given by

$$y-d = \frac{\pm x}{\sqrt{M^2-1}}$$

where d is a constant and U is in the direction x. The characteristics form a net of straight lines.

If we suppose that a small two dimensional body perturbs uniform flow when U > C then there would be an undisturbed region in front of the body which could not be reached by disturbances of the vorticity traveling at velocity C. In a linear theory the first changes in the vorticity would occur across the leading characteristics which form an angle like the Mach angle of gas dynamics (see Fig. 11.1). We could call the undisturbed region of uniform flow a "region of silence".

No one has yet solved a problem of supercritical flow over a body, even in the linearized case and no one has looked for a region of silence in experiments.

An air bubble rising in a liquid will reach a steady terminal velocity. The larger the bubble the larger is the terminal speed. Astarita and Apuzzo (1965) were the first to notice that the terminal velocity of the bubble is not a smooth function of the bubble volume. There is a critical volume which is associated with a jump in the terminal velocity, as in Fig. 11.2.

This phenomenon has also been studied by Calderbank, Johnson & Loudon (1970). The fluids used in these experiments are water-based polymers. The lower value of the critical terminal velocity ranges from 0.1 to 10 cm per sec. These values are not unlike values which we think are typical for wave speeds in these polymers (see Section 4). The reasons for the abrupt rise in the terminal velocity are not understood. Maybe the abrupt change is associated with a change of type. Then we might expect to see an abrupt change near M = 1. For Maxwell models M = RW where R is the Reynolds number and W is the Weissenberg number. Zana and Leal (1978) give values of RW which vary by decades at criticality. Their method of calculating was not clear. They seemed to have used some normal stress data which could give very inaccurate values for W. If they are nearly right it will be hard to support the idea that the abrupt rise is associated with a change of type. The problem is open. To settle this problem we need to have accurate values for the wave velocity used in experiments.

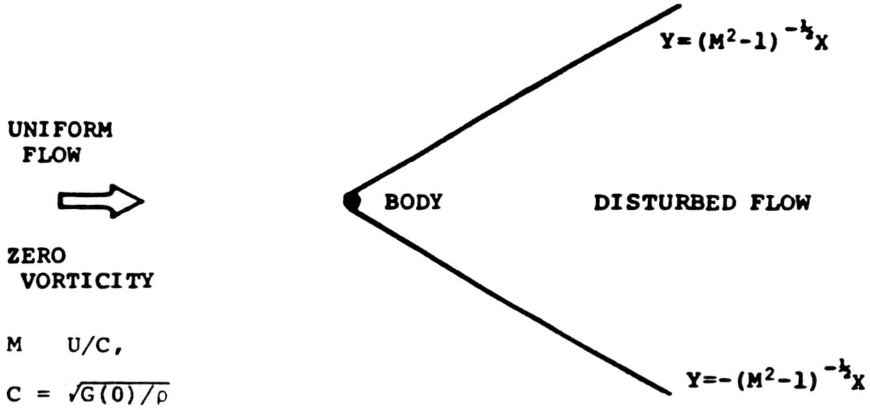

Fig. 11.1. Leading characteristics for vorticity in a plane uniform flow.

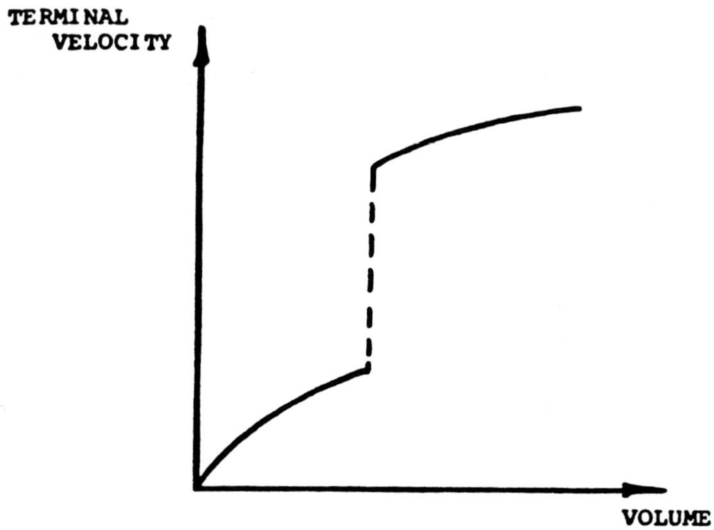

Fig. 11.2. The jump in the terminal velocity for an air bubble rising in a liquid. The jump can be large, say a five-fold increase.

12. FLOW OVER A FLAT PLATE

The problem of flow around a flat plate could be thought to be closer than flow around a body to a perturbation of uniform flow. In fact, the theory of slender bodies in aerodynamics is a perturbation of uniform flow which is perturbed less by slender bodies than by fat ones. The aerodynamic theory works well, but only because of the flow following condition in which the fluid is required to slip along the slender body. In this chapter I am going to give an exact supercritical theory, valid for all fluids with instantaneous elasticity, such that the vorticity is prescribed arbitrarily along the flat plate, with a vanishing normal component of velocity on the along the flat plate, with a vanishing normal component of velocity on the plate surface. Presumably the no slip condition on the plate can also be satisfied by choosing the correct vorticity distribution on the plate. In the nonlinear problem there will be a diffusive subcritical region near the plate and a "transonic" surface outside where the viscoelastic Mach number passes through one. This type of important problem has yet to be considered and solved in the theory of flow of viscoelastic fluids.

We now consider two dimensional flow past a flat plate. The plate is on the half line $y = 0,\ x \geqq 0$. The velocity components corresponding to (x,y) are (u,v) and there is only one component

$$\zeta = \frac{\partial v}{\partial x} - \frac{\partial u}{\partial y} \tag{12.1}$$

of vorticity. The velocity components

$$(u,v) = \left[-\frac{\partial \Psi}{\partial y}, \frac{\partial \Psi}{\partial x}\right] \tag{12.2}$$

may be obtained from a stream function Ψ,. We suppose that the velocity (u,v) vanishes at infinity and that

$$(u,v) = (-U,0) \text{ on } y = 0,\ x \geqq 0. \tag{12.3}$$

The vorticity of the perturbed flow satisfies

$$(M^2-1)\,\frac{\partial^2\zeta}{\partial x^2} - \frac{G'(0)}{UG(0)}\,\frac{\partial\zeta}{\partial x} = \frac{\partial^2\zeta}{\partial y^2} + \frac{G''(0)}{U^2G(0)}\,\zeta$$

$$-\ \frac{1}{UG(0)}\int_{-\infty}^{x}\left[G'\left[\frac{x-\chi}{U}\right]\frac{\partial^2}{\partial y^2} + \frac{1}{U}\,G'''\left[\frac{x-\chi}{U}\right]\right]\zeta(\chi)\,d\chi. \tag{12.4}$$

For the Maxwell model (11.4) we get a telegraph equation

$$(M^2-1)\,\frac{\partial^2\zeta}{\partial x^2} - \frac{\partial^2\zeta}{\partial y^2} + \frac{\rho U}{\eta}\,\frac{\partial\zeta}{\partial x} = 0.$$

We may write this telegraph equation in dimensionless form

$$\frac{\partial^2\zeta}{\partial x^2} + \frac{M^2}{\beta^2 W}\,\frac{\partial\zeta}{\partial x} - \frac{1}{\beta^2}\,\frac{\partial^2\zeta}{\partial y^2} = 0 \tag{12.5}$$

where $(x,y) \sim (x/\ell, y/\ell)$, ℓ is a scale length

$$\beta^2 = M^2 - 1 > 0$$

and

$$W = U\lambda/\ell$$

is a Weissenberg number based on ℓ.

Equation (12.5) can be approximated by a wave equation without damping when $W \to \infty$ and $M > 1$ is fixed. For this same M we may expect rapid damping of the backward heat equation when W is very small.

The other prescribed conditions are that the vorticity must vanish far from the plate and since $v = 0$ on the plate,

$$\zeta = -\frac{\partial u}{\partial y} \text{ on } y = 0,\ x \geq 0 \tag{12.6}$$

We need vorticity fields satisfying (12.4) or (12.5) over R^2 and (12.6) and (12.3) on the plate.

The stream function and vorticity are related by a second order equation

$$\left[\frac{\partial^2}{\partial x^2} + \frac{\partial^2}{\partial y^2}\right]\Psi = -\zeta(y,x). \tag{12.7}$$

We may therefore expect that corresponding to any good field $\zeta(x,y)$ we may solve (12.7) subject to the conditions that Ψ vanishes far from the plate and

$$v = \frac{\partial\Psi}{\partial x} = 0 \text{ on } y = 0,\ x \geqq 0. \tag{12.8}$$

In general, we cannot expect that for any field $\zeta(y,x)$

$$U = \frac{\partial\Psi}{\partial y} \text{ on } y = 0,\ x \geqq 0. \tag{12.9}$$

It follows that the existence of a solution of the linearized flat plate problem, in both the subcritical $M < 1$ and supercritical $M > 1$ case depends on finding a special prescription of the vorticity ζ on the plate (12.6) such that (12.9) is verified.

I wish now to consider the supercritical case and to indicate the method by which our linearized problem may be resolved. It will be convenient to start this discussion with an analysis of the telegraph equation (12.5) for $M > 1$. To put this problem into canonical form we change variables setting

$$(x,y) = \left[\frac{\beta^2 W}{M^2}t, \frac{\beta W}{M^2}z\right] \tag{12.10}$$

and find that $\tilde{\zeta}(z,t) = \zeta(y,x)$ satisfies

$$\frac{\partial^2\tilde{\zeta}}{\partial t^2} + \frac{\partial\tilde{\zeta}}{\partial t} - \frac{\partial^2\tilde{\zeta}}{\partial z^2} = 0 \tag{12.11}$$

We may expect that this hyperbolic problem will not allow the plate to influence the flow upstream. We therefore seek solutions $\tilde{\zeta}(z,t)$ such that the upstream vorticity vanishes;

$$\tilde{\zeta}(z,t) = 0 \text{ for } t < 0,\ -\infty < z < \infty$$

This implies that "initially"

$$\tilde{\zeta}(z,0) = \tilde{\zeta}_t(z,0) = 0 \tag{12.12}$$

whereas

$$\tilde{\zeta}(0,t) = g(t) \tag{12.13}$$

is prescribed for $t > 0$. Asymptotically, for large $|z|$ we require that $\tilde{\zeta} \to 0$.

The problem (12.11), (12.12) and (12.13) can be solved by Laplace transform techniques. A well known solution of this problem with uniform vorticity on the plate

$$f(t) = H(t) = \begin{matrix} 0, t<0 \\ 1, t>0 \end{matrix} \tag{12.14}$$

has been given by Carlslaw and Jaeger (1963) in the form

$$\tilde{\zeta}(z,t) = \left[e^{-z/2} + \frac{z}{2} \int_z^t \frac{e^{-\sigma/2}}{\sqrt{\sigma^2 - z^2}} I_1 \left(\frac{1}{2}\sqrt{\sigma^2 - z^2}\right) d\sigma \right] H(t-z) \tag{12.15}$$

where I_1 is the modified Bessel function of the first kind. The vorticity $\tilde{\zeta}(-z,t)$ on the bottom of the plate is given by (12.15) with $-z$ replacing z. This solution for flow over a flat plate was given by B. Caswell (1976) in a study of the effects of a leading edge singularity.

We next note that the field $\zeta(y,x) = \tilde{\zeta}(z,t)$ can now be inserted into (12.7), which is an elliptic problem leading to a nearly everywhere differentiable Ψ, even when ζ has simple discontinuities, as in Figure 1. We solve (12.7) subject to (12.8), but the solution will not satisfy (12.9). We can hope to satisfy (12.8) and (12.9) simultaneously by prescribing the perfect vorticity distribution $f(t)$ on the plate.

To get the $\tilde{\zeta}(t,z,g)$ corresponding to different prescriptions (12.13) of the vorticity $f(t)$ on the plate, we could use the method of Duhamel type integrals introduced for start up problems by Narain and Joseph (1983). For these integrals we need to superpose using the fundamental singular solution of (12.11), (12.12) and (12.13) when

$$g(t) = \delta(t-\tau) \tag{12.16}$$

is a Dirac function. It is easy to see and not hard to prove that the required solution is the time derivative of the step function problem just derived when

$$\zeta(0,t) = \begin{matrix} 1 \text{ for } t > \tau \\ 0 \text{ for } t \leq \tau \end{matrix}. \tag{12.17}$$

Using the aforementioned method I find that the solution of (12.11), (12,12) and (12.13) is

$$\tilde{\zeta}(t,z) = \int_0^{t-z} g(\tau) \frac{\partial f}{\partial t}(z,t-\tau)d\tau + e^{-z/2} g(t-z) \tag{12.18}$$

where

$$f(z,t) = e^{-z/2} + \frac{z}{2} \int_z^t e^{-\sigma/2} \frac{I_1(\frac{1}{2}\sqrt{\sigma^2-z^2})}{\sqrt{\sigma^2-z^2}} d\sigma .$$

The solution $\zeta(y,x)$of

$$\frac{\partial^2\zeta}{\partial x^2} + \frac{M^2}{\beta^2 W}\frac{\partial\zeta}{\partial x} - \frac{1}{\beta^2}\frac{\partial^2\zeta}{\partial y^2} = 0 ,$$

$$\zeta(0,x) = g(x), \tag{12.19}$$

$$\zeta(y,0) = 0$$

is

$$\zeta(y,x) = \int_0^{x-\beta y} g(\tau) \frac{\partial f}{\partial x}(y,x-\tau)d\tau + \exp\left[-\frac{yM^2}{2\beta W}\right] g(x-\beta y) \tag{12.20}$$

where

$$\frac{\partial f}{\partial x} = \frac{yM^2}{2\beta W}\,\frac{e^{-x/2}I_1\left[\frac{1}{2}\left[x^2 - \frac{M^4y^2}{4\beta^2W^2}\right]^{1/2}\right]}{\left[x^2 - \frac{M^4y^2}{4\beta^2W^2}\right]^{1/2}}$$

The amplitude of ζ on $x = \beta y$ is given by $\exp[-\frac{yM^2}{2\beta W}]$. The amplitude decays rapidly when W is small (see Fig. 12.1).

Turning next to the stream function, we find that the solution of (12.7) with $\zeta(y,x)$ given by (12.20) and

$$\psi(x,0) = 0 \qquad \text{for } x > 0 \tag{12.21}$$

is

$$\psi = -\int\int_{R_-} G(y,y_0|x,x_0)\zeta(y_0,x_0)dy_0\,dx_0 \tag{12.22}$$

where R_- is the complement to the line $x > 0$, $y = 0$. The Green function for this domain, vanishing on the plate is given by (10.1.40) on p. 1208 of Morse & Feshbach (1953). We have to divide their solution by 4π.

$$G(y,y_0|x,x_0) = \overline{G}(r,r_0|\phi,\phi_0)$$

$$= \frac{1}{\pi}\sum_{n=1}^{\infty}\frac{1}{n}\sin\left[\frac{n\phi_0}{2}\right]\sin\left[\frac{n\phi}{2}\right]\left[\frac{r}{r_0}\right]^{n/2}, \qquad r < r_0$$

$$= \frac{1}{\pi}\sum_{n=1}^{\infty}\frac{1}{n}\sin\left[\frac{n\phi_0}{2}\right]\sin\left[\frac{n\phi}{2}\right]\left[\frac{r_0}{r}\right]^{n/2}, \qquad r > r_0 \tag{12.23}$$

where $(x,y) = r(\cos\phi,\sin\phi)$

The solution (12.7), satisfying (12.21) is given by (12.22) with $\zeta(y,x)$ given by (2.20). The fluid will not slip on the plate if $U = \frac{\partial\psi}{\partial y}(x,0)$ when $x > 0$; that is,

$$-U = \int\int_{R_-} \frac{\partial}{\partial y}\bigg|_{\substack{y=0\\x>0}} G(y,y_0|x,x_0)\zeta(y_0,x_0)dy_0\, dx_0 \tag{12.24}$$

We have to choose the prescribed plate vorticity g(x) to satisfy (12.24). I have no guarantee that there is a g(x) satisfying (12.24). Edmond O'Donovan is trying to find a numerical approximation for the g(x) which satisfies (12.24).

It is probable that the approach which I have taken in this problem can be generalized to fluids satisfying (12.4). This equation applies to all fluids with instantaneous elasticity. Such a study would show how the flow depends on the kernel G(s). The Laplace transform methods used by Narain and Joseph (1982, 1983) to study the linearized dynamics of shearing motions perturbing rest in viscoelastic fluids are appropriate for this study too. As in their work I get the following problem for the transform

$$\bar{\zeta}(y,\omega) = \int_0^\infty e^{\omega x}\ \zeta(x,y)\ dx \tag{12.25}$$

where $\bar{\zeta}(y,\omega)$ vanishes for large y and satisfies

$$k(\omega)\bar{\zeta} = \bar{g}(\omega)\ \frac{d^2\bar{\zeta}}{dy^2}$$

$$\bar{\zeta}(0,\omega) = \int_0^\infty e^{-\omega x}\ f(x)\ dx \tag{12.26}$$

and

$$k(\omega) = \omega[\rho U^2 - G(0)] - \frac{G'(0)}{U} - \frac{\bar{h}(\omega)}{U},$$

$$\bar{h}(\omega) = \int_0^\infty e^{-\omega s} G''\left[\frac{s}{U}\right] ds,$$

$$\bar{g}(\omega) = \int_0^\infty e^{-\omega s} G\left[\frac{s}{U}\right] ds.$$

One finds that

$$\bar{h}(\omega) = \frac{G''(0)}{\omega} + O\left[\frac{1}{\omega}\right],$$

$$\bar{g}(\omega) = \frac{G(0)}{\omega} + \frac{G'(0)}{\omega^2 U} + O\left[\frac{1}{\omega^2}\right]$$

if G" is integrable. The character of the solutions is determined by the symbol of the operator. We should therefore look at the problem which arises at large ω

$$\left[\omega[\rho U^2 - G(0)] - \frac{G'(0)}{U}\right]\bar{\zeta} = \left[\frac{G(0)}{\omega} + \frac{G'(0)}{\omega^2 U}\right] \frac{d^2\bar{\zeta}}{dy^2}$$

or

$$\left[\beta^2\omega^2 - \frac{\rho U G'}{G^2(0)}\ \omega\right]\bar{\zeta} = \frac{d^2\bar{\zeta}}{dy^2}$$

which is the transform of the equation

$$\beta^2\zeta_{xx} - \frac{\rho U G'(0)}{G^2(0)}\ \zeta_x = \zeta_{xx}. \tag{12.27}$$

From this equation we expect that the solution of the present problem is a simple wave as in Figure 12.1 with $\zeta = 0$ to the left of the line $x = \beta y$ on the top of the plate, and $x = -\beta y$ on the bottom plate. The amplitude of the wave on $x = \beta y$ is given by

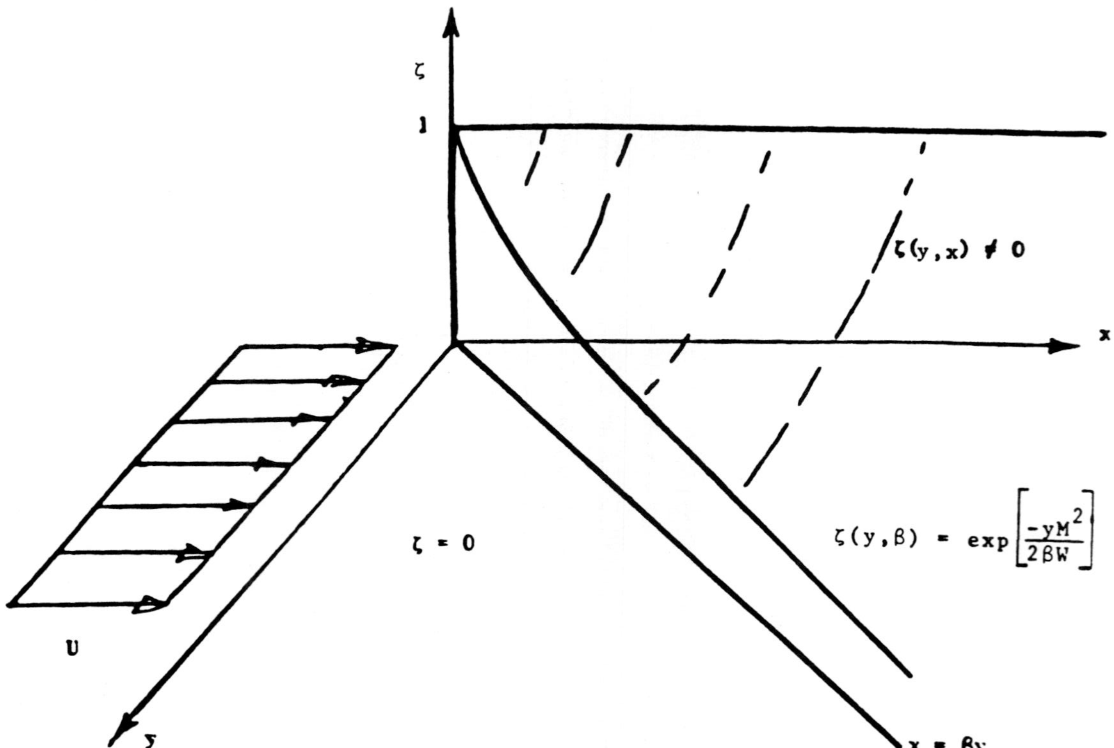

Fig. 12.1. Solution (12.20) for the problem (12.19) for a unit jump of vorticity. The solution decays rapidly if $1/W$ is large.

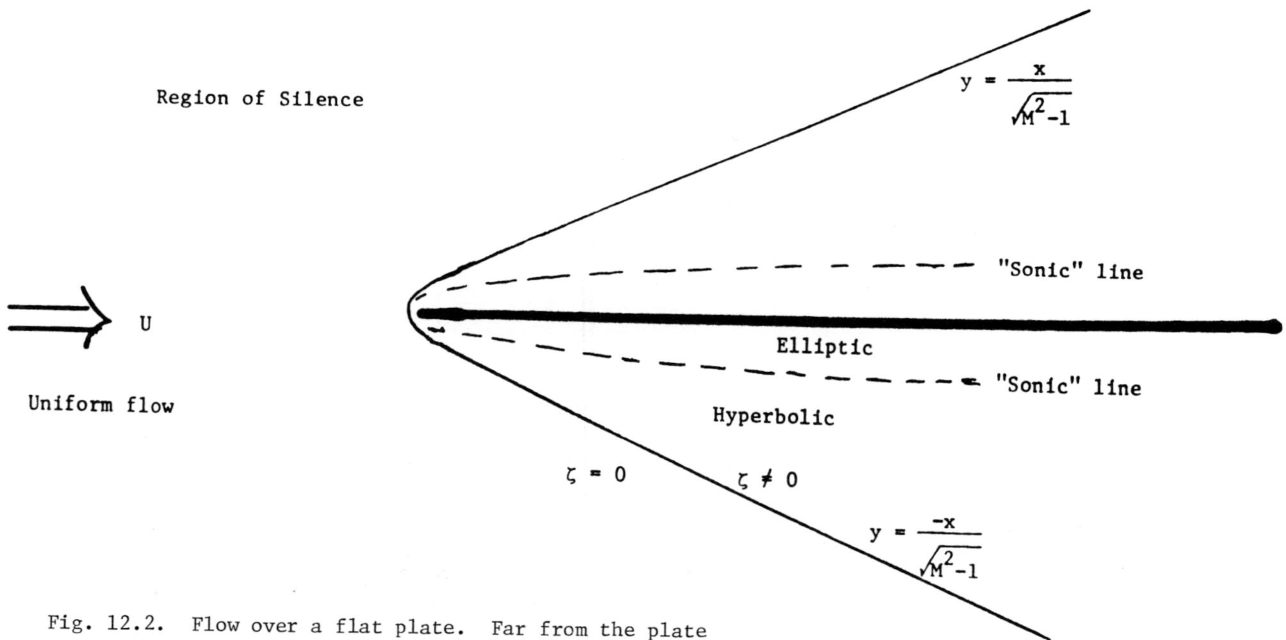

Fig. 12.2. Flow over a flat plate. Far from the plate the linear theory is valid and the vorticity will be confined to the region behind the "Mach" cone. Near the plate the flow must be subcritical and the details will depend on the constitutive equation.

$$\exp\left[\frac{x\rho U G'(0)}{\beta^2 G^2(0)}\right]$$

This amplitude tends to zero as $G'(0) \to -\infty$. When $G(0) > 0$ and $G'(0) = -\infty$ we have the case of singular kernels. Renardy (1982) showed that for some special singular kernels shear waves would propagate into a fluid at rest with the wave velocity $C = \sqrt{G(0)/\rho}$, with a zero amplitude at the front and C^∞ connection with non-zero solution across the front. These properties, as well as the analytic smoothing of sharp fronts with small viscous terms (small retardation times) occur in the present steady problem as well as in the theory of propagation of waves.

Supercritical flow past a flat plate cannot be considered to be a perturbation of uniform flow. If it is supercritical, the free stream velocity is finite, perhaps large, but the fluid must come to rest on the plate. Near the plate, the velocity will be small and the local "Mach" number less than one even when $M_\infty^2 = \rho U^2/G(0) > 1$. The governing problem near the plate is therefore elliptic, or at least not hyperbolic. It may be true that when M is large the solution of the nonlinear problem is close to the linear solution (if it exists) except in regions immediately near the plate. This might suggest boundary layers, but that thought should be eschewed. We should instead think of a narrow subcritical region near the plate which goes supercritical at the "sonic" line which could also be close to the plate when $M >> 1$. We have therefore to consider a "transonic" type of problem and not a boundary layer type of problem. We know almost nothing about such problems for viscoelastic fluids.

13. NONLINEAR WAVE PROPAGATION AND SHOCKS

It is well to motivate this chapter by reminding the reader of the huge differences between linear and nonlinear theories of gas dynamics. In the linearized theory there are no essential differences between rarefaction and compression waves. These waves propagate according to the wave equation without change of form. In the nonlinear theory an impulsive rarefaction will be smoothed by nonlinear effects and a smooth compression will shock up.

Compression waves are impossible in incompressible fluids. Instead we may perhaps speak of waves of shear or of waves of vorticity. Very little is known about the nonlinear effects in the flow of viscoelastic fluids. We should like to know the answers to the following questions. We are given a constitutive equation in the class with instantaneous elasticity: Different subclasses in this class can lead to different results. We want some classification of results.

(i) Suppose that we are given smooth data. Is there a shock up?

(ii) When there is a shock, what variables (vorticity, velocity, displacement) become discontinuous?

(iii) We are given impulsive data. Is it possible that the nonlinear terms smooth discontinuities?

We first consider the problem of wave propagation in fluids undergoing rectilinear shear flow. The formulation, due to Coleman and Noll, is embodied in the representations for shear flow shown in (1.2) of the Appendix, by M. Slemrod to this paper. Coleman and Gurtin (1968), following earlier work on longitudinal acceleration waves in compressible materials (1965), showed that the amplitude of a jump discontinuity in the fluid acceleration satisfies a simple nonlinear differential equation of Bernoulli type with coefficients determined by the instantaneous value of the relaxation kernel at the wave and second order instantaneous modulus evaluated at the wave. The instantaneous value of the relaxation kernel is designated as

$$G(\kappa,0)$$

where κ is the shear rate at the wave and $G(\kappa,s)$ arises as the kernel of the integral representation implied by the Riesz theorem for first functional derivatives of the functional $t\,(\cdot)$ of (1.2) in Appendix A evaluated in the weighted $L^2_h[0,\infty]$ spaces of Coleman and Noll. This kernel reduces to $G(s)$ when the wave advances into a region at rest. The speed of the wave

$$C = \sqrt{\frac{G(\kappa,0)}{\rho}}$$

depends on the rate of shear where the first derivative is evaluated. In general problems the speed of waves depends on the motion. The second order general problems the speed of waves depends on the motion. The second order modulus is an instantaneous evaluation the second Frechet derivative at the wave.

The amplitude equation of Coleman and Gurtin is notable because it is simple, general, rigorous and implies interesting physical results. They showed that an initial jump discontinuity may either decay or grow depending on the sign and magnitude of the initial discontinuity. The assumed jump discontinuity may blow up in finite time. The cause of this blow up is associated with the nonlinearity. They showed that an acceleration wave entering into a region at rest would always decay. This is a type of nonlinear result of category (ii) which shows that nonlinear terms can lead to blow up. It is generally assumed that the loss of C^n regularity (blow up) implies the formation of a jump in the n-1 derivative of velocity (shock up). The decay of acceleration waves is a result of category (iii) which shows that nonlinear terms can force the decay of initial discontinuities.

It was not clear how discontinuities in acceleration, which are equivalent to shocks of the vorticity, would appear in the fluid. The results discussed in the appendix by M. Slemrod help to clarify this issue.

Slemrod (1978) showed that the equation of motion for the shearing perturbation $\hat{v}(x,t)$ of a shearing motion will admit a differentiable (in x,t) solution for only a finite time for appropriate smooth initial velocity histories when the constitutive relation is

$$T^{\langle xy\rangle} = \sigma\left(\int_0^\infty e^{-\tau/\lambda}\, v_x(x,t-\tau)d\tau\right). \tag{13.1}$$

where σ is a nonlinear odd function. The loss of differentiability is assumed to imply the appearance of a discontinuity in v, as shock up of the velocity, a vortex sheet. This result is like the one proved by Coleman and Gurtin, but implies the shock of smooth data.

In the appendix of this paper Slemrod shows that when the constitutive relation is

$$T^{\langle xy\rangle} = -\int_0^\infty e^{\tau/\lambda}\, \sigma(v_x(x,t-\tau)d\tau \tag{13.2}$$

for a nonlinear $\sigma(\cdot)$, the second derivatives of v(x,t) can blow-up when the smooth data is given in a certain way. The shock up assumption which one is obliged to associate with this blow-up of second derivatives is the appearance of a jump in v_x, a vortex shock or an acceleration discontinuity. It is interesting to note that $T^{\langle xy\rangle}$ given by (13.2) satisfies a rate equation

$$\frac{d}{dt} T^{\langle xy\rangle} + \frac{1}{\lambda} T^{\langle xy\rangle} = \sigma(v_x(x,t)) \tag{13.3}$$

vaguely resembling some popular models.

The results just reviewed show the remarkable effects of the choice of constitutive relations.

It is possible to entertain the notion of successive shock ups from smooth data. First we get a vortex shock from nonlinear effects associated with (13.2). This gives an acceleration discontinuity which will lead to blow up of the vorticity if the amplitude of the vortex shock is larger than the critical amplitude of Coleman & Gurtin. In reservation, I wish to note that recent calculations with popular models show that the critical amplitude can be infinite.

I think it would be interesting to see what sort of blow up results could be obtained for shear flows of Oldroyd models with instantaneous elasticity.

The shear flow of the Oldroyd models (6.2) are governed by the system of first order quasilinear equations.

$$\sigma_t - (a+1)\tau v_x + \frac{\sigma}{\lambda} = 0,$$

$$\tau_t + \left[(1-a)\sigma - \frac{\eta}{\lambda}\right] v_x + \frac{\tau}{\lambda} = 0,$$

$$\rho v_t - \tau_x = 0 \tag{13.4}$$

where $v(x,t)$ is the rectilinear velocity in the direction y and $(\sigma, \gamma, \tau) = (T^{\langle yy\rangle}, T^{\langle xx\rangle}, T^{\langle xy\rangle})$. This can be written as

$$\underline{A}\,\underline{q}_t + \underline{B}\,\underline{q}_x + \underline{D} = 0$$

$$\underline{q} = \begin{bmatrix} \sigma \\ \tau \\ v \end{bmatrix}, \quad \underline{A} = \begin{bmatrix} 1 & 0 & 0 \\ 0 & 1 & 0 \\ 0 & 0 & \rho \end{bmatrix}, \quad \underline{D} = \begin{bmatrix} \sigma/\lambda \\ \tau/\lambda \\ 0 \end{bmatrix}$$

$$\underline{B} = \begin{bmatrix} 0 & 0 & -(a+1)\tau \\ 0 & 0 & (1-a)\sigma-\eta/\lambda \\ 0 & -1 & 0 \end{bmatrix}$$

The characteristic directions are given by

$$\det\left|\underline{A}\dot{x} - \underline{B}\dot{x}\right| = \left[\rho^2\dot{x}^2 + (1-a)\sigma - \eta/\lambda\right] = 0$$

where $\overset{\circ}{x}$ is the wave speed. The streamlines are characteristic; $\dot{x} = 0$, because streamlines don't move. The two waves

$$\overset{\circ}{x} = \pm\sqrt{\frac{(a-1)\sigma + \eta/\lambda}{\rho}}$$

can be identified with waves of vorticity. It is clear that the speed of the wave depends on how the liquid is stressed.

We should like to know if smooth initial data, given to (13.4) can blow up. This depends on the constitutive equation, as we have already seen. For example, the problem for the upper and lower convected Maxwell models ($a^2=1$) cannot lead to blow up of smooth data because the governing problems are linear.

Johnson and Segalman (1977) have shown that $(13.4)_1$ and $(13.4)_2$ can be integrated once:

$$\tau = \frac{\eta}{\lambda}\int_{-\infty}^{t} e^{-(t-s)/\lambda}\,\kappa(s)\cos\left(\sqrt{1-a^2}\int_{s}^{t}\kappa(s')ds'\right)ds,$$

$$\sigma = \frac{\eta}{\lambda}\sqrt{\frac{1+a}{1-a}}\int_{-\infty}^{t} e^{-(t-s)/\lambda}\,\kappa(s)\sin\left(\sqrt{1-a^2}\int_{s}^{t}\kappa(s')ds'\right)ds$$

where

$$\kappa(s) = v_x(x,s)$$

is the shear rate at the layer x at time $s \leq t$. It follows that

$$\tau_t - \frac{\eta}{\lambda}\kappa(t)\left[1 - \sqrt{1-a^2}\int_{-\infty}^{t} e^{-(t-s)/\lambda}\,\kappa(s)\sin\left(\sqrt{1-a^2}\int_{s}^{t}\kappa(s')ds'\right)ds\right] = -\frac{\tau}{\lambda}.$$

We may consider fluids with short memories, small λ. After integrating by parts we find that

$$\tau_t - \frac{\eta}{\lambda}\kappa(t)[1 - (1 - a^2)\lambda^2\kappa^2(t) + O(\lambda^3)] + \frac{\tau}{\lambda} = 0 \tag{13.5}$$

This is in the form

$$\frac{d}{dt}T^{\langle xy\rangle} + \frac{1}{\lambda}T^{\langle xy\rangle} = \frac{\eta}{\lambda}v_x(x,t)\left[1 - (1 - a^2)\lambda^2 v_x^2(x,t) + O(\lambda^3)\right]$$

which, up to terms $O(\lambda^2)$ is like (13.3), leading to the blow up of vorticity when $1 - a^2 < 0$.

We turn next to a still more complicated flow involving wave propagation, which is originally set in two space dimensions.

N. Phan Thien and R. I. Tanner (1983), hereafter called PT&T, have considered the problem of flow of an upper convected Maxwell model which is induced by squeezing the fluid between infinite parallel planes. They do not linearize or neglect terms. They call their solution exact because they find a separable solution (13.6) which reduces the governing three dimensional quasilinear system to a nonlinear two-dimensional system which they integrate numerically. They have impulsive initial data in the following sense. Initially an irrotational plug flow is prescribed. The velocity v normal to the plates increases linearly from zero at the bottom to the prescribed value at the top. At this same initial instant the velocity u parallel to the plate is squeezed out as a plug flow. Since u must vanish at the plates, this component is prescribed as discontinuous. In linear problems this discontinuity would propagate and decay along characteristics. PT&T do note that their numerical work shows wave propagation but their graphs are confusing; it is hard to tell what is propagating and how it is propagating.

I have analyzed their problem using some elementary hyperbolic theory. I work with their reduced equations. The reduced equations given by (13.10) are a consequence of representation.

$$[u,v] = V[-xf_y(y,t), f(y,t)] \tag{13.6}$$

Here, subscripts denote differentiation. We treat the rest of the reduction later. For now, consider Eq. (6.4) with a = 1 for the vorticity

$$\rho \frac{\partial^2 \zeta}{\partial t^2} + \rho u \frac{\partial^2 \zeta}{\partial x \partial t} + 2\rho v \frac{\partial^2 \zeta}{\partial y \partial t} + (\rho u^2 - \sigma - \frac{\eta}{\lambda}) \frac{\partial^2 \zeta}{\partial x^2}$$

$$+ 2(\rho u v + \tau) \frac{\partial^2 \zeta}{\partial x \partial y} + (\rho v^2 - \gamma - \frac{\eta}{\lambda}) \frac{\partial^2 \zeta}{\partial y^2} = \text{L.O.T.} \tag{13.7}$$

The vorticity compatible with (13.6) satisfies

$$\zeta = -\frac{1}{2} \frac{\partial u}{\partial y} = \frac{V}{2} x f_{yy} \overset{\text{def}}{=} x\omega,$$

$$\zeta_x = \omega, \qquad \zeta_{xx} = 0.$$

These relations imply that

$$\left[\frac{\partial^2 \zeta}{\partial x \partial t}, \frac{\partial^2 \zeta}{\partial x \partial y}\right] = \left[\frac{\partial \omega}{\partial t}, \frac{\partial \omega}{\partial y}\right]$$

are first derivatives of ω and hence of lower order. We may then write (13.7) as

$$A \frac{\partial^2 \zeta}{\partial t^2} + 2B \frac{\partial^2 \zeta}{\partial y \partial t} + C \frac{\partial^2 \zeta}{\partial y^2} = \text{L.O.T.}$$

where $A = \rho$, $B = \rho v$, $C = \rho v^2 - \gamma - \frac{\eta}{\lambda}$. The characteristic equation for this second order problem are given by

$$\frac{dy}{dt} = \frac{B}{A} \pm\sqrt{\frac{B^2-AC}{A}}$$

$$= v \pm\sqrt{\frac{\Upsilon}{\rho} + C^2} \qquad (13.8)$$

where $C = \sqrt{\eta/\lambda\rho}$ is the speed of small amplitude vorticity waves into regions of uniform motion. The wavespeed formula (13.8) shows how the speed of vorticity for solutions of the type (13.6) depends on the pointwise values of the normal component of velocity and the tensile stress $\Upsilon = \tau_{yy}$. Every problem which can be represented in (13.6) will have wave speeds for the vorticity which will satisfy (13.8). Numerical results of PT&T suggest that for their problem $\Upsilon/\rho + C^2 > 0$.

In principle, it should be possible to determine some of the effects of nonlinearity on propagation by more careful studies of the problem treated by PT&T.

Let h_0 be the initial distance between the plates and let v be the constant squeezing velocity

$$(y,t) = \frac{1}{h_0}(y,Vt)$$

are dimensionless variables, also called y and t. The configuration of flow is shown as Fig. 13.1. The dimensionless velocity components are given by

$$(u,v) = (xf_y\,(y,t),f(y,t)$$

and have been made dimensionless with V. The stresses are made dimensionless with $\eta V/h_0$.

The dimensionless stress and pressure π are then represented by

$$[\tau^{\langle xx\rangle},\tau^{\langle yy\rangle},T^{\langle xy\rangle},P] = \left[X_1 + x^2X_2,xT,Y,\ \frac{1}{2}x^2p_1 + p_2\right] \qquad (13.9)$$

where X_1,X_2,T,Y depend on (y,t) and p_2 and

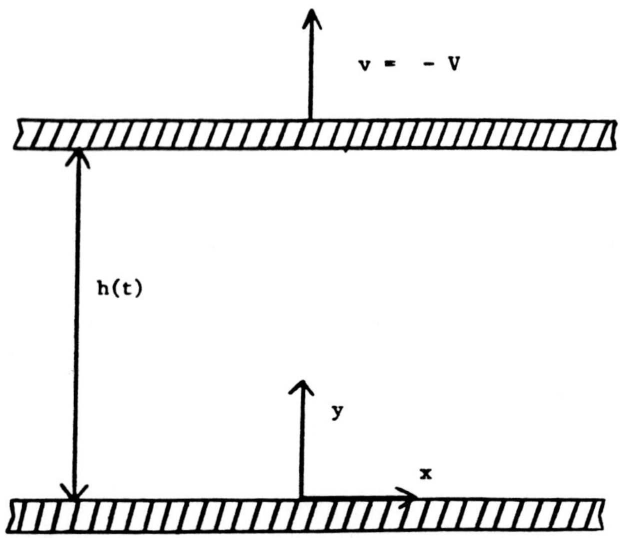

Fig. 13.1 Squeeze Flow, $V > 0$, $h(0) = h_o$.

$$p_1(t) = 2X_2 + T_y + R\left[f_{ty} + f_y^2 + ff_{yy}\right]$$

depend only on t, and R is the Reynolds number.

We may write the reduced problem as

$$\underline{A}\,\underline{q}_y + \underline{B}\,\underline{q}_t = \underline{D} \tag{13.10}$$

where $\underline{q}$ is the column vector whose components are $[f,p,X_1,Y,X_2,T,g]$ where $g = f_y$,

$$\underline{A} = \begin{bmatrix} 1 & 0 & 0 & 0 & 0 & 0 & 0 \\ 0 & 1 & 0 & 0 & 0 & 0 & 0 \\ 0 & 0 & fW & 0 & 0 & 0 & 0 \\ 0 & 0 & 0 & fW & 0 & 0 & 0 \\ 0 & 0 & 0 & 0 & fW & 0 & 2WT \\ 0 & 0 & 0 & 0 & 0 & fW & 1+WY \\ 0 & 0 & 0 & 0 & 0 & 1 & fR \end{bmatrix}$$

$W = V\lambda/h_0$ is the Weissenberg number, $\underline{B}$ is diagonal with entries $[0,0,W,W,W,W,R]$ and $\underline{D}$ depends on $\underline{q}$ but not on its derivatives. The wave speeds for this system $\overset{\circ}{y} = dy/dt$ are given as the real roots of

$$\det[\underline{A} - \overset{\circ}{y}\underline{B}] = W^3(f-\overset{\circ}{y})^3[M^2(f-\overset{\circ}{y})^2 - 1 - WY] = 0$$

where $M^2 = RW$. The normal component of velocity $v = f$ is triply characteristic. The wave speeds for the vorticity are given by the dimensionless form of (13.8).

$$\overset{\circ}{y} = f \pm \frac{1}{M}(1+WY)^{1/2}$$

Numerical results of PT&T show that $Y \to 0$ as $W \to \infty$ in such a way that $1 + WY > 0$. There is no Hadamard instability.

An initial boundary value problem for the reduced quasilinear system requires that one set of boundary and initial conditions. At the boundaries $y = 0$ and $y = H(t)$ we have adherence

$$u = v = 0 \quad \text{at } y = 0,$$

$$v = 0, \quad v = -V \quad \text{at } y = H(t).$$

To specify initial conditions it is enough to give the fields u and v, that is, $f(y,t)$ at $t = 0$. PT&T set $f(y,0) = -y$. Hence

$$(u,v)\Big|_{t=0} = (x,-y) \tag{13.11}$$

which is an irrotational squeeze flow with vanishing stresses and an outflow independent of y at each x. The wave speeds (13.8) are

$$dy/dt\Big|_{t=0} = C \quad \text{at the bottom plate}$$

and

$$dy/dt\Big|_{t=0} = -1 - C \quad \text{at the top plate.}$$

The velocity $v = -y$ at $t = 0$ is compatible with the prescribed boundary conditions but $u = 0$ on $y = 0$ and on $y = H(t)$ is incompatible with $u = x$. The velocity v and all the stresses are continuous as $t \to 0$ but the velocity $u = 0$ at the walls is incompatible with (13.9). In a linear theory we could expect this initial discontinuity to propagate into the interior with a decay in the magnitude of the jump in u. This does not appear to be true in the results presented by PT&T.

In Fig. 13.2 I have reproduced their results for the evolution of $-f_y(y,t) = u/x$ for $R = 1$ and increasing W. The graphs of the evolution of $f = v$ do not show interesting features and are not presented here.

I cannot find evidence for propagating discontinuities of u/x near the top wall. The wave in u/x propagating from the top plate seems to have been smoothed. On the other hand the points near the bottom plate which could show the development of discontinuous u/x for large time and large W. Perhaps it is these points at which we may see the development of shocks of u/x are only what PT&T call "unavoidable noise." It is possible to find pictures, say $\tau = 0.5$, $W = 1$, in which one can imagine a point of discontinuity of the slope of u/x which can be identified with a shock in the vorticity.

I think that the results shown support the notion, already explored in Sections 10 and 12, that the Weissenberg number is a measure of the fluids elasticity and in hyperbolic problems with damping, the damping is large when the Weissenberg number is small and vice versa. The effect of the waves is almost immediately damped in the case $W = 0.1$ and is extraordinarily persistent when $W = 500$.

It would be interesting to see if and what discontinuities develop in the squeeze flow problem when all of the prescribed data is smooth at $t = 0$.

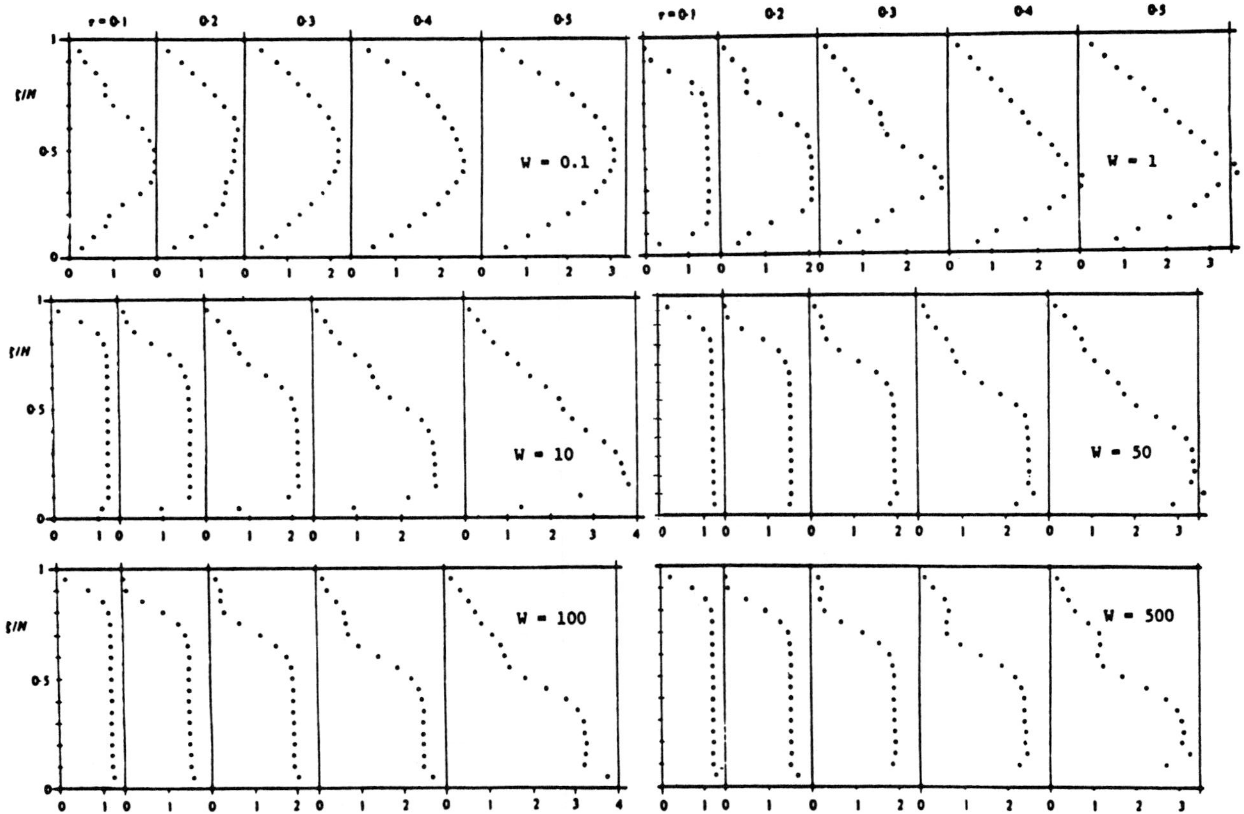

Fig. 13.2. (After Phan Thien and Tanner (1983)). Evolution of the vertical component of velocity $u_x - f_y = u/x$ is plotted, in squeeze flow. The Reynolds number is unity. W is the Weissenberg number.

REFERENCES

A. Ambari, C. Deslouis, B. Tibollet, Coil-stretch transition of macromolecules in laminar flow around a small cylinder. Chem. Eng. Comm. (to appear).

G. Astarita and G. Apuzzo, Mition of gas bubbles in Non-Newtonian liquids, A.I.Ch.E.J., 11, 815-816 (1965).

E. Brenschede and J. Klein, Druckverbuste und instabiles Fliessen elastischer Flussigheiter. Hochdruck Kappillarviskometer. Rheol. Acta 9, 130-139 (1970).

P.H. Calderbank, , D.S.L. Johnson and J. Loudon, Mechanics and mass transfer of single bubbles in free rise through some Newtonian and non-Newtonian liquids. Chem. Eng. Science, 25, 235-256 (1970).

B. Caswell, The stress singularity in a linear viscoelastic fluid at the leading edge of a flat plate. Proceedings of the VIIth International Congress son Rheology. Chalmers University of Technology, 432-33, 1976.

B.D. Coleman, M.E. Gurtin and I.R. Herrera, Waves in materials with memory. I. Arch. Rational Mech. Anal. 19, 1-19 (1965).

B.D. Coleman and M.E. Gurtin, On the stability against shear waves of steady flows of non-linear viscoelastic fluids, J. Fluid Mech. 33, 65-18 (1968)

A.G. Dodson, P. Townsend and K. Walters, On the flow of Newtonian and non-Newtonian liquids through corrugated pipes, Rheol. Acta 10, 508-516 (1971).

A.G. Fabula, An experimental study of grid turbulence in dilute high-polymer solutions, Ph.D. Thesis, Pennsylvania State University, University Park, 1966.

M. W. Johnson and D. Segalman, A model for viscoelastic fluid behavior which allows non-affine deformation. J. Non Newtonian Fluid Mech. 2, 255-270 (1977).

A.B. Metzner, E.A. Uebler and C.F.C.M. Fong, Converging flows of viscoelastic materials, AICME J. 15, 5 750-578 (1969).

P.M. Morse and H. Feshbach, Methods of Theoretical Physics, McGraw-Hill, New York, 1953.

A. Narain and D.D. Joseph, Linearized dynamics for step jumps of velocity and displacement of shearing flows of a simple fluid, Rheol. Acta 21, 228-250 (1982).

A. Narain and D.D. Joseph, Remarks about the interpretation of impulse experiments in shear flow of viscoelastic liquids. Rheologica Acta, 22, 528-538 (1983A).

A. Narain and D.D. Joseph, Linearized dynamics of shearing deformation perturbing rest in viscoelastic materials, transactions of the twenty-eighth conference of Army Mathematicians ARO Report 83-1 (1983). Also in Springer Lecture Notes in Mathematics Equadiff 82, Vol. 1017, 1983B.

A.C. Papanatasiou, L.E. Scriven and C.W. Macosko. An integral constitutive equation for mixed flows: Viscoelastic Characterization. J. Rheology, 27, 387-410 (1983).

H. Giesekus, A simple constitutive equation for polymer fluids based on the concept of deformation-dependent tensorial mobility. J. of Non Newtonian Fluid Mechanics. 11, 69-126 (1982).

H. Giesekus, Zur Formulierung der Randbedingungin in Stromungen viskoelastischer Flussigheiten mit Injektion und Absaugang and den Wanden, Rheol. Acta 9, 474-487 (1970).

H. Giesekus, Verschiedene Phanomene in Stromungen viskoelasticher Flussigkeiten durch Dusen, Rheol. Acta. 8 411-421 (1968).

D.F. James, Laminar flow of dilute polymer solutions around circular cylinders, Ph.D. Thesis, California Institute of Technology, Pasadena, 1967.

D.D. Joseph, M. Renardy, J.C. Saut, Hyperbolicity and change of type in the flow of viscoelastic fluid, to appear in the Arch. Rational Mech. Anal.

H.M. Laun, Description of the non-linear shear behavior of a low density polyethylene melt by means of an experimentally determined strain dependent memory function, Rheol. Acta 17, 1-15 (1978).

G. Leal, J. Skoog and A. Acrivos, On the motion of gas bubbles in viscoelastic liquid, Can. J. of Chem. Eng. 49, 569-575 (1971).

E.H. Lieb. Ph.D. thesis. University of Delaware, 1975.

M. Luskin, On the classification of some model equations for viscoelasticity, J. Non-Newtonian Fluid Mech. 16, 3-11 (1984).

I.M. Rutkevich, Some general properties of the equation of viscoelastic incompressible fluid dynamics. J. Appl. Math. Mech. 33 42-51 (1969).

I.M. Rutkevich, The propagation of small perturbations in a viscoelastic fluid. J. Appl. Math. Mech. 34, 35-50 (1970).

I.M. Rutkevich, On the thermodynamic interpretation of the evolutionary conditions of the equations of the mechanics of finitely deformable viscoelastic media of Maxwell type, J. Appl. Math. Mech. 36, 283-295 (1972).

J.C. Saut and D.D. Joseph, Fading Memory, Arch. Rational Mech. Anal. 81, 53-95 (1983).

M. Slemrod, Instability of steady shearing flows in a nonlinear viscoelastic fluid, Arch. Rational Mech. Anal. 68, 211-255 (1978).

Phan Thien and R. Tanner, Viscoelastic squeeze film flow-Maxwell fluids, J. Fluid Mech. 129, 165-281 (1983).

J.S. Ultman and M.M. Denn, Anomalous heat transfer and a wave phenomenon in dilute polymer solutions, Trans. Soc. Rheology 14, 307-317 (1970).

J.S. Ultman and M.M. Denn, Slow viscoelastic flow past submerged objects. Chem. Eng. J. 2, 81-89 (1971).

G.B. Whitham, Linear and nonlinear waves. Wiley Interscience, New York, 1974.

J. Yoo, M. Ahrens and D.D. Joseph, Hyperbolicity and change of type in sink flow. J. Fluid Mech. (to appear).

J. Yoo and D.D. Joseph, Hyperbolicity and change of type in the flow through channels. Submitted to J. Fluid Mech.

E. Zana and L.G. Leal. The dynamics and dissolution of gas bubbles in a viscoelastic liquid. J. Multiphase Flow, 4, 237-262 (1978).

Sponsored by the U.S. Army, ARO (Math), DAAG-29-82-K005) and the Fluid Mechanics Branch of The National Science Foundation.

APPENDIX A: BREAKDOWN OF SMOOTH SHEARING FLOW IN VISCOELASTIC FLUIDS FOR TWO CONSTITUTIVE RELATIONS: THE VORTEX SHEET VS. THE VORTEX SHOCK.

A0. INTRODUCTION

The purpose of this note is to study the effect of choice of constitutive relations on shearing perturbations of steady shearing flows in a non-linear, isotropic, incompressible, viscoelastic fluid. In an earlier paper [1], I showed that the simple constitutive relation relating the shearing stress to the shear rate v_x

$$T^{xy}(t) = \sigma\left(\int_0^\infty e^{-\alpha\tau} v_x(x,t-\tau)d\tau\right). \tag{0.1}$$

yields the following result:

The equation of motion for the shearing perturbation $\hat{v}(x,t)$ of a non-trivial steady shearing motion will admit a C^1 in (x,t) solution for only a finite time for appropriately chosen smooth initial velocity histories. Here σ is a non-linear, odd, real analytic function, $\sigma' > 0$, α is a positive constant.

The proof of the result hinged on the fact that equations for the perturbed fluid motion $\hat{v}(x,t)$ can be written as a non-linear hyperbolic conservation law with linear damping. In this form the equations of motion are amenable to study via the use of Riemann invariants and an argument of Lax [2]. As a side benefit from this formulation continuation theory for hyperbolic equations shows that loss of C^1 regularity implies $|\hat{v}_t| + |\hat{v}_x| \to \infty$ in finite time. ; This suggests the appearance of a jump discontinuity in $\hat{v}$, i.e. formation of a vortex sheet.

Sometime after appearance of my result C.J.S. Petrie remarked to me that a similar result could be obtained for the choice of constitutive relation

$$T^{xy}(t) = \int_0^\infty e^{-\alpha\tau}\sigma(v_x(x,t-\tau))d\tau \tag{0.2}$$

for σ,α as above. Upon working out Petrie's idea I noticed the results are similar but not the same. Specifically the above mentioned result holds with C^1 replaced by C^2. D. D. Joseph observed that this may have important physical consequences as now the appearance of jump discontinuity in $\hat{v}_x$ may be expected, i.e. formation of a vortex shock.

At the request of D. D. Joseph I have prepared this short note analyzing the different effects of (1.1) and (1.2). The technical details of the breakdown proof are omitted and may be found [1].

A1. RECTILINEAR SHEARING FLOWS

If in a fixed Cartesian co-ordinate system x,y,z, the velocity fields of a flowing fluid body has the form

$$v^x = 0,\ v^y = v(x,t),\ v^z = 0 \tag{1.1}$$

we say that the motion is a rectilinear shearing flow. For such a flow the condition of incompressibility div $\underline{v} = 0$ is automatically satisfied. Coleman & Noll [3] have shown that if the fluid is a simple fluid, then the components of stress obey the relations

$$T^{xy}(t) = \mathop{\mathbf{t}}_{s=0}^{\infty} (\Lambda^t(s)) ,$$

$$T^{xx}(t) - T^{zz}(t) = \mathop{\mathbf{s}_1}_{s=0}^{\infty} (\Lambda^t(s)) ,$$

$$T^{yy}(t) - T^{zz}(t) = \mathop{\mathbf{s}_2}_{s=0}^{\infty} (\Lambda^t(s)) ,$$

$$T^{xz} = T^{yz} = 0 , \tag{1.2}$$

where Λ^t is the relative shearing history defined by

$$\Lambda^t(s) = -\int_{t-s}^{t} v_x(x,\tau)d\tau \qquad (0 \leqq s \leqq \infty) ,$$

and $\mathbf{t},\mathbf{s}_1,\mathbf{s}_2$ are real valued functionals obeying the identities

$$\mathop{\mathbf{t}}_{s=0}^{\infty}(-\Lambda^t(s)) = -\mathop{\mathbf{t}}_{s=0}^{\infty}(\Lambda^t(s)),$$

$$\mathop{\mathbf{s_i}}_{s=0}^{\infty}(-\Lambda^t(s)) = \mathop{\mathbf{s_i}}_{s=0}^{\infty}(\Lambda^t(s)), \quad (i = 1,2). \tag{1.3}$$

Here subscripts x,t denote partial derivatives with respect to x and t respectively.

A2. CONSTITUTIVE ASSUMPTIONS

To proceed further it is necessary to make some mathematical assumptions as to the nature of the functional **t**. For this analysis we assume **t** has two particularly simple forms, i.e.,

$$\mathop{\mathbf{t}}_{s=0}^{\infty}(\Lambda_t(s)) = \sigma\left[\int_0^\infty e^{-\alpha s}v_x(x,t-s)ds\right], \tag{2.1}$$

and

$$\mathop{\mathbf{t}}_{s=0}^{\infty}(\Lambda_t(s)) = \int_0^\infty e^{-\alpha s}\sigma(v_x(x,t-s))ds, \tag{2.2}$$

where σ is a real valued, odd, analytic function defined on the real line and α is a positive constant. In (2.1) we see **t** is a (generally) nonlinear function of the linear functional

$$\int_0^\infty e^{-\alpha s}v_x(x,t-s)ds .$$

In (2.2) we see that **t** is a linear functional of the nonlinear function $\sigma(v_x(x,t-s))$.

A motivation for the choice (2.1) may be based on the multiple integral expansions for **t** originally presented by Green & Rivlin [4] and Chacon & Rivlin [5] for general viscoelastic materials. In [5] the authors showed that for **t** continuous on an appropriate function space, **t** may be approximated by a finite series of multiple integrals:

$$\underset{s=0}{\overset{\infty}{\mathbf{t}}} (\Lambda^t(s)) \approx \sum_{n \text{ odd}} \int_0^\infty \cdots \int_0^\infty K_n (s_1,..,s_n) \Lambda^t(s_1)\ldots \Lambda^t(s_n)ds_1\ldots ds_n \tag{2.3}$$

(The restriction to odd order follows from the isotropy condition (1.3).) If we assume more is true, namely that in a manner similar to the Taylor series expansion for an analytic function we take (2.3) to be an equality where the sum may be infinite. Choose

$$K_n (s_1,\ldots,s_n) = -\alpha^n \sigma_n e^{-\alpha(s_1+\ldots+s_n)} \quad n = 1,3,5\ldots$$

where α is a positive constant. (This is consistent with fluid of fading memory type [3], [6].) Finally if we define

$$\alpha(\xi) = \sum_{n \text{ odd}} \sigma_n \xi^n$$

and integrate (2.3) by parts we obtain (2.1).

Constitutive equation (2.2) is equivalent to the choice

$$\underset{s=0}{\overset{\infty}{\mathbf{t}}} (\Lambda^t(s)) = -\int_0^\infty e^{-\alpha s} \sigma((\Lambda^t(s))_s)ds .$$

An alternative motivation arises from consideration of a material of rate type where $T^{xy}(t)$ satisfies the ordinary differential equation

$$\frac{d}{dt} T^{xy}(t) + \alpha T^{xy}(t) = \sigma(v_x(x,t)) \tag{2.4}$$

for all t, $-\infty < t < \infty$, and all x; $0 < x < h$, where α as before is a positive constant. Integration of (2.4) in the usual manner for first order ordinary differential equations yields (2.2).

A3. SHEARING PERTURBATION OF A STEADY SHEARING FLOW

Let us assume a viscoelastic fluid satisfies either the constitutive relation (2.1) or (2.2). Consider the problem when the fluid is confined between two parallel walls of infinite extent at $x = 0$ and $x = h$. The top wall at $x = h$ moves with velocity V. In the absence of body and driving forces the equation of conservation of linear momentum (see Coleman & Gurtin [7]) becomes

$$\rho v_t(x,t) = \sigma\left(\int_0^\infty e^{-\alpha s} v_x(x,t-s))ds\right)_x \tag{3.1}$$

for constitutive relation (2.1) and

$$\rho v_t(x,t) = \int_0^\infty e^{-\alpha s}\sigma(v_x(x,t-s))_x \, ds \tag{3.2}$$

for constitutive relation (2.2). Here ρ is an (assumed) constant mass density, $\rho > 0$. Also we consider the case of no-slip boundary conditions

$$v(0,t) = 0 , \qquad v(h,t) = V . \tag{3.3}$$

Systems (3.1), (3.3) and (3.2), (3.3) admit the steady rectilinear flow ($v_t(x,t) = 0$) solution

$$v(x) = \frac{V_x}{h} .$$

To study stability of the flow against shearing perturbations we set

$$\hat{v}(x,t) = v(x,t) - \frac{Vx}{h} .$$

We observe that for relation (2.1) this implies

$$\rho\hat{v}_t(x,t) = \sigma\left(\int_0^\infty e^{-\alpha s}\hat{v}_x(x,t-s)ds + \frac{V}{\alpha h}\right)_x \tag{3.4}$$

with boundary condition

$$\hat{v}(0,t) = \hat{v}(h,t) = 0 \tag{3.5}$$

On the other hand for relation (2.2) we see that the perturbation $\hat{v}(x,t)$ satisfies

$$\rho\hat{v}_t(x,t) = \int_0^\infty e^{-\alpha s}\sigma\left(\hat{v}_x(x,t-s) + \frac{V}{h}\right)_x ds \tag{3.6}$$

along with boundary condition (3.5).

In either case we prescribe a smooth velocity history

$$\hat{v}(x,t) = \hat{v}_0(x,t) \quad -\infty < \tau \leqq 0$$

consistent with either (3.4), (3.5) or (3.5), (3.6). Thus the two fluid cases are governed by non-linear boundary initial history value problems.

A4. ANALYSIS OF FLOW WITH FIRST CONSTITUTIVE ASSUMPTION: STRESS NON-LINEAR FUNCTION OF A LINEAR FUNCTIONAL OF SHEAR RATE.

We consider the perturbed flow governed by constitutive relations (2.1). In this case the time evolution is governed by (3.4), (3.5). To simplify matters we write

$$\hat{\sigma}(\xi) = \frac{1}{\rho}\left[\sigma\left(\xi + \frac{V}{\alpha h}\right) - \sigma\left(\frac{V}{\alpha h}\right)\right],$$

$$u(x,t) = \int_0^\infty e^{-\alpha s}\,\hat{v}_t(x,t-s)ds,$$

$$w(x,t) = \int_0^\infty e^{-\alpha s}\,\hat{v}_x(x,t-s)ds.$$

An integration by parts shows

$$u(x,t) = \hat{v}(x,t) - \alpha\int_0^\infty e^{-\alpha s}\,\hat{v}(x,t-s)ds$$

and hence

$$u_t = \hat{v}_t - \alpha u \; . \tag{4.1}$$

If we combine (4.1) with (3.4) we find u,w satisfy the system

$$w_t = u_x \; ,$$

$$u_t = \hat{\sigma}(w)_x - \alpha u \; , \tag{4.2}$$

with boundary conditions

$$u(o,t) = u(h,t) = 0 \; , \tag{4.3}$$

and initial conditions

$$u(x,0) = u_0(x),$$

$$w(x,0) = w_0(x), \qquad 0 \leqq x \leqq h \; .$$

The values of $u_0(x)$ and $w_0(x)$ are obtained from their respective definitions by insertion of the given velocity history $\hat{v}_0(x,\tau)$, $-\infty < \tau \leqq 0$.

In order that the constitutive relation(2.1) be truly non-linear we must have $\sigma''(\xi^x) \neq 0$ for some real ξ^x . Hence when the speed of top wall is given

$$V = \alpha h\;\xi^k$$

we see $\hat{\sigma}$ satisfies

$$\hat{\sigma}(0) = 0 \,,$$

$$\hat{\sigma}''(0) \neq 0 \,. \tag{4.4}$$

We further impose the condition that (4.2) be strictly hyperbolic, i.e. the matrix

$$\begin{bmatrix} 0 & \hat{\sigma}' \\ 1 & 0 \end{bmatrix}$$

possesses real distinct eigenvalues. Strict hyperbolicity is easily seen to be equivalent to the condition $\sigma' > 0$.

A5. ANALYSIS OF FLOW WITH SECOND CONSTITUTIVE ASSUMPTION: STRESS LINEAR FUNCTIONAL OF A NON-LINEAR FUNCTION OF SHEAR RATE.

We consider the perturbed flow governed by constitutive relation (2.2). In this case the time evolution is governed by (3.5), (3.6). Again to simplify matters we write

$$\sigma(\xi) = \frac{1}{\rho}\left[\sigma(\xi + \frac{V}{h}) - \sigma\left[\frac{V}{h}\right]\right] \,,$$

$$\bar{u} = \hat{v}_t \,,$$

$$\bar{w} = \hat{v}_x \,.$$

We note (3.6) can be written as

$$\hat{v}_t = \int_{-\infty}^{t} e^{-\alpha(t-\tau)} \, \sigma(\hat{v}_x(x,\tau))_x \, d\tau \tag{5.1}$$

which upon differentiating with respect to t yields

$$\hat{v}_{tt} = -\alpha \int_{-\infty}^{t} e^{-\alpha(t-z)} \, \sigma(\hat{v}_x \, (x,\tau))_x \, d\tau + \sigma(v_x(x,t))_x \tag{5.2}$$

i.e.

$$\hat{v}_{tt} + \alpha\hat{v}_t = \sigma(\hat{v}_x(x,t))_x \tag{5.3}$$

where we have used (5.1).

From the definitions of $\overline{u}$, $\overline{w}$ above we see

$$\overline{w}_t = \overline{u}_x \,,$$

$$\overline{u}_t = \sigma(\overline{w})_x - \alpha\overline{u} \,, \tag{5.4}$$

with boundary conditions

$$\overline{u}(0,t) = \overline{u}(h,t) = 0 \,, \tag{5.5}$$

and initial conditions

$$\overline{u}(x,0) = \overline{u}_0(x),$$

$$\overline{w}(x,0) = \overline{w}_0(x), \qquad 0 < x < h \,.$$

Again the values of $\overline{u}_0(x)$, $\overline{w}_0(x)$ can be obtained from the respective definition of $\overline{u},\overline{w}$ in terms of the given velocity history $\hat{v}_0(x,\tau)$, $-\infty < \tau \leqq 0$.

Also we note that if $\sigma''(\xi) \neq 0$ for some ζ real and when the speed of top wall is given by

$$V = h\zeta$$

then

$$\sigma(0) = 0 \,,$$

$$\sigma''(0) \neq 0 \,. \tag{5.6}$$

As before the condition $\sigma' > 0$ implies (5.4) is strictly hyperbolic.

A6. A BREAKDOWN RESULT.

We have shown in sections 4 and 5 that evolution of perturbed flow is governed by a system of the form

$$
\begin{aligned}
W_t &= U_x \\
U_t &= K(W)_x - \alpha U \,, \qquad (6.1)
\end{aligned}
$$

with

$$
\begin{aligned}
&U(0,t) = U(h,t) = 0, \\
&U(x,0) = U_0(x), \qquad W(x,0) = W_0(x) \qquad (6.2)
\end{aligned}
$$

and

$$K' > 0, \quad \kappa(0) = 0, \quad K''(0) \neq 0.$$

Analysis of (6.1), (6.2) has been given in [1]. We shall not repeat that analysis but only state the relevant breakdown result.

Define Riemann invariants for (6.1) by

$$\begin{matrix} r \\ s \end{matrix} = U \pm \emptyset(W) \qquad (6.3)$$

where

$$\emptyset(W) = \int_0^W \sqrt{K'(s)}\, ds.$$

The transformation given by (6.3) from (U,W) R x R to (r,s) R x R is one-one. Also we assume the initial data $r(0,x) = r_0(x)$, $s(0,x) = s_0(x)$ to be smooth functions.

Our main breakdown result is as follows.

Theorem 6.1: Suppose $|r_0|$, $|s_0|$ are sufficiently small and $K'(0) > 0$, $K''(0) > 0$. If $r_{0,x}$ or $s_{0,x}$ is positive and sufficiently large at any point x, then (6.1), (6.2) has a solution (W,U) in $C^1[0,h] \times C^1[0,h]$ for only a finite time. A similar result holds if $K''(0) < 0$ and $r_{0,x}$ or $s_{0,x}$ is sufficiently negative at any point x.

Proof: See the proof of Thm. 3.1 in [1].

Standard existence theorems (see for example Section 1.8 of [8] and Chaps. 2 and 3 of [9]) imply that under the hypothesis of Theorem 6.1 we have $|U_t| + |U_x| + |W_t| + |W_x| \to \infty$ in finite time. This suggests but does not prove the occurrence of a jump discontinuity in U and W, i.e. the formation of a shock.

Also we note that since

$$\begin{matrix} r_x \\ s_x \end{matrix} = U_x \pm \phi'(W)W_x$$

we will have r_{o_x} large if U_{o_x} or W_{o_x} is large, s_{o_x} large if U_{o_x} or $-W_{o_x}$ is large, $-r_{o_x}$ large if $-U_{o_x}$ or $-W_{o_x}$ large, $-s_{o_x}$ large if $-U_{o_x}$ or W_{o_x} large.

7. PHYSICAL IMPLICATIONS OF BREAKDOWN OF SMOOTH SOLUTIONS

We examine the implications of the breakdown result of Section 6 with respect to our two constitutive relations.

First we consider constitutive relation (2.1) In this case we see that if $\sigma''(V/\alpha h) \neq 0$ and $u_{o_x}(x)$ or $w_{o_x}(x)$ is appropriately sufficiently large of sufficiently negative u_{o_x} (depending on the sign of $\sigma''(V/\alpha h)$), Theorem 6.1 implies $|u_x| + |w_x| \to \infty$ in finite time.

Since we know from (3.6) that

$$\hat{u}_t = \hat{\sigma}(w)_x$$

and from the definition of u we have

$$\hat{u}_x = u_x + \alpha w,$$

$|u_x| + |w_x| \to \infty$ in finite time implies $|\hat{v}_t| + |\hat{v}_x| \to \infty$ in finite time. Again this suggests but doesn't prove that $\hat{v}$ and hence v forms a jump discontinuity in finite time. In this case the singular surface across which the discontinuity in v is called a vortex sheet.

Next we consider the case of constitutive relation (2.2). In this case we see that if $\bar{u}_{o_x}(x) = v_{tx}(x,0)$ or $\bar{w}_{o_x}(x) = v_{xx}(x,0)$ is appropriately sufficiently large or sufficiently negative (depending on the sign of $\sigma''(V/h)$) Theorem 6.12 implies $|v_{tx}| + |v_{xx}| \to \infty$ in finite time. Once more this suggests but doesn't prove that either v_t or v_x forms a jump discontinuity in finite time. However from the Rankine-Hugoniot jump condition for (5.4)

$$-(\frac{ds}{dt})[\bar{w}] = [\bar{u}]$$

$$-(\frac{ds}{dt})[\bar{u}] = [\sigma(\bar{w})] \tag{7.1}$$

we see

$$-(\frac{ds}{dt})[v_t] = \left[\sigma(v_x - \frac{V}{h})\right] = \frac{1}{\rho}[\sigma(v_x)] \tag{7.2}$$

where $x = s(t)$ denotes the surface across which the jump occurs. Hence if $[v_x] \neq 0$ then $[\sigma(v_x)] \neq 0$ (since $\sigma' > 0$) and by (7.2) $[v_t] \neq 0$. Conversely if $[v_t] \neq 0$ then from (7.1) we have

$$-(\frac{ds}{dt})[v_x] = [v_t] \tag{7.3}$$

and hence $[v_x] \neq 0$. Thus the appearance of a jump in v_t or v_x implies a jump in the other.

We define a propagating singular surface across which the acceleration v_t experiences a jump discontinuity as an <u>acceleration wave</u>. Similarly we define a propagating singular surface across which the vorticity $\underline{w} = \text{curl}(v^z, z^y, y^z) = v_x(x,t)\underline{e}_z$ experiences a jump discontinuity as a <u>vortex shock</u>. Our analysis shows that for our flow the vortex shock and acceleration waves are equivalent and can be expected to form in finite time if constitutive relation (2.2) holds.

We thus see the remarkable effect of choice of constitutive relation. In one case (2.1) appropriately chosen initial data appears to force the formation of a jump in v (a vortex sheet) while in the second case (2.2) appropriately chosen initial data suggests formation of a jump discontinuity in v_x (a vortex shock and equivalently in v_t (an acceleration wave).

REFERENCES

1. M. Slemrod, Instability of steady shearing flows in a non-linear viscoelastic fluid, Archive for Rational Mechanics and Analysis, 68, 211-225 (1978).

2. P.D. Lax, Development of singularities of solutions of nonlinear hyperbolic partial differential equations, J. Mathematical Physics 5, 611-613 (1964).

3. B.D. Coleman & W. Noll, Recent results in continuum theory of viscoelastic fluids, Ann. N.Y. Acad. Sci. 89, 672-714 (1961).

4. A.E. Green & R.S. Rivlin, Mechanics of nonlinear materials with memory, Arch. Rational Mechanics Analysis 1, 1-21 (1957).

5. R.V.S. Chacon & R.S. Rivlin, Representation theorems in the mechanics of materials with memory, ZAMP 15, 444-447 (1964)

6. B.D. Coleman & W. Noll, Foundations of linear viscoelasticity, Rev. Modern Physics 33, 239-249 (1961).

7. B.D. Coleman & M.E. Gurtin, On the stability against shear waves of steady flows in non-linear viscoelastic fluids, J. Fluid Mech. 33, 165-181 (1968).

8. B.L. Rozdestvenskii & N.N. Janeko, Systems of quasilinear equations and their applications to gas dynamics, Translations of Mathematical Monographs, American Mathematical Society, Vol. 55, Providence, R I. (1983).

9. A. Majda, Compressible fluid flow and systems of conservation laws in several space variables, Applied Mathematical Sciences 53, Springer-Verlag, New York-Berlin-Heidelberg-Tokyo (1984).

This research was sponsored in part by the Air Force Office of Scientific Research, Air Force Systems Command, USAF, Contract/Grant No. AFOSR-81-0172. The U.S. Government's right to retain a nonexclusive royalty free license in and to copyright this paper for governmental purposes is acknowledged.

Appendix A by
Marshall Slemrod
Department of Mathematical Sciences
Rensselaer Polytechnic Institute
Troy, New York 12180

ABSORBING BOUNDARIES FOR VISCOELASTICITY

R. C. MacCamy

Department of Mathematics
Carnegie-Mellon University
Pittsburgh, Pennsylvania

I. Introduction

The ideas discussed here fall into a general category which we term interface problems. Such problems deal with situations in which there are two or more regions in which distinct differential equations hold, with transition conditions across the interface. Typically, one of the regions is infinite in extent. A major goal in the study of such problems is the development of numerical approximations. Such procedures, to be effective, must be over finite regions and thus one must somehow eliminate the infinite part.

Very effective procedures have been developed for a wide range of problems in acoustic, electromagnetic and elastic scattering but almost all of them apply only to the special case of time periodic fields. A notable exception is the work of Engquist and Majda [1]. We describe their idea in a simple situation.

Suppose one wants to solve the wave equation in two dimensions in the exterior of an obstacle. The idea is to introduce an artificial boundary outside the obstacle and then solve the wave equation only in the annular region. There is an exact, theoretical boundary condition which must hold on the artificial boundary. This condition is termed absorbing because if it is imposed there will be no reflection back into the annulus. It is, however, very difficult to implement numerically. It is a pseudo-differential operator, that is, it is non-local, in both space and time, and is represented in terms of Fourier integrals.

The idea of Majda and Engquist was to replace the exact relations by approximate boundary conditions which are more nearly local but still retain at least some of the absorbing property.

Our goal is to present analogs of the absorbing boundary idea for viscoelasticity. While the absorbing idea becomes trivial for the wave equation in one dimension (see the next section) it is quite non-trivial for one-dimensional viscoelasticity. Our results are quite fragmentary and are primarily limited to the linear one-dimensional situation, although we make a few remarks on a nonlinear problem in section two and on two dimensional problems in section six. Our work relies very heavily on the techniques in [1].

We remark that the ideas presented here are closely related to the abstract notion of an environmental operator introduced by Noll [4].

Let us consider the problem of the distortion of a uniform bar. Let x denote the position of a plane section in a reference (unstrained) configuration and suppose the bar is semi-infinite, $x > 0$. Let $u(x,t)$ be the displacement of a plane section. Then one has the global balance law,

$$\frac{d}{dt}\int_{x_1}^{x_2} \rho u_t d\xi = \sigma(x_2,t) - \sigma(x_1,t), \quad \text{for any } x_1 < x_2. \tag{1.1}$$

Here ρ is the density and σ the stress. Where u is smooth we have the local balance law,

$$\rho u_{tt} = \sigma_x \tag{1.2}$$

and across curves $x = \zeta(t)$ of discontinuities in u_x and u_t we have the Rankine-Hugoniot relation,

$$\dot{\zeta}[\rho u_t] = [\sigma], \tag{1.3}$$

where the brackets indicate jumps across the curve.

We consider impact problems which means:

$$u(x,0) = u_t(x,0) \equiv 0 \quad \text{on} \quad x > 0 \tag{1.4}$$

$$\sigma(0,t) = g(t). \tag{1.5}$$

(The displacement impact problem $u(0,t) = g(t)$ has an analogous treatment.) We distinguish three cases in all of which (1.4) and (1.5) hold:

(1) Continuous bar

Here we assume there are no discontinuities in the bar and the displacement is to satisfy,

$$\rho u_{tt} = \sigma_x \qquad x > 0. \tag{1.6}$$

(2) Artificial boundary

Often one is not really interested in what happens over the whole bar but only near the end $x = 0$. For instance, one might want to know only $u(0,t)$ given $\sigma(0,t)$. A natural question then is whether one can solve the equation only over the interval $0 < x < L$ and replace the part for $x > L$ by an appropriate boundary condition at $x = L$. Thus we have, schematically,

$$\rho u_{tt} = \sigma_x \qquad 0 < x < L \qquad ? \text{ at } x = L. \tag{1.7}$$

(3) Composite bar

Here we imagine that the bar consists of two distinct materials, one for $0 < x < L$ and one for $x > L$. We require that the displacement be continuous across the interface but the velocity will jump. Since the discontinuity is stationary the relation (1.3) shows that the stress will be continuous. Thus if we denote by $-$ and $+$ values in $x < L$, $x > L$, we have the problem,

$$\begin{aligned} &\rho^- u_{tt} = \sigma^-_x, \quad 0 < x < L; \qquad \rho^+ u_{tt} = \sigma^+_x, \quad x > L \\ &u^-(L,t) = u^+(L,t); \qquad\qquad \sigma^-(L,t) = \sigma^+(L,t). \end{aligned} \tag{1.8}$$

In the following sections we consider these problems under various constitutive assumptions on σ. In the last section we make a few remarks on the extension to two dimensions.

II. Elastic Bars

We consider four special cases:

A. Linear, homogeneous continuous bar

Here we have,

$$\sigma(x,t) = E_0 u_x(x,t), \quad \rho(x) = \rho_0 \qquad x > 0, \tag{2.1}$$

in (1.6). The solution, with conditions (1.4) and (1.5), is easy to obtain. If we set $C_0^2 = E_0/\rho_0$ it is,

$$u(x,t) = G(t - \frac{x}{C_0}), \qquad G(\xi) = C_0 \int_0^{\xi} g(\eta)\,d\eta. \tag{2.2}$$

This solution is <u>outgoing</u>, that is u is identically zero for $x > C_0 t$. It follows that we have,

$$\sigma(x,t) = -\frac{E_0}{C_0} u_t(x,t), \tag{2.3}$$

for all x and t.

B. <u>Linear, homogeneous artificial boundary</u>

The formula (2.3) shows what to use as a boundary condition in (1.7); that is if we solve

$$u_{tt} = C_0^2 u_{xx} \quad 0 < x < L; \qquad \sigma(L,t) = -\frac{E_0}{C_0} u_t(L,t), \tag{2.4}$$

together with (1.4) and (1.5), we will have the exact solution in $0 < x < L$.

C. <u>Linear inhomogeneous-linear homogeneous composite bar</u>

Let us suppose the bar is a homogeneous linear material on $x > L$, that is, (2.1) holds but that on $0 < x < L$ we have $\rho(x)$ and

$$\sigma(x,t) = E(x) u_x(x,t). \tag{2.5}$$

This is a prototype problem in interface theory. We observe that (2.3) holds on $x > L$ and hence, using (1.8) we can reduce the problem to,

$$\begin{aligned} \rho(x) u_{tt} &= \frac{\partial}{\partial x}(E(x) u_x(x,t)) \qquad 0 < x < L \\ \sigma(L,t) &= -\frac{E_0}{C_0} u_t(L,t), \end{aligned} \tag{2.6}$$

together with (1.4) and (1.5).

The value of the above observation is that once the infinite region $x > L$ is eliminated we can treat (2.6) by variational methods. If we multiply $(2.6)_1$ by a test function v, integrate by parts and use (1.5) and $(2.6)_2$ we obtain the variational formulation,

$$\int_0^L \rho u_{tt} v\,dx + \int_0^L E u_x v_x dx + \frac{E_0}{C_0} u_t^-(L,t) v(L) = -g(t) v(0), \tag{2.7}$$

with (1.5) still holding. If one introduces finite dimensional approximate spaces (say, finite elements) (2.7) reduces to a set of ordinary differential equations which gives us a way of obtaining approximate solutions.

Equation (2.7) also yields an energy estimate. If we put $v = u_t$ in (2.7) and integrate with respect to t we obtain,

$$\frac{1}{2}\int_0^L \rho(x)u_t^2(x,T)\,dx + \frac{1}{2}\int_0^L E(x)u_x^2(x,T)\,dx + \frac{E_0}{C_0}\int_0^T u_t^2(L,t)\,dt = -\int_0^T g(t)u_t(0,t)\,dt. \qquad (2.8)$$

If one assumes that g and $u_t(0,t)$ belong to $L_2(0,\infty)$ then (2.8) shows that we have u_t and u_x bounded in $L_\infty(0,T:L_2(0,L))$ and $u_t(L,\cdot)$ bounded in $L_2(0,\infty)$. Notice that the sign in $(2.6)_2$ is such that it contributes a positive term in the energy estimate. For this reason we call it _absorbing_.

D. _Non-linear homogeneous-linear homogeneous composite bar_

Suppose that the bar again satisfies (2.1) on $x > L$ but that this time we have,

$$\sigma(x,t) = E(u_x(x,t)), \qquad \rho(x) \equiv \rho_1 \qquad 0 < x < L \qquad (2.9)$$

where, as usual, one assumes $E(\xi)$ is monotone increasing with $E(0) = 0$. We can again reduce to a problem on $(0,L)$, namely,

$$\begin{aligned} \rho_1 u_{tt} &= \frac{\partial}{\partial x}E(u_x) \qquad 0 < x < L \\ \sigma(L,t) &= -\frac{E_0}{C_0}u_t(L,t), \end{aligned} \qquad (2.10)$$

with (1.4) and (1.5).

Greenberg and Tsien [2] considered the problem

$$\begin{aligned} u_{tt} &= \frac{\partial}{\partial x}E(u_x) \qquad 0 < x < L \\ u(0,t) &\equiv 0, \qquad \sigma(L,t) = -\lambda u_t(L,t) \\ u(x,0) &= u_0(x), \qquad u_t(x,0) = u_1(x). \end{aligned} \qquad (2.10')$$

They established the following result.

Theorem 2.1. _For any positive_ λ _if_ u_0 _and_ u_1 _and their first three derivatives are sufficiently small_ (2.10') _has a unique global solution._

This is quite a remarkable theorem since the same problem with the boundary condition $u(L,t) = 0$ is known to have the property that solutions break down in finite time. Thus the boundary condition (2.6) has a stabilizing effect. We believe the same result would hold for (2.10) under smoothness assumptions on g. Thus one can stabilize a nonlinear bar by connecting it to a linear bar.

III. Linear, Homogeneous, Viscoelastic Bars

We use the standard model,

$$\sigma(x,t) = f_t(x,t); \quad f(x,t) = a(\cdot) * u_x(x,\cdot)(t), \tag{3.1}$$

where $*$ denotes convolution and the kernel a satisfies,

$$a \in C^{(2)}[0,\infty), \quad a(t) = a_\infty + b(t), \quad a_\infty > 0,$$
$$b^{(j)} \in L_1(0,\infty) \quad j \leq 3.$$
$$(-1)^k a^{(k)}(t) > 0 \quad k \leq 2. \tag{3.2}$$

For ease of writing we assume $a(0)$ and ρ are both one.

We can integrate the local balance law and the boundary condition (1.5) and obtain the first order problem,

$$u_t = f_x, \; u(x,0) = 0 \; \text{ on } \; x > 0; \quad f(0,t) = G(t) = \int_0^t g(\tau)\,d\tau. \tag{3.3}$$

Because of the form (3.1) this problem is conveniently treated by Laplace transforms. If $\hat{h}(s)$, $s = \xi + i\eta$, denotes the Laplace transform of h then (3.3) yields,

$$s\hat{u} = \hat{a}\hat{u}_{xx}, \; x > 0; \quad \hat{a}(s)u_x(0,s) = \hat{G}(s). \tag{3.4}$$

We observe that once $\hat{u}$ is found one can formally recover u by the complex inversion integral,

$$u(x,t) = (2\pi)^{-1}\int_{-\infty}^{+\infty} e^{i\eta t}\hat{u}(x,i\eta)\,d\eta. \tag{3.5}$$

1. Continuous bar. Here we have to solve (3.4) for $x \in (0,\infty)$. We set,

$$\hat{\alpha}(s) = \sqrt{s\hat{a}(s)}, \quad \hat{\beta}(s) = \sqrt{s/\hat{a}(s)}. \tag{3.6}$$

We choose the solution of (3.4) which remains bounded as $x \to \infty$. (This gives an outgoing wave.) We have,

$$\hat{u}(x,s) = -\frac{\hat{G}(s)}{\hat{\alpha}(s)} e^{-\hat{\beta}(s)x} \tag{3.7}$$

and then, by (3.1),

$$\hat{f}(x,s) = -\hat{\alpha}(s)\hat{u}(x,s). \tag{3.8}$$

In order to proceed we need some well known facts about transforms (see, for instance [3]). It follows from $(3.2)_1$ that $\hat{a}(s)$ has the form,

$$\hat{a}(s) = a_\infty s^{-1} + \hat{b}(s) \tag{3.9}$$

where $\hat{b}$ is continuous in $\xi \geq 0$ and analytic in $\xi > 0$. Moreover, $(3.2)_2$ yields,

$$\operatorname{Re}\hat{a}(i\eta) > 0 \quad \text{for all} \quad \eta. \tag{3.10}$$

For large s we have, with $a(0) = 1$,

$$\hat{a}(s) = s^{-1} + \dot{a}(0)s^{-2} + o(s^{-2}) \text{ as } |s| \to \infty \text{ in } \xi \geq 0, \tag{3.11}$$

and we have from (3.6),

$$\hat{\alpha}(s) = 1 + \frac{1}{2}\dot{a}(0)s^{-1} + o(s^{-1}); \quad \hat{\beta}(s) = s + \frac{1}{2}\dot{a}(0) + o(1) \quad \text{as} \quad s \to 0. \tag{3.12}$$

For small s we obtain,

$$\hat{\alpha}(s) = \sqrt{a_\infty} + o(1), \quad \hat{\beta}(s) = \sqrt{a_\infty}\, s + o(s) \quad \text{as} \quad s \to 0. \tag{3.13}$$

From (3.10)-(3.13) one derives the following conclusions.

Lemma 3.1. (i) *$\hat{\alpha}$ and $\hat{\beta}$ are continuous in $\xi \geq 0$ and analytic in $\xi > 0$.*

(ii) $\hat{\alpha}(s) \neq 0$ in $\xi \geq 0$.

(iii) *There exists a $\delta > 0$ such that $\operatorname{Re}\hat{\alpha}(i\eta) \geq \delta$ for all η.*

We note that under suitable integrability conditions of g and its derivatives we will have $\hat{G}(s) = G_\infty s^{-1} + \hat{H}(s)$, $G_\infty = \int_0^\infty g(\tau)d\tau$ with $\hat{H}$ regular in $\xi \geq 0$. It follows that the right side of (3.7) is analytic in $\xi > 0$ and continuous in $\xi \geq 0$ except at $s = 0$ where it has a pole. Moreover, (3.12) shows that $\hat{u}$ decays exponentially as $s \to \infty$ in $|\arg s| \leq \pi/2 - \epsilon$, $\epsilon > 0$. With some further technical conditions on a and g it is then possible to show that (3.8) and (3.7) yield a solution of (3.4).

Let us consider the condition (3.8). By (3.12) we can write

$$\hat{\alpha}(s) = 1 + \hat{r}(s), \tag{3.14}$$

and thus (3.8) translates back into the time domain as a convolution. For a function $v \in C[0,\infty)$ we define the <u>history</u> v^t of v by $v^t(\tau) = v(t-\tau)$, $0 \le \tau \le t$. Now define a linear functional $\mathfrak{F}$ on these histories by the formula,

$$\mathfrak{F}(v^t) = v(t) + (r(\cdot) * v(\cdot))(t). \tag{3.15}$$

Then (3.8) implies,

$$f(x,t) = \mathfrak{F}(u^t(x,\cdot)). \tag{3.16}$$

This should be compared with (2.3). Notice that (3.16) is non-local in time.

2. <u>Artificial boundary</u>. If we want to solve (3.3) only on $(0,L)$ and we want to obtain the exact solution then the problem to be solved is,

$$u_t = f_x,\ u(x,0) = 0 \text{ on } 0 < x < L; \quad f(0,t) = G(t);$$
$$f(L,t) = \mathfrak{F}(u^t(L,\cdot)). \tag{3.17}$$

Once again let us give a variational formulation. For an arbitrary test function (3.17) yields

$$\int_0^L u_t\, v\, dx + \int_0^L (a * u_x) v_x dx + \mathfrak{F}(u^t(L,\cdot))v(L) = -G(t)v(0). \tag{3.18}$$

Once again we can obtain an energy estimate. Put $v = u$ in (3.18) and integrate from $t = 0$ to T to obtain,

$$\frac{1}{2}\int_0^L u^2(x,T)\,dx + \int_0^T\int_0^L (a * u_x)u_x dx dt + \int_0^T \mathfrak{F}(u^t(L,\cdot)u(L,t))dt$$
$$\equiv \frac{1}{2}\int_0^L u^2(x,T)\,dx + I + II = -\int_0^T G(t)u(0,t)\,dt. \tag{3.19}$$

<u>Lemma</u> 3.2. (i) $I \ge 0$.

(ii) <u>There is a constant</u> δ, <u>independent of</u> T, <u>such that</u> $II \ge \delta\int_0^T u^2(t,t)\,dt$.

Proof: The result (i) is well known and both (i) and (ii) are consequences of the following fact. For any kernel k, any $v \in C[0,\infty)$ and any $T > 0$ put

$$Q_k[v : T] = \int_0^T v(t)(k*v)(t)\,dt.$$

Then ([3], Lemma (4.3)) we have,

$$Q_k[v : T] = \int_0^\infty \operatorname{Re}\hat{k}(i\eta)|\tilde{v}_T(\eta)|^2 d\eta \tag{3.20}$$

where $\tilde{v}_T$ is the Fourier transform of the function $v_T(t) = v(t)$ $0 \leq t < T$, $v_T(t) = 0$ for $t > T$. Then (i) follows immediately from (3.10). For (ii) we have, by (3.15), Plancherel's theorem and Lemma 3.1(iii),

$$\begin{aligned} II &= \int_0^\infty (1 + \operatorname{Re}\hat{r}(i\eta))|\tilde{v}_T(\eta)|^2 d\eta = \int_0^\infty \operatorname{Re}\hat{\alpha}(i\eta)|\tilde{v}_T(\eta)|^2 d\eta \\ &\geq \delta \int_0^\infty |\tilde{v}_T(\eta)|^2 d\eta = \delta \int_0^T v^2(t)\,dt. \end{aligned}$$

Equation (3.19) and Lemma 3.2 show that we can obtain estimates for u in $L_\infty(0,\infty: L_2(0,L))$ and $u(L,\cdot)$ in $L_2(0,\infty)$ in terms of L_2 norms of G and $u(0,\cdot)$. (The estimate for u_x is lost here.) Thus we can again term $f(L,t) = \mathfrak{F}(u^t(L,\cdot))$ an absorbing boundary condition.

The case of a composite bar with an inhomogeneous region on $0 < x < L$ and a homogeneous region on $x > L$ can be treated in a similar way.

IV. Approximate Boundary Conditions

In the work of [1] the difficulty of non-local exact boundary conditions was encountered even for acoustic scattering (the wave equation) when one goes to two dimensions. The major idea in [1] was to introduce approximate boundary conditions which are more local. For our case the non-local nature of the exact boundary condition is not so drastic since the equation inside is already non-local in time. It is of interest, however, to search for approximate conditions which might be easier to implement.

What we seek are problems of the form,*

$$v_t = f_x, \quad v(x,0) = 0 \quad \text{on} \quad 0 < x < L;$$

$$f(0,t) = G(t), \quad f(L,t) = -\mathcal{J}(v^t(L,\cdot)) \tag{4.1}$$

such that the solution v approximates the solution u of (3.17). We require in addition that the approximate boundary condition be absorbing, that is

$$\int_0^T \mathcal{J}(v^t)v(t)\,dt \geq \gamma \int_0^T v^2(t)\,dt \quad \text{for some } \gamma > 0 \text{ and all } v. \tag{4.2}$$

We suppose that $\mathcal{J}$ can be described by its "transform", that is,

$$\widehat{\mathcal{J}v} = \hat{p}(s)\hat{v}, \tag{4.3}$$

as was $\mathfrak{F}$.

Our analysis borrows heavily from ideas in [ ]. We put $v = u + V$ so that we have,

$$V_t = f_x, \quad V(x,0) = 0 \quad \text{on} \quad 0 < x < L, \quad V_x(0,t) = 0$$
$$f(L,t) + \mathcal{J}(V^t(L,\cdot)) = (\mathfrak{F} - \mathcal{J})(u^t(L,\cdot)) = \phi(t). \tag{4.4}$$

We transform this problem and obtain from (3.7),

$$s\hat{V} = \hat{a}\hat{V}_{xx} \quad 0 < x < L, \quad \hat{V}_x(0,s) = 0$$
$$\hat{a}(s)\hat{V}_x(L,s) + \hat{p}(s)\hat{V}(L,s) = \hat{\phi}(s) = -\frac{\hat{G}(s)}{\hat{\alpha}(s)}(\hat{\alpha}(s) - \hat{p}(s))e^{-\hat{\beta}(s)L}. \tag{4.5}$$

The solution of (4.5) is,

$$\hat{V}(x,s) = \frac{\hat{\phi}(s)}{\hat{\Gamma}(s)} \cosh \hat{\beta}(s)x; \tag{4.6}$$

$$\hat{\Gamma}(s) = \hat{\alpha}(s) \sinh \hat{\beta}(s)L + \hat{p}(s) \cosh \hat{\beta}(s)L. \tag{4.7}$$

We make two hypotheses:

$$\hat{\alpha}(s) - \hat{p}(s) \to 0 \quad \text{as } |s| \to \infty \quad \text{in} \quad \xi \geq 0. \tag{H.1}$$

There exists a $\gamma > 0$ such that $\operatorname{Re}\hat{p}(i\eta) \geq \gamma$ for all η. (H.2)

---

*In (4.1) and what follows f indicates (3.1) formed for the function appearing.

If (H.1) and (H.2) are satisfied we say the condition in (4.1) is an _approximate absorbing boundary condition_. We observe that (H.2) will guarantee that (4.2) is satisfied. We proceed to show that under (H.1) and (H.2) v does approximate u. Our convergence results are analogous to those in [1] and are very weak.

The treatment of convergence becomes technically complicated unless we make the assumption that $\int_0^\infty g(\tau)\,d\tau = 0$ and we will assume this. More precisely, we assume that

$$G \in L_2(0,\infty). \tag{4.8}$$

Theorem 4.1. _Suppose_ (H.1), (H.2) _and_ (4.8) _hold_. _Then given any_ $\epsilon > 0$ _and any_ $L_0 > 0$ _there is an_ $L > 0$ _such that_,

$$\int_0^\infty \int_0^L (u(x,t)-v(x,t))^2 dx = \int_0^\infty \int_0^L V(x,t)^2 dx < \epsilon. \tag{4.9}$$

Note that the theorem indicates that to have approximate solutions "accurate on $(0,L_0)$" we may have to solve the approximate problem on a much larger interval. We sketch the proof of the theorem.

Our first observation is that, by Plancherel's theorem, it suffices to show that the solution $\hat{V}$ of (4.5) satisfies,

$$\int_0^\infty \int_0^L |\hat{V}(x,i\eta)|^2 dx\, d\eta < \epsilon. \tag{4.10}$$

We estimate the integral in (4.10) in three parts: $0 < \eta < \eta_0$, $\eta_0 < \eta < \eta_1$, $\eta_1 < \eta$. In order to do this we need estimates of $\int_0^L |\hat{V}(x,i\eta)^2 d\eta$ in the three regions.

(1) _Large_ η. We start with an energy estimate. Put,

$$D(\hat{V},s) = s\int_0^L |\hat{V}(x,s)|^2 dx + \hat{a}(s)\int_0^L |\hat{V}_x(x,s)|^2 dx. \tag{4.11}$$

Then if we multiply by $\bar{\hat{V}}$ in (4.5) and integrate with respect to x we obtain,

$$D(\hat{V},s) + \hat{p}(s)|\hat{V}(L,s)|^2 = \phi(s)\bar{\hat{V}}(L,s). \tag{4.12}$$

Lemma 4.1. _There exists an_ $N > 0$ _and a_ $\delta > 0$, _independent of_ L, _such that_

$$|D(\hat{W},s)| \geq \delta(\int_0^L |\hat{W}(x)|^2 dx) \tag{4.13}$$

for all s *with* $|s| \geq N$, $\xi \geq 0$ *and any* $\hat{W}$.

Proof: Consider first s in a wedge $|\arg s| \leq \pi/2 - \epsilon$ for some $\epsilon > 0$. By (3.10) we see that $\operatorname{Re} \hat{a}(s) > 0$ in the wedge. Thus if we take the real part of (4.11) we see that (4.13) holds in the wedge, for $|s|$ large, with a δ depending on ϵ.

It follows from the preceding argument that the only way the conclusion can fail is that there is a sequence $s_n = r_n e^{i\theta n}$, $r_n \to \infty$ $\theta_n \to \pi/2$ and a sequence $\hat{W}^n$ with $|\int_0^L |\hat{W}^n(x)|^2 dx = 1$ such that $|D(\hat{W}^n, s_n)| \to 0$ as $n \to \infty$. We show that this leads to a contradiction and hence cannot happen.

We write (3.11) as

$$\hat{a}(s) = s^{-1} + \dot{a}(0)s^{-2} + \hat{a}_2(s)s^{-2} \tag{4.14}$$

where $\hat{a}_2(s) \to 0$ as $|s| \to \infty$. Then for $s_n = r_n e^{i\theta n}$ we have,

$$\begin{aligned} \operatorname{Re} D(\hat{W}^n, s_n)| &= r_n \cos\theta_n(1+L_n) + \dot{a}(0)L_n \cos 2\theta_n + \operatorname{Re}, R_n \\ \operatorname{Im} D(\hat{W}^n, s_n)| &= r_n \sin\theta_n(1-L_n) - \dot{a}(0)L_n \sin 2\theta_n + \operatorname{Im}, R_n \end{aligned} \tag{4.15}$$

where

$$L_n = \frac{1}{r_n^2}\int_0^L |\hat{W}_x^n|^2 dx, \quad R_n = \frac{1}{s_n^2}\int_0^L |\hat{W}_x^n|^2 dx. \tag{4.16}$$

Now if $D(\hat{W}^n, s_n)$ tends to zero its real and imaginary parts must separately do so. Since $\sin\theta_n \to 1$ and $\dot{a}(0) < 0$ it follows from $(4.15)_2$ that $L_n \to 1$. But since $\cos 2\theta_n \to -1$ and $\dot{a}(0) < 0$, $(4.15)_1$ shows that $\operatorname{Re} D(\hat{W}^n, s_n)$ would not then tend to zero.

From Lemma 4.1, (H.1), (3.12) and (4.12) we obtain immediately the fact that there is an $N > 0$ and a δ' such that,

$$\int_0^L |\hat{V}(x,s)|^2 dx + |\hat{V}(L,s)|^2 \leq \delta|\hat{\phi}(s)|^2, \quad |s| \geq N. \tag{4.17}$$

From (3.12) we also conclude that for $|s|$ large $|\hat{\alpha}(s)^{-1}e^{-\hat{\beta}(s)L}|$ is bounded independently of L. Thus (4.17) yields,

$$\int_0^L |\hat{V}(x,i\eta)|^2dx + |\hat{V}(L,i\eta)|^2 \le \delta'|\hat{G}(i\eta)|^2|\hat{\alpha}(i\eta)-\hat{p}(i\eta)|. \tag{4.18}$$

(2) $\eta_0 \le \eta \le \eta_1$. Since $\operatorname{Re}\hat{a}(i\eta)$ is uniformly bounded away from zero in this region it follows from (3.6) that the same is true of both $\hat{\alpha}$ and $\hat{\beta}$, that is, there is a $\tau(\eta_0,\eta_1) > 0$ such that

$$\operatorname{Re}\hat{\alpha}(i\eta) \ge \tau, \quad \operatorname{Re}\hat{\beta}(i\eta) \ge \tau. \tag{4.19}$$

<u>Lemma</u> 4.2. <u>There exists</u> $L, \tau_1(\eta_0,\eta_1)$ <u>and</u> $\tau_2(\eta_0,\eta_1)$ <u>positive such that</u>,

$$|\Gamma(i\eta)| \ge \tau_1 \ell^{\tau_2 L} \quad \text{for} \quad \eta_0 \le \eta \le \eta_1, \quad L \ge L_0. \tag{4.20}$$

<u>Proof</u>: From (4.7) we have,

$$\Gamma(i\eta) = \frac{\ell^{\hat{\beta}(i\eta)L}}{2}\{(\hat{\alpha}(i\eta)+\hat{\beta}(i\eta)+(\hat{p}(i\eta)-\hat{\alpha}(i\eta))\ell^{-2\hat{\beta}(i\eta)L}\}. \tag{4.21}$$

By Lemma 3.1 and (H.1) we know that $\operatorname{Re}(\hat{\alpha}(i\eta)+\hat{p}(i\eta))$ is bounded below on $[\eta_0,\eta_1]$ by a positive constant. By (3.6) and (3.10) we conclude that the same is true of $\operatorname{Re}\hat{\beta}(i\eta)$. On the other hand, $\hat{p}(i\eta) - \hat{\alpha}(i\eta)$ is bounded above on the same interval. Hence for L sufficiently large the absolute value of the quantity in brackets in (4.21) is bounded below by a positive constant. Since $|\ell^{\hat{\beta}(i\eta)L}| = \ell^{\operatorname{Re}\hat{\beta}(i\eta)L}$ we obtain (4.20) from (4.21).

Now $\operatorname{Re}\hat{\beta}(i\eta)$ is also bounded above on the interval $[\eta_0,\eta_1]$. We can conclude then from (4.8) and Lemma 4.2 that there are constants $k(\eta_0,\eta_1) > 0$, $K(\eta_0,\eta_1 > 0$ such that,

$$\int_0^{L_0} |\hat{V}(x,i\eta)|^2dx \le K\ell^{-kL}|\hat{G}(i\eta)|^2 \quad \text{for} \quad \eta_0 \le \eta \le \eta_1. \tag{4.22}$$

(3) η small. We see from (3.12) that on an interval around zero we can no longer bound $\operatorname{Re}\hat{\beta}(s)$ below. Thus we lose the exponential factor in (4.22). From (4.9) and (3.12) we see that $\Gamma(0) \ne 0$. We conclude then from (4.8) that there is an

$\bar{\eta}_0$ and a K_1 such that,

$$\int_0^{L_0} |\hat{V}(x,i\eta)|^2 dx \leq K_1 |\hat{G}(i\eta)|. \tag{4.23}$$

<u>Proof of Theorem</u> 4.1: Given $\epsilon > 0$ we first choose $\eta_0 \leq \bar{\eta}_0$ such that $\int_0^{\eta_0} |\hat{G}(i\eta)|^2 d\eta < \frac{\epsilon}{3K_1}$. This is possible since $G \in L_2$ and $\hat{G}(i\eta)$ is the Fourier transform of the function which is G on $t > 0$ and zero on $t < 0$. Next we choose η_1 so large that

$$|\hat{\alpha}(i\eta) - \hat{p}(i\eta)| < \epsilon\{3\delta' \int_0^{\infty} |\hat{G}(i\eta)|^2 d\eta\}^{-1}.$$

Then by (4.18) we have

$$\int_{\eta_1}^{\infty} \int_0^{L} |\hat{V}(x,i\eta)|^2 dx \leq \frac{\epsilon}{3} \quad \text{independently of } L.$$

Finally, by (4.22) we can choose L so large that

$$\int_{\eta_0}^{\eta_1} |\hat{V}(x,i\eta)|^2 dx < \frac{\epsilon}{3}$$

and then (4.10), hence (4.9), will be satisfied.

We want to make an observation about the above convergence result. Suppose (H.1) is strengthened to,

$$|\hat{\alpha}(i\eta) - \hat{p}(i\eta)| \leq \frac{M}{|\eta|^N}. \tag{4.24}$$

We say then that the condition in (4.1) is an <u>Nth order</u> approximate condition. In this case (4.18) yields,

$$\int_{\eta_1}^{\infty} \int_0^{L} |\hat{V}(x,i\eta)|^2 dx \leq \frac{\delta' M}{\eta_1^N} \int_0^{\infty} |\hat{G}(i\eta)|^2 d\eta.$$

To make this less than $\epsilon/3$ then we can choose η_1 smaller according as N increases. But then one can see that the L can also be chosen smaller and the process becomes easier to implement.

The question of determining approximate absorbing boundary conditions is a difficult one. As a first observation we note that we have one very simple condition, namely,

$$f(L,t) = -u(L,t) \tag{4.25}$$

(more generally $f = -a(0)u$ if $a(0) \neq 1$). That this is an acceptable condition follows immediately from (3.12) and we see from that equation that it is first order.

One can approach the approximation property by using rational approximations, that is, $\hat{p}(s) = \hat{P}(s)/\hat{Q}(s)$ with $\hat{P}$ and $\hat{Q}$ polynomials with real coefficients. Such an approximation to $\mathfrak{F}$ corresponds to a differential approximate boundary condition,

$$\hat{Q}(\frac{d}{dt}) f(L,t) = \hat{P}(\frac{d}{dt}) u(L,t)$$

in the time domain. This is the kind of approximation used in [1]. The question to be considered here, for which we do not have an answer, is whether it is possible to find such approximations which are Nth order and which satisfy $\mathrm{Re}\ \hat{p}(i\eta) \geq \delta > 0$, that is, are absorbing.

We give one result which seems very interesting but probably not very practical. This is that we can obtain approximate absorbing conditions of arbitrarily high order by using Padé approximations. This is a result which closely parallels one in [1].

We begin with the observation that $s\hat{a}(s) = 1 + \hat{p}(s)$ where $\hat{p}(s)$ is the transform of $\dot{a}$. Following [1] we introduce the Padé approximations $P_N(Z)$ for $\sqrt{1+Z}$. These are defined by,

$$P_1 = 1 \qquad P_2 = 1 + \frac{Z}{2}, \ldots, P_{N+1} = 1 + \frac{Z}{1+P_N}. \tag{4.26}$$

One has the following results from [1].

$$P_N = 1 + \frac{T_N}{S_N}, \qquad T_N,\ S_N \ \text{polynomials}$$
$$\sqrt{1+Z} - P_N(Z) = O(Z^N). \tag{4.27}$$

This suggests that we take as approximate boundary conditions those corresponding to $\hat{p}_N(s) = P_N(\hat{\rho}(s))$. Since we have $\hat{\rho}(s) = O(s^{-1})$ we see by (4.27) that these will be Nth order. Also, by (4.27), they correspond in the time domain to convolution conditions of the form,

$$S_N(\mathbb{C}) f(L,\cdot) = -(S_N(\mathbb{C}) + T_N(\mathbb{C}))u(L,\cdot) \tag{4.28}$$

where $\mathcal{C}$ is convolution with $\dot{a}$, thus the conditions are very complicated. For the second order condition we have,

$$f(L,t) = -u(L,t) - \int_0^t \dot{a}(t-\tau)u(L,\tau)d\tau. \tag{4.29}$$

This has the same form as the exact condition,

$$f(L,t) = -u(L,t) - \int_0^t r(t-\tau)u(L,t)d\tau,$$

but at least in (4.29) we know the kernel explicitly while in the exact condition r would have to be computed by taking the inverse transform of $\sqrt{s\hat{a}(s)} - 1$.

The feature that makes the $\hat{p}_N$ interesting is that the conditions (4.28) are automatically absorbing for any N as we show now.

<u>Theorem</u> 4.2. <u>Suppose</u> $(3.2)_2$ <u>holds for</u> $k \leq 3$. <u>Then there exists a</u> $\delta > 0$ <u>such that for any</u> N $\hat{p}_N(i\eta) = P_N(\hat{\rho}(i\eta)) \geq \delta$ <u>for all</u> η.

<u>Proof</u>: We consider positive η. We have

$$\hat{\rho}(i\eta) = \int_0^\infty \dot{a}(t)\cos\eta\, t\, dt - i\int_0^\infty \dot{a}(t)\sin\eta\, t\, dt = R(\eta) + iI(\eta).$$

Since $(-\dot{a}(t))(-1)^k > 0$ $k \leq 2$ the result from [3] shows that $R(\eta) < 0$ for all η. There is a second result from [3] which states that if a function $m(t)$ satisfies $(-1)^k m^{(k)}(t) > 0$ then $\operatorname{Im}\hat{a}(i\eta) < 0$ for $\eta > 0$. It follows that $I(\eta) > 0$ for $\eta > 0$. We can estimate further. Since $\dot{a}(t)$ is always negative we have,

$$R(\eta) > \int_0^\infty \dot{a}(t)dt = -(1-a_\infty)$$

$$I(\eta) < -\int_0^t \dot{a}(t)dt < (1-a_\infty).$$

Thus if we put $\tau = (1-a_\infty)$ then $0 < \tau < 1$ and we have,

$$0 > R(\eta) > -\tau \quad \text{for all } \eta \qquad 0 < I(\eta) < \tau \quad \text{for } \eta > 0. \tag{4.30}$$

We have now

$$\hat{p}_2(i\eta) = 1 + \tfrac{1}{2}(R(\eta) + iI(\eta)) = P_2(\eta) + i\,Q_2(\eta),$$

where

$$P_2(\eta) > 1 - \frac{1}{2}\tau, \qquad 0 < Q_2(\eta). \tag{4.31}$$

We write $\hat{p}_N = P_N(\eta) + i\,Q_N(\eta)$ and we assert that the estimates (4.31) hold for all N. We argue by induction. We have,

$$P_{N+1}(\eta) + i\,Q_{N+1}(\eta) = 1 + \frac{1}{2}\,\frac{R(\eta)+iI(\eta)}{1+P_N(\eta)+i\,Q_N(\eta)}$$

$$= 1 + \frac{1}{2}\,\frac{R(1+P_N)+IQ_N}{(1+P_N)^2+Q_N^2} + \frac{1}{2}\,\frac{I(1+P_N)-RQ_N}{(1+P_N)^2+Q_N^2}\,i\,.$$

Thus

$$P_{N+1} \geq 1 + \frac{1}{2}\,\frac{R}{1+P_N} \geq 1 - \frac{\tau}{2}, \qquad Q_{N+1} > 0$$

and the proof of Theorem 4.2 is complete.

V. Inhomogeneous Elastic Bars

In this section we make a few remarks on problems for inhomogeneous continuous bars. We will discuss only the elastic case. Our findings here are very preliminary and we have no analysis yet for inhomogeneous viscoelastic bars.

We assume that (2.5) holds and we consider the problem

$$\rho(x)u_{tt} = (E(x)u_x)_x, \quad x > 0; \quad u(x,0) = 0;$$

$$E(0)u_x(0,t) = G(t). \tag{5.1}$$

Here E may vary throughout the length of the bar. We want to consider the artificial boundary problem, finding both the exact boundary condition to be used at $x = L$ and some approximation to it.

We transform (5.1) to obtain

$$s^2\rho\hat{u} = (E\hat{u}_x)_x \quad x > 0 \qquad E(0)\hat{u}_x(0,s) = \hat{G}(s). \tag{5.2}$$

To determine the exact boundary condition for $x = L$ we can proceed as follows. Solve

$$s\rho\hat{u} = (E\hat{u}_x)_x \quad \text{on} \quad x > L; \qquad \hat{u}(L,s) = \hat{\varphi}(s),$$

$$\hat{u} \ \text{bounded as} \ x \to \infty \tag{5.3}$$

in the form (Duhamel's principle)

$$\hat{u}(x,s) = \hat{U}(x,s)\hat{\varphi}(s), \qquad s\rho\hat{U} = (E\hat{U}_x)_x \quad \text{on} \quad x > L;$$
$$\hat{U}(x,L) = 1, \quad \hat{U} \text{ bounded as } x \to \infty. \tag{5.4}$$

Then the exact non-local boundary condition for $x = L$ is given in the transform plane by

$$E(L)\hat{u}_x(L,s) = \hat{\alpha}(s)\hat{u}(L,s) \qquad \hat{\alpha}(s) = E(L)\hat{U}_x(L,s). \tag{5.5}$$

The preceding result is very hard to implement since we would have to solve (5.4) for all $s = i\eta$ and then try to return to the time domain by forming the inverse transform of $\hat{\alpha}$. We try to replace this by an approximate condition.

We observe that the right side of (5.5), like that of (3.8), defines a pseudo-differential operator and $\hat{\alpha}(s)$ is called its <u>symbol</u>. The basic idea to approximate such operators is to approximate the symbol for large values of the transform variables. This was the idea of (H.1). We do this here for the $\hat{\alpha}$ of (5.5).

We use the idea of geometrical optics. We seek a solution $\hat{U}$ of (5.4) of the form,

$$\hat{U}(x,i\eta) = e^{i\eta\varphi(x)}\hat{V}(x,i\eta). \tag{5.6}$$

A little calculation shows that this will solve (5.4) if,

$$\begin{aligned} &\varphi'(x) = -\sqrt{\rho(x)/E(x)}^{\,*} \qquad \varphi(L) = 0 \\ &\qquad - i\eta[(\sqrt{\rho E})'V + 2\sqrt{\rho E}\,V_x] + (EV_x)_x = 0, \quad \hat{V}(L,s) = 1. \end{aligned} \tag{5.7}$$

We seek an asymptotic expansion of V of the form,

$$\hat{V}(x,i\eta) \sim \sum_{n=0}^{\infty} V_n(x)(i\eta)^{-n} \qquad V_0(L) = 1, \quad V_n(L) = 0 \quad n \geq 1. \tag{5.8}$$

We substitute (5.8) into $(5.7)_2$ and equate powers of η. This yields,

$$\begin{aligned} &(\sqrt{\rho E})'V_0 + 2\sqrt{\rho E}\,V_0' = 0 \qquad V_0(L) = 1 \\ &(\sqrt{\rho E})'V_{n+1} + 2\sqrt{\rho E}\,V_{n+1} = -(EV_n')' \quad V_n(L) = 0 \qquad n \geq 0. \end{aligned} \tag{5.9}$$

---

*The choice of the minus sign is to make the wave outgoing.

Now from (5.6) we have

$$\hat{U}_x(L,i\eta) \sim i\eta\varphi'(L) + \sum_{n=0}^{L} V_n'(L)(i\eta)^{-n}. \tag{5.10}$$

By using (5.9) one can successively compute $V_n'(L)$ $n=0,1,\ldots$ without actually having to solve the equations (5.9). We content ourselves with finding the first two terms. As a first approximation we have,

$$E(L)\hat{U}_x(L,i\eta) = -i\eta\sqrt{\rho(L)E(L)} = \hat{\alpha}_0(i\eta). \tag{5.11}$$

If we translate this back to the time domain we have the approximate boundary condition,

$$\sigma(L,t) = -\sqrt{\rho(L)E(L)}\, u_t(L,t). \tag{5.12}$$

If we compare this with the results in section two we see that this is the same condition we would have used for a composite bar in which $\rho(x) \equiv \rho(L)$ and $E(x) \equiv E(L)$ for $x > L$. This is called the <u>frozen coefficient</u> condition. To get the next term we observe that $V_0'(L) = -\frac{1}{2}\{(\sqrt{\rho E})'/\sqrt{\rho E}\}_{x=L}$. Thus we obtain a refined approximate boundary condition in the time domain;

$$\begin{aligned} \sigma(L,t) &= -\sqrt{\rho(L)E(L)}\, u_t(L,t) + \alpha u(L,t), \\ \alpha &= -\frac{1}{2}E\{(\sqrt{\rho E})'/\sqrt{\rho E}\}_{x=L}. \end{aligned} \tag{5.13}$$

We note that the boundary condition (5.13) will be absorbing if $\alpha \leq 0$ that is, if $\rho(x)E(x)$ is increasing at $x = L$.

The higher order conditions obtained by keeping more terms in (5.8) will give rise to boundary conditions of the form,

$$\begin{aligned} \sigma(L,t) &= -\sqrt{\rho(L)E(L)}\, u_t(L,t) + \alpha u(L,t) \\ &\quad + \int_0^t r_N(t-\tau)u(L,t)\,d\tau \end{aligned}$$

where r_N is a polynomial.

VI. Remarks on Two-Dimensional Problems

In this final section we want to give an indication of the difficulties that arise in two-dimensional problems. We consider the simplest situation, that of S - H waves. In these the displacement $\underset{\sim}{u}$ has the form,

$$\underset{\sim}{u} = u(x,y,t)\hat{k}. \tag{6.1}$$

We consider the displacement impact problem for a half-space which we take as $y > 0$. This means we consider the problem,

$$\begin{aligned} &\rho \underset{\sim}{u}_{tt} = \operatorname{div} \underset{\sim}{\sigma}, \quad u(x,y,0) = u_t(x,y,0) = 0 \\ &\qquad\qquad\qquad \text{in } -\infty < x < \infty, \quad y > 0 \\ &\qquad \underset{\sim}{u}(x,0,t) = g(x,t)\hat{k}. \end{aligned} \tag{6.2}$$

1. Homogeneous elastic region. Here we have,

$$\underset{\sim}{\sigma} = \begin{pmatrix} 0 & 0 & E_0 u_x \\ 0 & 0 & E_0 u_y \\ E_0 u_x & E_0 u_x & 0 \end{pmatrix} \qquad \rho = \rho_0 \qquad c_0^2 = E_0/\rho_0$$

and (6.2) becomes the wave equation problem,

$$\begin{aligned} &u_{tt} = c_0^2 \Delta u, \quad u(x,y,0) = u_t(x,y,0) = 0 \quad -\infty < x < +\infty, \ y > 0 \\ &\qquad u(x,0,t) = g(x,t). \end{aligned} \tag{6.3}$$

We take the Laplace transform with respect to t and the Fourier transform with respect to x. If $\tilde{u}(\omega,y,s)$ is the transform we have,

$$s^2\hat{u} = -\omega^2\hat{u} + \hat{u}_{yy} \quad y > 0, \quad \hat{u}(\omega,0,s) = \hat{g}(\omega,s).$$

We can solve this problem and obtain (as the outgoing solution),

$$\hat{u}(\omega,y,s) = \hat{g}(\omega,s)e^{-\sqrt{s^2+\omega^2}\,y}$$

and hence,

$$\hat{u}_y(\omega,L,s) = -\sqrt{s^2+\omega^2}\ \hat{u}(\omega,L,s) \text{ for any } L. \tag{6.4}$$

Equation (6.4) describes the exact boundary condition which should be used if one wishes to introduce an artificial boundary at $y = L$. It represents a pseudo-differential operator and if transformed back to the time-x domain gives a boundary condition of the form,

$$u_y(x,L,t) = \mathfrak{F}(u^t(\cdot,L,\cdot)) \tag{6.5}$$

which is a functional of the time history of $u(\cdot,y,\cdot)$. Thus it is non-local both with respect to time and space.

The above is the problem considered by Majda and Engquist. They discussed approximate boundary conditions.

We see that we can write,

$$\sqrt{-\eta^2+\omega^2} = -i\eta\sqrt{1-\frac{\omega^2}{\eta^2}}. \tag{6.6}$$

What was done in [1] was to expand the square root in (6.6) using the Padé approximations. The idea is to approximate the symbol of the operator for large $i\eta$. The difficulty is the presence of the other variable ω, for even if η is large the ratio ω/η will not always be small. This causes considerable technical difficulty.

2. Homogeneous viscoelastic bar. Here we have,

$$\underset{\sim}{\sigma} = \frac{\partial}{\partial t}\begin{pmatrix} 0 & 0 & a*u_x \\ 0 & 0 & a*u_x \\ a*u_x & a*u_x & 0 \end{pmatrix}$$

and our problem can again be integrated into a one-dimensional one. We simply write down the transformed version of the resulting problem (with $\rho \equiv 1$). It is,

$$s\hat{u} = \hat{a}(-\omega^2\hat{u} + \hat{u}_{yy}), \qquad \hat{u}(\omega,0,s) = \hat{g}(\omega,s).$$

In this case we have for the exact condition at $x = L$,

$$\hat{u}_x(\omega,L,s) = -\sqrt{\frac{s}{\hat{a}(s)} + \omega^2}\;\hat{u}(\omega,L,s). \tag{6.7}$$

Once again (6.7) translates into a condition of the form (6.8). The question of appropriate approximate boundary conditions in this case is not yet resolved.

REFERENCES

[1] Engquist, B. and Majda, A., Radiation boundary conditions for wave calculations, Comm. on Pure and Appl. Math. 32 (1979), 314-358.

[2] Greenberg, J.M. and Tsien, L.T., The effect of boundary damping for the quasi-linear wave equation, (Preprint).

[3] MacCamy, R.C., Remarks on frequency domain methods, J. Math. Anal. and Appl. 55 (1976), 555-575.

[4] Noll, W., A general framework for problems in the statics of finite electricity, Contemporary Developments in Continuum Mechanics and Partial Differential Equations, North Holland Mathematics Studies 30 (1978).

The author's work was supported by the National Science Foundation under Grant DMS-8400986.

RECENT DEVELOPMENTS AND OPEN PROBLEMS IN THE MATHEMATICAL THEORY OF VISCOELASTICITY

Michael Renardy

Mathematics Research Center
and Mathematics Department
University of Wisconsin-Madison
Madison, Wisconsin

1. EXISTENCE RESULTS FOR INITIAL VALUE PROBLEMS

The existence of solutions for systems of ordinary differential equations

$$\dot{u} = f(u), \; u(0) = u_0 \; , \tag{1.1}$$

can be established by using the iteration

$$\dot{u}^{n+1} = f(u^n), \; u^{n+1}(0) = u_0 \; , \tag{1.2}$$

which can be shown to converge for small enough $t > 0$. For partial differential equations such as

$$u_t = \phi(u_x)_x, \; u(t = 0,x) = u_0(x) \; , \tag{1.3}$$

the iteration (1.2) cannot be used. This is because at each step of the iteration we would be taking two x-derivatives and, if data are sufficiently oscillatory, these x-derivatives will grow larger and larger. An iteration that can be used is the following

$$u_t^{n+1} = \phi'(u_x^n)u_{xx}^{n+1}, \; u^{n+1}(t = 0,x) = u_0(x) \; . \tag{1.4}$$

This example illustrates a general strategy used for so-called "quasilinear" equations. In such equations the highest derivatives occur only linearly, and nonlinearities contain only lower-order derivatives. One then puts an $n + 1$ on the highest derivatives (the "principal terms" in the equation) and an n on lower order derivatives. Two things must then be proved: first that the linear problems to be solved at each step are well-posed and secondly, that the iteration converges.

In order to apply these ideas to problems in viscoelasticity, we have to know that the equations can be written in some quasilinear form and we have to know what the principal terms are. In the following, we illustrate this for a class of one-dimensional model equations, whose structure is typical of problems arising in viscoelasticity. Let us consider equations of the form

$$u_{tt} = \mu f(u_x, u_{xt})_x + g(u_x)_x + \int_{-\infty}^{t} m(t-\tau) h(u_x(t), u_x(\tau))_x d\tau \qquad (1.5)$$

with appropriately smooth boundary and initial data. For $\mu > 0$, the term of highest differential order on the right hand side is the one involving u_{xxt}. If the derivative of f with respect to the second argument is positive, (1.5) can be regarded as a parabolic equation and the iteration to be used is

$$u_{tt}^{n+1} = \mu D_2 f(u_x^n, u_{xt}^n) u_{xxt}^{n+1} + \mu D_1 f(u_x^n, u_{xt}^n) u_{xx}^n + g(u_x^n)_x + \int_{-\infty}^{t} m(t-\tau) h(u_x^n(t), u_x^n(\tau))_x d\tau \; . \qquad (1.6)$$

Under appropriate smoothness assumptions it can be shown that such a scheme is well-defined and converges. This is accomplished by adapting a result due to Sobolevskii [36] for equations without integral terms. At each step of the iteration, we have to solve a linear parabolic equation with time-dependent coefficients. In Sobolevskii's approach, this equation is regarded as a perturbation of the problem with time-independent coefficients, evaluated at the initial time $t = 0$, and a convergent iteration based on the variation of constants formula can be formulated. Finally, it is proved that the mapping $u^n \to u^{n+1}$ given by (1.6) is a contraction in an appropriate Banach space.

One-dimensional model problems of parabolic type are discussed e.g. in [11], [29]. In [30], the same ideas are extended to three-dimensional flows of incompressible fluids under both displacement and traction boundary conditions.

If $\mu = 0$, m, g and h are smooth, and m, g' and D_1h are positive, then (1.5) can be regarded as a hyperbolic equation. The appropriate iteration method is

$$u_{tt}^{n+1} = g'(u_x^n)u_{xx}^{n+1} + \int_{-\infty}^{t} m(t-\tau)D_1h(u_x^n(t),u_x^n(\tau))u_{xx}^{n+1}(t) + D_2h(u_x^n(t),u_x^n(\tau))u_{xx}^n(\tau)d\tau \ . \qquad (1.7)$$

It looks surprising at first that we can put an n on the $u_{xx}(\tau)$-term, since this term appears to be of the same differential order as the $u_{xx}(t)$-term. However, when we differentiate the equation with respect to t, then $u_{xx}(t)$ becomes u_{xxt}, but $u_{xx}(\tau)$ remains $u_{xx}(\tau)$. This explains, on a heuristic level, why this term can be regarded as being of lower order.

At each step of the iteration, one now has to solve a linear hyperbolic equation with time-dependent coefficients. Existence of solutions for such problems can be shown e.g. by the implicit Euler scheme. Again, a contraction argument is used to show convergence of the iteration. This argument consists essentially of two steps: First, it is shown that the iterates remain bounded in a certain norm (e.g. that all third derivatives of u^n have bounds independent of n). Then one uses this fact to show that the iterates converge in a weaker norm (i.e. one proves that second, rather than third, derivaties of u^n converge). An abstract formulation of these ideas is given in the work of Kato [17], [19], [20].

Applications to problems in viscoelasticity are due to Mac Camy [23], Staffans [37], Dafermos and Nohel [5], [6], Hrusa [13], Kim [21], Heard [12] and the author [31], [32]. Three-dimensional problems with incompressibility have been studied [21], [32], but only for the pure Cauchy problem posed on all of space, not for initial-boundary value problems.

There is some interest in relaxing the assumption that m is smooth. The first term on the right hand side of (1.5) can be thought of as arising from setting m equal to the derivative of the δ-function. It is only reasonable to think that m may also have weaker singularities as $t - \tau \to 0$. If m is unbounded at 0, the iteration (1.7) will not work. Instead, one will have to put $n + 1$ on the $u_{xx}(\tau)$-term as well. This means that the "quasilinearized" problems to be solved at each iteration are now history dependent. (In both the parabolic and hyperbolic cases the linear problems involved no history dependence at all, i.e. history dependence was basically treated as a perturbation.) In a recent paper [15], W. Hrusa and the author show that such an iteration converges for the special problem

$$u_{tt} = \phi(u_x)_x - \int_{-\infty}^{t} m(t - \tau)\psi(u_x(\tau))_x d\tau \ , \qquad (1.8)$$

where ϕ', ψ' and $\phi' - \psi' \int_0^\infty m(\tau)d\tau$ are positive, m is positive and monotone decreasing and has an integrable singularity at 0. The method used is based on approximating the singular kernel by smooth kernels and showing that it is possible to pass to the limit.

Thus far, we have talked about existence results of local type. This means that we have initial data given at a certain time, and we prove that solutions exist on some time interval thereafter, which may be short. For large times, smooth solutions may not exist as we shall see in the next section. However, there are results on existence of smooth solutions for all times if one considers smooth and small data perturbing a stable state of rest. Dafermos and Nohel [5], [6] have established such a result for the problem

$$u_{tt} = \phi(u_x)_x - \int_{-\infty}^{t} m(t - \tau)\psi(u_x(\tau))_x d\tau + f(x,t) \ , \qquad (1.9)$$

posed on a finite interval with boundary conditions of Dirichlet or Neumann type. The proof is based on energy estimates which guarantee that small solutions will remain small for all times.

Hrusa [13] has generalized this method to a large class of one-dimensional problems. Hrusa and Nohel [14] study the Cauchy problem on all of space for (1.9). This problem is more difficult than the initial-boundary value problem, because an estimate on x-derivatives of a function no longer implies an estimate on the function itself. This makes the required energy estimates far more complicated. In three dimensions, Kim [21] has shown a global existence result on all of space for a covariant quasilinear model fluid. It is to be expected that Kim's proof can be generalized to K-BKZ fluids.

2. PROPAGATION AND DEVELOPMENT OF SINGULARITIES

While parabolic equations, like the heat equation, instantly smooth out discontinuous initial data, hyperbolic equations like the wave equation will propagate discontinuities and, if nonlinearities are introduced, discontinuities can develop from initially smooth data. As we shall see, the intermediate nature of model equations for viscoelastic materials leads to interesting possibilities.

One-dimensional linear wave propagation in a viscoelastic material is described by the equation

$$u_{tt} = bu_{xx} + \mu u_{xxt} + \int_{-\infty}^{t} m(t - s)(u_{xx}(t) - u_{xx}(s))ds \ . \tag{2.1}$$

Here b and μ are non-negative constants and m is a non-negative, non-increasing function on $[0,\infty)$ such that m is integrable at ∞ and $tm(t)$ is integrable at 0. For the following, we are interested only in the case $\mu = 0$.

In recent joint work with W. Hrusa [16], we study (2.1) for $x \in R$ with the following initial data

$$\begin{aligned} &u = 0, \quad t < 0 \\ &u(x,t = 0) = \tfrac{1}{2}\,\alpha \operatorname{sgn} x \ , \\ &u_t(x,t = 0) = \tfrac{1}{2}\,\beta \operatorname{sgn} x \ . \end{aligned} \tag{2.2}$$

The problem (2.1), (2.2) can easily be solved by Laplace transforms. We are interested in the regularity of the solution and the location of singularities. It is easy to see that these are determined by the asymptotic behavior at infinity of the Laplace transform of u, which is related to the transform of m.

If m is a smooth function, then, for $\lambda \to \infty$, $\mathrm{Re}\,\lambda \geq 0$, the transform has the asymptotic behavior

$$\hat{m}(\lambda) = \frac{1}{\lambda} m(0) + O\left(\frac{1}{\lambda^2}\right) . \tag{2.3}$$

In this case, it is well known [1] that the singularity propagates with the wave speed $\sqrt{b + \hat{m}(0)}$ and decays exponentially with a factor $\exp(-m(0)t/2(b + \hat{m}(0)))$. In addition, we show in [16] that, except in a very special case, there is a discontinuity in u_{xx} across $x = 0$. Away from the lines $x = \pm t\sqrt{b + \hat{m}(0)}$, $x = 0$, the solution is smooth.

The presence of a singularity in m alters the asymptotic behavior of $\hat{m}(\lambda)$. Of particular physical interest are kernels m which are smooth away from 0, but become infinite at 0. Such a possibility has been suggested by certain molecular theories [7], [34], [39] and by experiments [22].

As is expected from the formula for the wave speed above, there remains a finite wave speed as long as m is integrable. However, the exponential decay factor becomes zero if $m(0)$ is infinite, and we can no longer expect propagation of a discontinuity. In fact, Hannsgen and Wheeler [10] show that, for any singular kernel, there must be some kind of "smoothing"; that is, the solution at any positive time will lie in a more restricted function space than the space admitted for the initial data. How much smoothing there is, depends, not surprisingly, on the nature of the singularity in m. In [16], [28], the following examples are discussed:

1. $m(t) = \sum_{n=1}^{\infty} e^{-n^{\alpha}t}$, $\alpha > \frac{1}{2}$.

This kernel behaves like $t^{-1/\alpha}$ as $t \to 0$, and is integrable if $\alpha > 1$. The solution is C^{∞} across the wave and analytic behind the wave, i.e. for $0 < x < t\sqrt{b + \hat{m}(0)}$ (the latter is in fact true for any completely monotone kernel). If $\alpha \leqslant 1$, m is not integrable. The wave speed is infinite, and the solution is analytic everywhere in $t > 0$, $x \neq 0$.

2. $m(t) = \sum_{n=1}^{\infty} \exp(-e^{n}t)$.

This kernel behaves like $|\ell n\ t|$ as $t \to 0$. The number of derivatives of the solution which exist across the wave is increasing proportional with time.

3. $m(t) = \int_0^{\infty} \exp(-e^{e^{\nu}}t)d\nu$.

This behaves like $\ell n|\ell n\ t|$ as $t \to 0$. If regularity is measured by the number of existing derivatives, the gain of regularity in the solution is infinitesimal. However, the initial discontinuity disappears and u is continuous across the wave.

We believe that continuity across the wave should hold under very mild assumptions on m (provided of course that m is singular), but we have not been able to prove a general result. It would be enough to show that the total variation remains finite.

As this discussion shows, singular kernels lead to completely new qualitative features of wave propagation. The coexistence of finite speed and smoothing may be of interest in modeling phenomena in other fields of physics, such as heat flow. It would be extremely interesting to know how singular kernels affect the solutions of nonlinear problems, but no results exist so far. In the following, I shall review some results for nonlinear problems with smooth kernels

Coleman and Gurtin [3], partly in joint work with Herrera [2] have studied the evolution of acceleration waves. They find that, if an acceleration wave propagates into a medium at rest, then its amplitude satisfies a Riccati equation of the form

$$\frac{da}{dt} = -\gamma a + \frac{\gamma}{\lambda} a^2 . \tag{2.4}$$

Here, in our notation from above, $\gamma = \frac{m(0)}{2(b + \hat{m}(0))}$, and λ is a second constant related to nonlinear properties of the material. From (2.4) it is clear that a will decay exponentially if its initial value is small, but will blow up in finite time if its initial value is large and has the right sign.

This result shows that weak singularities can grow in amplitude and lead to stronger singularities. This, and known results on hyperbolic conservation laws, have motivated the search for results establishing the development of singularities from smooth data. For certain specific model equations, such results were obtained by Slemrod [35], Gripenberg [8], Hattori [9], Malek-Madani and Nohel [24] and Malek-Madani, NOhel and John [25]. We do not present these proofs here, but the following, somewhat sketchy argument will illustrate some of the main ideas. Let us consider the model problem studied by Malek-Madani and Nohel [24]

$$u_t + \phi(u)_x - \int_0^t m(t - s)\psi(u(s))_x ds = 0 . \tag{2.5}$$

Here ϕ and ψ are strictly monotone functions: $\phi', \psi' > 0$ If we set

$$z = - \int_0^t m(t - s)\psi(u(s))ds ,$$

then

$$z_t = -m(0)\psi(u) - \int_0^t m'(t - s)\psi(u(s))ds . \tag{2.7}$$

If t is small and u is bounded, then the integral in (2.7) is small. We can then regard (2.5) as a perturbation of the problem

$$\begin{aligned} u_t + \phi(u)_x + z_x &= 0 , \\ z_t &= -m(0)\psi(u) , \end{aligned} \tag{2.8}$$

and we expect that the solution of (2.5) will develop a discontinuity in u if the solution to (2.8) develops such a discontinuity sufficiently fast, (and in fact, it is not difficult to include the integral as a perturbation in the following argument). In (2.8), we introduce Riemann invariants $r = \phi(u) + z$, $s = z$, and we define the characteristic derivative $\frac{d}{dt} = \frac{\partial}{\partial t} + \phi'(u)\frac{\partial}{\partial x}$. We then obtain

$$\begin{aligned} \frac{dr}{dt} &= -m(0)\psi(u) , \\ \frac{\partial s}{\partial t} &= -m(0)\psi(u) . \end{aligned} \tag{2.9}$$

We differentiate with respect to x, and use the fact that $\phi(u) = r - s$, to obtain

$$\begin{aligned} \frac{d}{dt} r_x &= -\frac{\phi''(u)}{\phi'(u)}(r_x - s_x)r_x - m(0)\frac{\psi'(u)}{\phi'(u)}(r_x - s_x) \\ \frac{\partial}{\partial t} s_x &= -m(0)\frac{\psi'(u)}{\phi'(u)}(r_x - s_x) \end{aligned} \tag{2.10}$$

If we now let $\rho = \sup_x |r_x|$, and $\sigma = \sup_x |s_x|$, then, under appropriate hypotheses on ϕ, ψ and the initial data (2.10) leads to differential inequalities of the form

$$\begin{aligned} \left(\frac{d}{dt}\right)_+ \rho &\geqslant \alpha\rho^2 - \beta\rho\sigma - \gamma(\rho + \sigma) \\ \left(\frac{d}{dt}\right)_+ \sigma &\leqslant \delta(\rho + \sigma) \end{aligned} \tag{2.11}$$

It is not hard to show that we can take data such that ρ tends to ∞ in a short time.

The analogy with hyperbolic conservation laws and the numerical evidence [26] suggest that this blow-up of smooth solutions means the development of shocks (i.e. weak solution with discontinuous u for (2.5) or solutions with discontinuous u_x and u_t for second order equations). No such result has been proved. Existence theorems for solutions with initial data of small total variation were established by Dafermos and Hsiao, but their work is still unpublished.

3. STEADY FLOWS OF VISCOELASTIC FLUIDS

Existence results for steady flows of viscoelastic liquids are lacking, even for slow flows perturbing rest. Formally, such slow flows have been analyzed by perturbation methods. Let the stress be given by a smooth functional of the history of the relative Cauchy strain

$$\underline{T} = \underset{s=-\infty}{\overset{t}{\mathbf{F}}} [\underline{\underline{C}}_t(s)] \ . \tag{3.1}$$

We assume $\underline{\underline{C}}_t(s)$ is a smooth function of s, expandable in a Taylor series about $s = t$:

$$\underline{\underline{C}}_t(s) = \underline{\underline{1}} + \sum_{i=1}^{n} \frac{(s-t)^i}{i!} \underline{\underline{A}}_i(t) + O(|s-t|^{n+1}) \ . \tag{3.2}$$

The Rivlin-Ericksen tensors $\underline{\underline{A}}_i$ satisfy the following recursion relation in a steady flow

$$\begin{aligned} \underline{\underline{A}}_{i+1} &= (\underline{u}\cdot\nabla)\underline{\underline{A}}_i + \underline{\underline{A}}_i\nabla\underline{u} + (\nabla\underline{u})^T\underline{\underline{A}}_i \ , \\ \underline{\underline{A}}_1 &= \nabla\underline{u} + (\nabla\underline{u})^T \ . \end{aligned} \tag{3.3}$$

We can therefore regard (3.2) as an expansion of $\underline{\underline{C}}_t(s)$ in increasing powers of a small and smooth velocity field $\underline{u}$. Suppose, e.g., that we want to solve the equations of motion

$$\begin{aligned} \rho(\underline{u}\cdot\nabla)\underline{u} &= \operatorname{div} \underline{T} - \nabla p + \varepsilon\underline{f} \ , \\ \operatorname{div} \underline{u} &= 0 \ , \\ \underline{u}\big|_{\partial\Omega} &= \varepsilon\underline{u}_0 \ , \end{aligned} \tag{3.4}$$

in a bounded domain Ω, where ε is a small parameter. Under appropriate conditions on the functional $\mathbf{F}$ (see Coleman and Noll [4]), we can formally expand $\underline{u}$ in powers of ε and equate coefficients of the same powers of ε in the equations. This leads to an infinite sequence of inhomogeneous Stokes problems from which an approximate "solution" can be determined to any order in ε.

It has not been proved that such a formal expansion converges or that it yields an asymptotic approximation to a solution of the full equation. Because of the presence of the term $(\underline{u}\cdot\nabla)\ \underline{\underline{A}}_i$ in (3.3), the expansion of $\underline{\underline{C}}_t(s)$ involves derivatives of $\underline{u}$ of arbitrary order. It is therefore clear that we can expect convergence of the expansion only if we require a high degree of smoothness for the data of the problem. It is possible that, under appropriate assumptions, convergence could be shown in certain spaces of analytic functions. An effort in this direction was made by Niggemann [27]. He studies one-dimensional model problems of the form

$$u'' = f + \sum_{N=1}^{\infty} \sum_{j_1,\ldots,j_N=1}^{\infty} a_{j_1\ldots j_N} u u^{(j_1)} \ldots u^{(j_N)} \tag{3.5}$$

His analysis requires certain assumptions on the coefficients $a_{j_1\ldots j_N}$, for which no physical motivation has been given.

An alternative approach to the problem is the use of iterative schemes, by which it may be possible to prove existence of solutions without recourse to small parameter expansions. The asymptotic validity of such expansions can then be establihsed a posteriori, by estimating derivatives of the solution with respect to ε. In recent work by the author [33], existence of solutions to (3.4) is proved in this way, provided that $\underline{u}_0$ is tangential to $\partial\Omega$, and $\underline{\underline{T}}$ is determined by a differential constitutive law of Maxwell or Jeffreys type, with an arbitrary number of relaxation modes. For simplicity, we hall here consider only the special case of the upper-convected Maxwell model

$$(\underline{u}\cdot\nabla)\underline{\underline{T}} - (\nabla\underline{u})\underline{\underline{T}} - \underline{\underline{T}}(\nabla\underline{u})^T + \lambda\underline{\underline{T}} = \eta\lambda(\nabla\underline{u} + (\nabla\underline{u})^T) \tag{3.6}$$

By taking the divergence of this equation, we obtain

$$(\underline{u}\cdot\nabla)\operatorname{div}\underline{\underline{T}} - (\nabla\underline{u})\operatorname{div}\underline{\underline{T}} + \lambda\operatorname{div}\underline{\underline{T}} = \underline{\underline{T}} : \partial^2\underline{u} + \eta\lambda\Delta\underline{u}\ , \tag{3.7}$$

where $\underline{\underline{T}} : \partial^2$ stands for $\sum_{j,k} T_{jk} \frac{\partial^2}{\partial x_j \partial x_k}$. We substitute $\operatorname{div}\underline{\underline{T}} = \rho(\underline{u}\cdot\nabla)\underline{u} + \nabla p - \underline{f}$ from (3.4) (we set $\varepsilon = 1$), introduce the notation $q = (\underline{u}\cdot\nabla)p + \lambda p$, and solve the equations by the following iteration scheme.

$$\underline{u}_0 = \underline{0},\quad p^0 = q^0 = 0,\quad \underline{\underline{T}}^0 = \underline{\underline{0}} \tag{3.8}$$

$$\underline{\underline{T}}^n : \partial^2 \underline{u}^{n+1} + \eta\lambda\Delta\underline{u}^{n+1} - \rho(\underline{u}^n\cdot\nabla)(\underline{u}^n\cdot\nabla)\underline{u}^{n+1} - \nabla q^{n+1}$$
$$= -[\nabla\underline{u}^n + (\nabla\underline{u}^n)^T]\nabla p^n - (\underline{u}^n\cdot\nabla)\underline{f} + (\nabla\underline{u}^n)\underline{f} - \lambda\underline{f}$$
$$-\rho(\nabla\underline{u}^n)(\underline{u}^n\cdot\nabla)\underline{u}^n + \lambda\rho(\underline{u}^n\cdot\nabla)\underline{u}^n \tag{3.9}$$
$$\text{div } \underline{u}^{n+1} = 0, \quad \underline{u}^{n+1}\big|_{\partial\Omega} = \underline{u}_0, \quad \iiint_\Omega q^{n+1} = 0$$

$$(\underline{u}^{n+1}\cdot\nabla)p^{n+1} + \lambda p^{n+1} = q^{n+1} \tag{3.10}$$

$$(\underline{u}^{n+1}\cdot\nabla)\underline{\underline{T}}^{n+1} - (\nabla\underline{u}^{n+1})\underline{\underline{T}}^{n+1} - \underline{\underline{T}}^{n+1}(\nabla\underline{u}^{n+1})^T$$
$$+ \lambda\underline{\underline{T}}^{n+1} = \eta\lambda(\nabla\underline{u}^{n+1} + (\nabla\underline{u}^{n+1})^T) \tag{3.11}$$

Equation (3.9) is basically a perturbation of the Stokes problem, while (3.10) and (3.11) are hyperbolic problems which can be solved by the method of characteristics. The proof for convergence of (3.8)-(3.11) is similar to existence proofs for initial value problems for quasilinear hyperbolic systems. First bounds for all iterates in a sufficiently strong norm are obtained, and then these bounds are used to obtain a contraction estimate in a weaker norm. Iterations similar to (3.8)-(3.11) have been used in numerical calculations (see e.g. [40]).

The following questions remain open:

1. Can similar results be obtained for integral type constitutive models?

2. What happens if $\underline{u}_0$ is not tangential to $\partial\Omega$? In this case we expect from the analysis of characteristics [18] that extra boundary conditions need to be imposed at inflow boundaries. What are the right conditions to impose and what compatibility conditions are required to avoid singularities of the solution where an inflow boundary joins with other parts of the boundary?

3. Formal expansions for small parameters require no extra boundary conditions at inflow. How are solutions approximated by these expansions distinguished from others?

4. Can one say anything about steady flows when $\underline{f}$ and $\underline{u}_0$ are not small? In this case it is possible that a change of type occurs in the equations governing steady flow [18]. However, attempts to compute steady flows have

encountered difficulties even in situations, such as creeping flows of an upper convected Maxwell model, where there should be no change of type. Although the existence of steady flows in the Newtonian case is well-known [38], even at high Reynolds number, we have no reason to believe that steady flows of viscoelastic fluids should always exist. It is possible that the numerical difficulties result from non-existence.

REFERENCES

1. Chu, B. T., Stress waves in isotropic linear viscoelastic materials, J. Mecanique 1 (1962), 439-446.
2. Coleman, B. D., M. E. Gurtin and I. R. Herrera, Waves in materials with memory I, Arch. Rat. Mech. Anal. 19 (1965), 1-19.
3 Coleman, B. D. and M. E. Gurtin, Waves in materials with memory II, Arch. Rat. Mech. Anal. 19 (1965), 239-265
4. Coleman, B. D. and W. Noll, An approximation theorem for functionals, with applications in continuum mechanics, Arch. Rat. Mech. Anal. 6 (1960), 355-370.
5. Dafermos, C. M. and J. A. Nohel, Energy methods for nonlinear hyperbolic Volterra integrodifferential equations, Comm. PDE 4 (1979), 219-278.
6. Dafermos, C. M. and J. A. Nohel, A nonlinear hyperbolic Volterra integrodifferential equation in viscoelasticity, Amer. J. Math. Supplement (1981), 87-116.
7. Doi, M. and S. F. Edwards, Dynamics of concentrated polymer systems, J. Chem. Soc. Faraday 74 (1978), 1789-1832 and 75 (1979), 38-54.
8. Gripenberg, G., Nonexistence of smooth solutions for shearing flows in a nonlinear viscoelastic fluid, SIAM J. Math. Anal. 13 (1982), 954-961.
9. Hattori, H., Breakdown of smooth solutions in dissipative nonlinear hyperbolic equations, Quart. Appl. Math. 40 (1982/83), 113-127.
10. Hannsgen, K. B. and R. L. Wheeler, Behavior of the solutions of a Volterra equation as a parameter tends to infinity, J. Integral Eq., to appear.

11. Heard, M. L., An abstract parabolic Volterra integro-differential equation, SIAM J. Math. Anal. 13 (1982), 81-105.

12. Heard, M. L., A class of hyperbolic Volterra integrodifferential equations, Nonlinear Analysis 8 (1984), 79-93.

13. Hrusa, W. J., A nonlinear functional differential equation in Banach space with applications to materials with fading memory, Arch. Rat. Mech. Anal. 84 (1983), 99-137.

14. Hrusa, W. J. and J. A. Nohel, The Cauchy problem in one-dimensional nonlinear viscoelasticity, J. Diff. Eq., to appear.

15. Hrusa, W. J. and M. Renardy, On a class of quasilinear partial integrodifferential equations with singular kernels, submitted to J. Diff. Eq.

16. Hrusa, W. J. and M. Renardy, On wave propagation in linear viscoelasticity, submitted to Quart. Appl. Math.

17. Hughes, T. J. R., T. Kato and J. E. Marsden, Well-posed quasi-linear second-order hyperbolic systems with applications to nonlinear elastodynamics and general relativity, Arch. Rat. Mech. Anal. 63 (1976), 273-284.

18. Joseph, D. D., M. Renardy and J. C. Saut, Hyperbolicity and change of type in the flow of viscoelastic fluids, Arch. Rat. Mech. Anal., to appear.

19. Kato, T., Quasi-linear equations of evolution with application to partial differential equations, in: W. N. Everitt (ed.), Spectral Theory of Differential Equations, Springer Lecture NOtes in Mathematics 448 (1975), 25-70.

20. Kato, T., Linear and quasilinear equations of evolution of hyperbolic type, in: G. da Prato and G. Geymonat (eds.), Hyperbolicity, Centro Internazionale Matematico Estivo, II ciclo, Cortona 1976, 125-191.

21. Kim, J. U., Global smooth solutions for the equations of motion of a nonlinear fluid with fading memory, Arch. Rat. Mech. Anal. 79 (1982), 97-130.

22. Laun, H. M., Description of the non-linear shear behaviour of a low density polyethylene melt by means of an experimentally determined strain dependent memory function, Rheol. Acta 17 (1978), 1-15.
23. R. C. MacCamy, A model for one-dimensional nonlinear viscoelasticity, Quart. Appl. Math. 35 (1977), 21-33.
24. Malek-Madani, R. and J. A. Nohel, Formation of singularities for a conservation law with memory, SIAM J. Math. Anal., to appear.
25. Malek-Madani, R., J. A. Nohel and F. John, to appear.
26. Markowich, P. and M. Renardy, Lax-Wendroff methods for hyperbolic history value problems, SIAM J. Num. Anal. 21 (1984), 24-51.
27. Niggemann, M., A model equation for non-Newtonian fluids, Math. Meth. Appl. Sci. 3 (1981), 200-217.
28. Renardy, M., Some remarks on the propagation and non-propagation of discontinuities in a linearly viscoelastic liquid, Rheol. Acta 21 (1982), 251-254.
29. Renardy, M., A class of quasilinear parabolic equations with infinite delay and application to a problem of viscoelasticity, J. Diff. Eq. 48 (1983), 280-292.
30. Renardy, M., Local existence theorems for the first and second initial-boundary value problems for a weakly non-Newtonian fluid, Arch. Rat. Mech. Anal. 83 (1983), 229-244.
31. Renardy, M., Singularly perturbed hyperbolic evolution problems with infinite delay and an application to polymer rheology, SIAM J. Math. Anal. 15 (1984), 333-349.
32. Renardy, M., A local existence and uniqueness theorem for a K-BKZ fluid, Arch. Rat. Mech. Anal., to appear.
33. Renardy, M., Existence of slow steady flows of viscoelastic fluids with differential constitutive equations, submitted to Z. angew. Math. Mech.
34. Rouse, P. E., A theory of the linear viscoelastic properties of dilute solutions of coiling polymers, J. Chem. Phys. 21 (1953), 1271-1280.

35. Slemrod, M., Instability of steady shearing flows in a nonlinear viscoelastic fluid, Arch. Rat. Mech. Anal. 68 (1978), 211-225.

36. Sobolevskii, P. E., Equations of parabolic type in a Banach space, Amer. Math. Soc. Transl. 49 (1966), 1-62.

37. Staffans, O., On a nonlinear hyperbolic Volterra equation, SIAM J. Math. Anal. 11 (1980), 793-812.

38. Teman, R., Navier-Stokes Equations, North Holland 1979.

39. Zimm, B. H., Dynamics of polymer molecules in dilute solutions: viscoelasticity, flow birefringence and dielectric loss, J. Chem. Phys. 24 (1956), 269-278.

40. Crochet, M. J. and K. Walters, Numerical methods in non-Newtonian fluid mechanics, Ann. Rev. Fluid Mech. 15 (1983), 241-260.

EVALUATION OF CONSTITUTIVE EQUATIONS: MATERIAL FUNCTIONS AND COMPLEX FLOWS OF VISCOELASTIC FLUIDS

Robert C. Armstrong, R. A. Brown,
A. N. Beris, J. V. Lawler, and S. J. Muller

Department of Chemical Engineering
Massachusetts Institute of Technology
Cambridge, Massachusetts

I. Introduction

In this paper we describe two methods for evaluating constitutive equations for viscoelastic fluids: the use of material functions as determined by rheometry and the comparison of measured and computed velocity and stress fields in complex flows. In traditional rheometry, the rheological properties of materials are determined by subjecting them to either well-defined kinematics or stresses and then measuring the other. Typically the measurements have been made mechanically, although optical methods for measuring both the stresses [1] and velocities [2] are becoming more common. Extensive characterization of viscoelastic materials can be done in shearing flow and to a lesser extent in shear-free flows. Among the serious limitations of these standard rheological characterization techniques are

(i) The lack of extensive shear-free flow methods for polymer solutions and many melts and

(ii) The inability to evaluate fluid behavior in complex kinematics that are mixtures of shear and shear-free flow.

These well defined experiments of course have important features that make them useful. Many of the simple flow fields used are homogeneous, so that the constitutive equation is identically satisfied. This means it is possible to interpret the results without having to make any constitutive assumptions.

This constitutive-free interpretation of experimental results can be contrasted with the study of complex flows where we specify boundary conditions. We then determine the velocity and stress fields that are consistent with these boundary conditions together with the equations of motion and a constitutive equation that we must assume to be valid for the material being studied.

In this paper we show how complex flows give useful insight into the proper constitutive description of materials and to the associated evaluation of constitutive equations. We suggest that the study of complex flows in the context of rheometry can be useful for three reasons:

(i) Information may be provided as to which of the host of possible material functions are most important in determining flow fields in general classes of flows.

(ii) New methods for measuring material functions may be suggested.

(iii) The ability of a constitutive equation to 'interpolate' between kinematics such as simple shear and simple elongational flow can be assessed.

In the next section we will introduce six relatively simple constitutive equations that we will use for the purpose of illustrating evaluation. Then in Section 3 we describe some of the most commonly measured material functions and give sample data that illustrates these. Comparisons of the constitutive equation predictions with these material functions is given in Section 4. Section 5 describes the behavior predicted by the six constitutive equations in a complex flow, flow between eccentric rotating cylinders, and compares these predictions with experimental data.

2. Six Differential Constitutive Equations

We will use six standard and rather simple differential constitutive equations to illustrate the process of constitutive evaluation. We give these equations in terms of the extra stress $\boldsymbol{\tau}$ [3]

Second-Order Fluid (SOF):

$$\boldsymbol{\tau} = -\eta_0\dot{\boldsymbol{\gamma}} + \frac{1}{2}\Psi_{10}\dot{\boldsymbol{\gamma}}_{(1)} - \Psi_{20}\{\dot{\boldsymbol{\gamma}}\cdot\dot{\boldsymbol{\gamma}}\} \tag{2.1}$$

Criminale-Ericksen-Filbey (CEF) Equation:

$$\boldsymbol{\tau} = -\eta\dot{\boldsymbol{\gamma}} + \frac{1}{2}\Psi_1\dot{\boldsymbol{\gamma}}_{(1)} - \Psi_2\{\dot{\boldsymbol{\gamma}}\cdot\dot{\boldsymbol{\gamma}}\} \tag{2.2}$$

Convected Maxwell Model (CM):

$$\boldsymbol{\tau} + \lambda_1\boldsymbol{\tau}_{(1)} = -\eta_0\dot{\boldsymbol{\gamma}} \tag{2.3}$$

White-Metzner Model (WM):

$$\boldsymbol{\tau} + \lambda\boldsymbol{\tau}_{(1)} = -\eta\dot{\boldsymbol{\gamma}} \tag{2.4}$$

Oldroyd Fluid-B (OLDB):

$$\boldsymbol{\tau} + \lambda_1\boldsymbol{\tau}_{(1)} = -\eta_0\left[\dot{\boldsymbol{\gamma}} + \lambda_2\dot{\boldsymbol{\gamma}}_{(1)}\right] \tag{2.5}$$

Giesekus: [4]

$$\boldsymbol{\tau} = \boldsymbol{\tau}_s + \boldsymbol{\tau}_p \tag{2.6a}$$

$$\boldsymbol{\tau}_s = -\eta_s \dot{\boldsymbol{\gamma}} \tag{2.6b}$$

$$\left(\boldsymbol{\delta} - \frac{\alpha}{\eta_p} \lambda_1 \boldsymbol{\tau}_p\right) \cdot \boldsymbol{\tau}_p + \lambda_1 \boldsymbol{\tau}_{p(1)} = -\eta_p \, \dot{\boldsymbol{\gamma}} \tag{2.6c}$$

or equivalently

$$\left(\boldsymbol{\delta} - \frac{a}{\eta_0} \lambda_1 \boldsymbol{\tau}\right) \cdot \boldsymbol{\tau} + \lambda_1 \boldsymbol{\tau}_{(1)} - a\lambda_2\{\boldsymbol{\tau}\cdot\dot{\boldsymbol{\gamma}} + \dot{\boldsymbol{\gamma}}\cdot\boldsymbol{\tau}\} = -\eta_0 \left[\dot{\boldsymbol{\gamma}} + \lambda_2 \dot{\boldsymbol{\gamma}}_{(1)} - \frac{a\lambda_2^2}{\lambda_1}\{\dot{\boldsymbol{\gamma}}\cdot\dot{\boldsymbol{\gamma}}\}\right] \tag{2.6d}$$

where in Eq. 2.6d η_0, λ_2, and a are defined in terms of "solvent" and "polymer" contributions to the viscosity, η_s and η_p, by

$$\eta_0 = \eta_s + \eta_p \tag{2.7}$$

$$\lambda_2 = \lambda_1 \frac{\eta_s}{(\eta_s + \eta_p)} \tag{2.8}$$

$$a = \frac{\alpha}{1 - (\lambda_2/\lambda_1)} \tag{2.9}$$

In the constitutive equations $\dot{\boldsymbol{\gamma}}$ is the rate-of-strain tensor, λ_1 is the relaxation time, λ_2 is the retardation time, and α is Giesekus's mobility parameter. The subscript $_{(1)}$ denotes the convected derivative of the contravariant components of a tensor, for example

$$\boldsymbol{\Lambda}_{(1)} \equiv \frac{D}{Dt}\boldsymbol{\Lambda} - [\{\nabla \mathbf{v}\}^{\dagger} \cdot \boldsymbol{\Lambda} + \boldsymbol{\Lambda} \cdot \{\nabla \mathbf{v}\}] \tag{2.10}$$

where D/Dt is the substantial derivative. In the White-Metzner model the relaxation time λ and the viscosity η are functions of the magnitude of the rate-of-strain tensor $\dot{\gamma} = (\dot{\boldsymbol{\gamma}}:\dot{\boldsymbol{\gamma}}/2)^{1/2}$; and $\lambda = \eta/G$ where G is a constant modulus. Similarly η, Ψ_1, and Ψ_2 in the CEF equation are functions of $\dot{\gamma}$.

The second-order is chosen because it gives the deviation from Newtonian behavior for a wide class of fluids and analytical results for the flow problem we consider later are available. The Giesekus model has been chosen because it represents in three parameters, the relaxation and retardation times and a mobility factor, most of the steady shear and elongation rheological behavior present in other differential models. In limits of these parameters the Giesekus model reduces to the Newtonian, second-order fluid, convected Maxwell, three constant Oldroyd-B, and a Leonov-like constitutive equation. The viscometric functions for these models are reviewed in Section 4. The limits of the three parameters $(\lambda_1, \lambda_2, \alpha)$ that reduce the Giesekus model to these six constitutive equations are summarized in Table 1. The Newtonian, convected Maxwell, second-order fluid and Oldroyd-B models are obtained exactly. The Leonov-like limit duplicates the Leonov model [5] only in steady simple shear. As indicated in Table 1, we use two forms of the Giesekus model with $\alpha = 0.5$: the one with $\lambda_2 = 0$ we refer to as the Leonov-like model (LEL) and for $\lambda_2 \neq 0$, we denote the constitutive equation as the LER model.

3. Shear and Shear-Free Flow Material Functions

The two standard types of flows in which viscoelastic fluids are normally characterized are shear and shear-free flows. A shear flow is locally of the form

$$v_x = \dot{\gamma}_{yx} y \; ; \qquad v_y = 0 \; ; \qquad v_z = 0 \tag{3.1}$$

in which the velocity gradient $\dot{\gamma}_{yx}$ can be a function of time. The absolute value of $\dot{\gamma}_{yx}$ is called the <u>shear rate</u> $\dot{\gamma}$ and is equal to the magnitude of $\dot{\boldsymbol{\gamma}}$ for this flow. The dependence of $\dot{\gamma}$ on time specifies the kind of shear flow and the shear flow material function to be measured. The most commonly used shear flow experiments are listed in Table 2 along with definitions of the relevant material functions. Of particular interest are the steady state or viscometric properties: the viscosity η and the first and second normal stress coefficients Ψ_1 and Ψ_2. These are illustrated in Figs. 1 - 3 for three polymer solutions. The shear thinning seen in these is typical of a diverse range of viscoelastic fluids. Note that for Newtonian fluids the viscosity is constant and the normal stress coefficients are both zero. Thus a measure of the 'elasticity' of the fluid can be given by the <u>stress ratio</u>

$$SR = (\tau_{xx} - \tau_{yy})/\tau_{yx} \tag{3.2}$$

Values of the stress ratio as a function of shear rate are shown in Fig. 4 for a low density polyethylene and for two polymer solutions. In the zero-shear-rate region where the normal stress coefficient and viscosity are constant, the stress ratio increases as $\dot{\gamma}$, whereas at higher shear rates the rate of increase of SR is less. SR appears to be a monotone increasing function of $\dot{\gamma}$, although if an infinite-shear-rate viscosity η_∞ were encountered with Ψ_1 still decreasing, then a maximum in SR would result.

Illustrative data for the many other shear flow experiments can be found in Ref. 3.

Shear-free flows are locally of the form

$$v_x = -\frac{1}{2}\dot{\varepsilon}\,(1+b)\,x; \qquad v_y = -\frac{1}{2}\dot{\varepsilon}\,(1-b)\,y; \qquad v_z = +\dot{\varepsilon}\,z \tag{3.3}$$

where $0 \leq b \leq 1$ and $\dot{\varepsilon}$ is the <u>elongation rate</u> which can depend on time. Several special shear-free flows are obtained for particular choices of the parameter b:

Table 1. Models Included in the Giesekus Model

Name of Model	Values of Constants			Remarks
	λ_1	λ_2	α	
Newtonian	0, λ_1	0, λ_1	0	
Second-order fluid	0	$\lambda_2 < 0$	0	
Convected Maxwell	$\lambda_1 > 0$	0	0	
Oldroyd Fluid B	$\lambda_1 > 0$	$\lambda_2 > 0$	0	
Leonov-like (LEL)	$\lambda_1 > 0$	0	1/2	Reduces to Leonov [5] model for steady shear flow
Leonov-like (LER)	$\lambda_1 > 0$	$\lambda_2 > 0$	1/2	Leonov-like model with retardation time

TABLE 2. Definitions of Material Functions in Simple Shearing Flows $v_x = \dot{\gamma}y$, $v_y = v_z = 0$.

Flow	Material Function	Defining Equation
a. Steady shear flow $\dot{\gamma}$ = constant	$\eta(\dot{\gamma})$ $\Psi_1(\dot{\gamma})$ $\Psi_2(\dot{\gamma})$	$\tau_{yx} = -\eta\,\dot{\gamma}$ $\tau_{xx}-\tau_{yy} = -\Psi_1\,\dot{\gamma}^2$ $\tau_{yy}-\tau_{zz} = -\Psi_2\,\dot{\gamma}^2$
b. Small amplitude shear $\dot{\gamma}=\dot{\gamma}^0\omega\cos\omega t$	$\eta'(\omega)$ $\eta''(\omega)$ $G'(\omega)$ $G''(\omega)$	$\tau_{yx} = -\eta'\gamma^0\omega\cos\omega t - \eta''\gamma^0\omega\sin\omega t$ $G' = \eta''\omega$ $G'' = \eta'\omega$

Table 2 continued

TABLE 2 (continued)

Flow	Material Function	Defining Equation
c. Stress growth upon inception of steady shear flow $\dot{\gamma} = \begin{cases} 0 & t<0 \\ \dot{\gamma}_0 & t\geq 0 \end{cases}$	$\eta^+(t;\dot{\gamma}_0)$ $\Psi_1^+(t;\dot{\gamma}_0)$ $\Psi_2^+(t;\dot{\gamma}_0)$	$\tau_{yx} = -\eta^+ \dot{\gamma}_0$ $\tau_{xx}-\tau_{yy} = -\Psi_1^+ \dot{\gamma}_0^2$ $\tau_{yy}-\tau_{zz} = -\Psi_2^+ \dot{\gamma}_0^2$
d. Stress relaxation after cessation of steady shear flow $\dot{\gamma} = \begin{cases} \dot{\gamma}_0 & t<0 \\ 0 & t\geq 0 \end{cases}$	$\eta^-(t;\dot{\gamma}_0)$ $\Psi_1^-(t;\dot{\gamma}_0)$ $\Psi_2^-(t;\dot{\gamma}_0)$	$\tau_{yx} = -\eta^- \dot{\gamma}_0$ $\tau_{xx}-\tau_{yy} = -\Psi_1^- \dot{\gamma}_0^2$ $\tau_{yy}-\tau_{zz} = -\Psi_2^- \dot{\gamma}_0^2$
e. Stress relaxation after a sudden shearing displacement $\dot{\gamma} = \gamma_0\ \delta(t)$	$G(t;\gamma_0)$ $G_{\Psi_1}(t;\gamma_0)$	$\tau_{yx} = -G\ \gamma_0$ $\tau_{xx}-\tau_{yy} = -G_{\Psi_1}\ \gamma_0^2$
f. Creep $\tau_{yx} = \begin{cases} 0 & t<0 \\ \tau_0 & t\geq 0 \end{cases}$	$J(t,\tau_0)$	$\gamma_{yx}(0,t) = -J\tau_0$
g. Constrained recoil after steady shear flow $\tau_{yx} = \begin{cases} \tau_0 & t<0 \\ 0 & t\geq 0 \end{cases}$	$\gamma_r(0,t,\tau_0)$ $\gamma_\infty(\tau_0)$ $J_e^0(\tau_0)$	$\gamma_r = \int_0^t \dot{\gamma}_{yx}(t')dt'$ $\gamma_\infty = \lim_{t\to\infty} \gamma_r$ $\gamma_\infty = -J_e^0\tau_0$

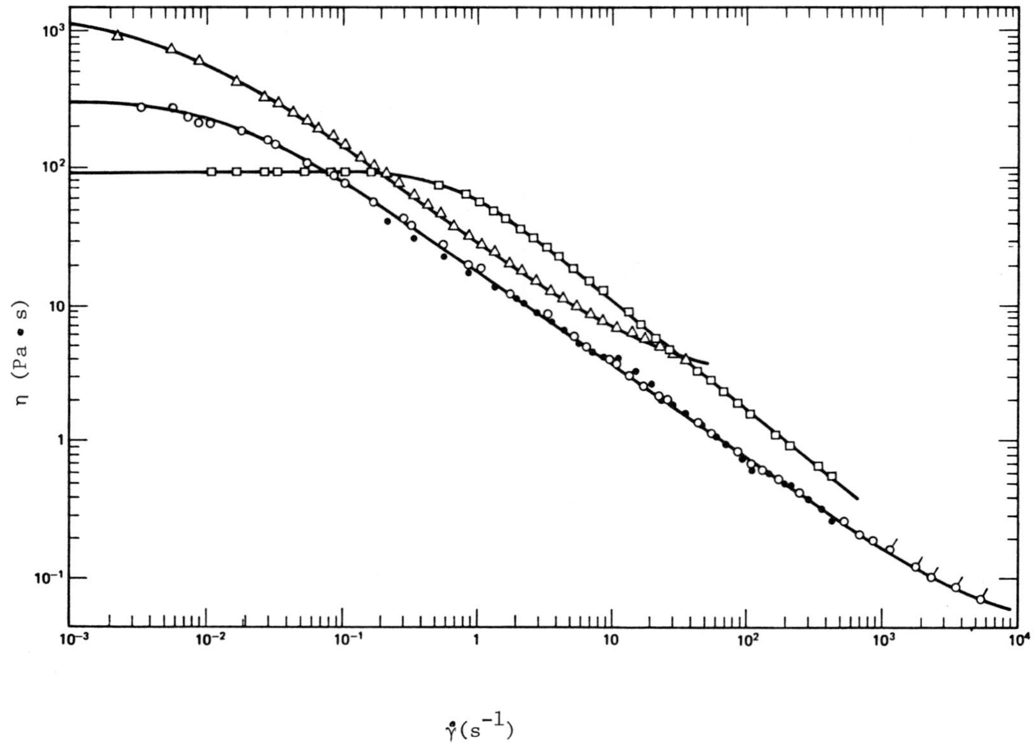

FIGURE 1. Dependence of viscosity on shear rate for two polymer solutions and an aluminum soap solution: o 1.5% polyacrylamide (Separan AP30) in a 50/50 mixture by weight of water and glycerin, Δ 2.0% polyisobutylene in Primol, and □ 7% aluminum laurate in a mixture of decalin and m-cresol. All data were taken at 298 K. [Data of J.D. Huppler, E. Ashare, and L.A. Holmes, Trans. Soc. Rheol., **11**, 159-179 (1967).]

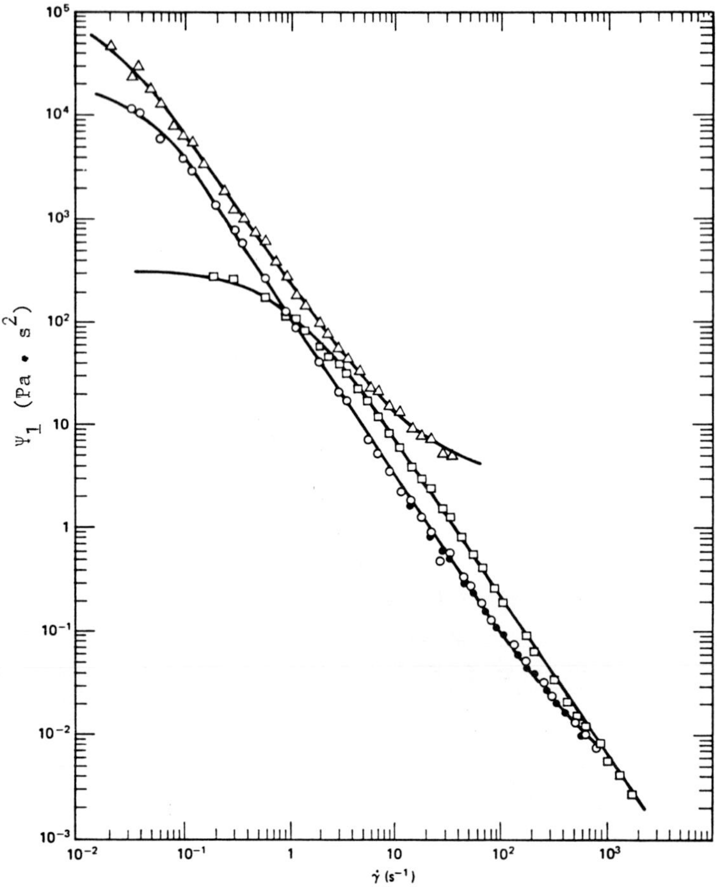

FIGURE 2. Dependence of the first normal stress coefficient on shear rate for the same fluids and temperature as Fig. 1. [Data replotted from J.D. Huppler, E. Ashare, and L.A. Holmes, Trans. Soc. Rheol., **11**, 159-179 (1967).]

Elongational flow ($b = 0$, $\dot{\varepsilon} > 0$)

Biaxial stretching ($b = 0$, $\dot{\varepsilon} < 0$)

Planar elongational flow ($b = 1$)

A host of material functions analogous to those for shear flows can be defined for shear-free flows; these will depend on the parameter b that specifies the type of flow as well as the elongation rate $\dot{\varepsilon}$. For steady simple shear-free flows we define two viscosity functions $\bar{\eta}_1$ and $\bar{\eta}_2$ to

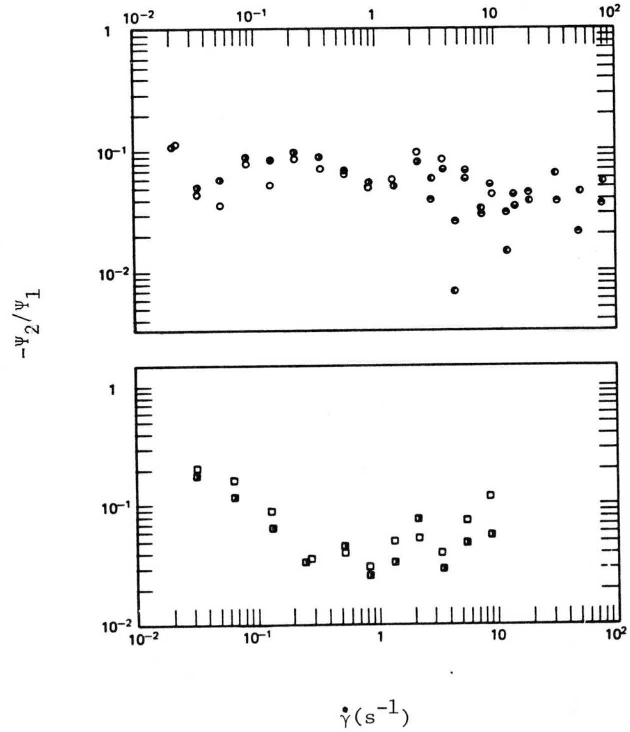

FIGURE 3. Ratio of the second normal stress coefficient to the first normal stress coefficient for two polymer solutions: the circles are for a 2.5% solution of polyacrylamide in a 50/50 mixture of water and glycerin, and the squares are for a 3% solution of poly(ethylene oxide) in a 57/38/5 mixture of water, glycerin, and isopropyl alcohol. [E.B.-Christiansen and W.R. Leppard, Trans. Soc. Rheol., **18**, 65-86 (1974)]

---

describe the two normal stress differences :

$$\begin{aligned} \tau_{zz} - \tau_{xx} &= -\bar{\eta}_1(\dot{\varepsilon},b)\,\dot{\varepsilon} \\ \tau_{zz} - \tau_{yy} &= -\bar{\eta}_2(\dot{\varepsilon},b)\,\dot{\varepsilon} \end{aligned} \tag{3.4}$$

For the special shear-free flows where b=0, the two normal stress differences

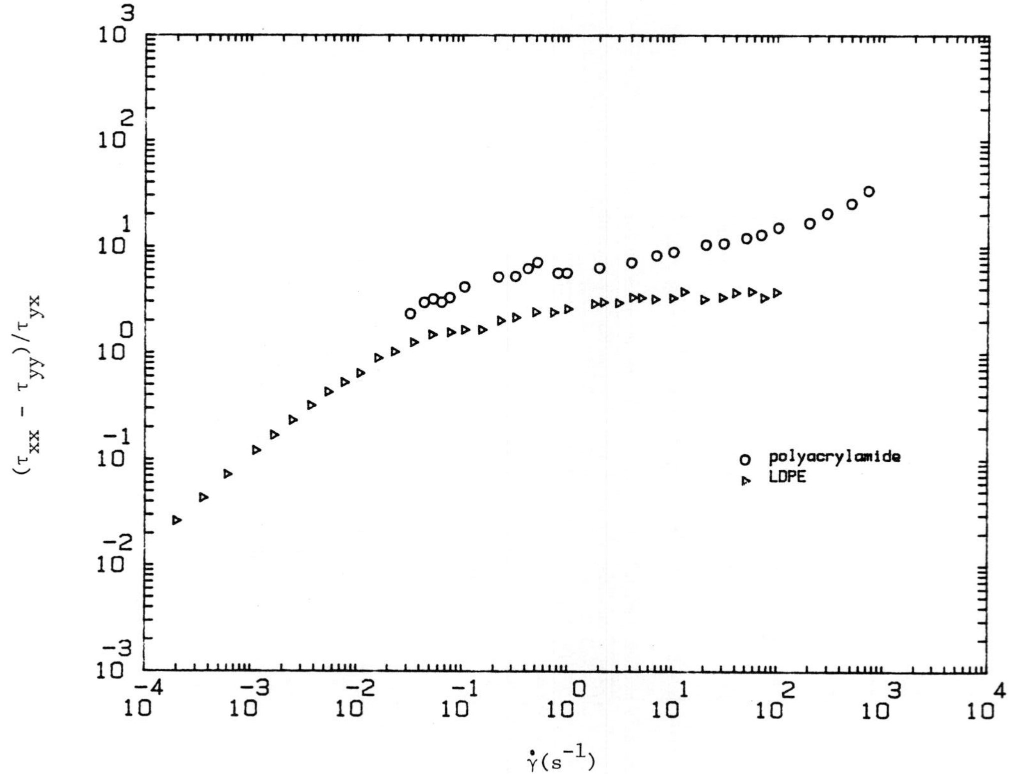

FIGURE 4. The stress ratio $(\tau_{xx}-\tau_{yy})/\tau_{yx}$ for a low density polyethylene melt [Melt I data from H.M. Laun, Rheol. Acta, **17**, 1-15 (1978)] and for the polyacrylamide solution of Fig. 1.

are the same, and the two functions $\bar{\eta}_1$ and $\bar{\eta}_2$ are both equal to the elongational viscosity $\bar{\eta}$:

$$\bar{\eta}(\dot{\varepsilon}) = \bar{\eta}_1(\dot{\varepsilon},0) = \bar{\eta}_2(\dot{\varepsilon},0) \tag{3.5}$$

For $\dot{\varepsilon}>0$, η gives the stress in elongational flow, and for $\dot{\varepsilon}<0$, it gives the stress for biaxial stretching. For start up of steady elongational flow, we define a transient property $\bar{\eta}^+$ analogously to η^+.

Data for a low density polyethylene melt [6] are shown in Fig. 5 for $\bar{\eta}^+$. At least for this polymer melt there appears to be no steady elongational viscosity. The curves for $\bar{\eta}^+$ are seen to go through a maximum and are still decreasing at the largest strains achieved. It should be emphasized that these strains are enormous; the maximum strains of $\varepsilon = 7$ shown here correspond to extension ratios of about 1100. Strain hardening in the material is seen to occur at Hencky strains $\varepsilon = \dot{\varepsilon}t$ of order unity. Data on high density polyethylene [7] and polystyrene [8] show that for these materials $\bar{\eta}$ is equal to $3\eta_0$ at low elongation rates, increases slightly with increasing elongation rate, and then decreases at high elongation rates.

There are presently no methods for obtaining these material functions on mobile liquid systems.

4. Material Functions for the Six Constitutive Equations

The viscometric functions for the six models are summarized in Fig. 6 and in Table 3. Shear thinning of the viscosity and the first normal stress difference is represented by the CEF, WM, LEL and LER models. For the LEL fluid, the viscosity is well approximated by a power law model (cf. Eq. 5.1) with n = 0 at large shear rates. This amount of shear thinning is excessive, as it gives a shear stress that is independent of shear rate. The small retardation time added to the LER fluid keeps the viscosity from reaching this limiting slope. The Giesekus model [9] gives a second normal stress difference N_2 which ranges between zero and $-(1/2)N_1$ depending on the values of α and $(\lambda_1-\lambda_2)$. The CM, OLDB and SOF limits of the Giesekus model all have $N_2 = 0$. The stress ratio is a monotone increasing function of shear rate for the SOF, CM, WM, and LEL models; for the OLDB and LER models the leveling off in viscosity at high shear rates causes the stress ratio to go through a maximum. For small values of λ_2/λ_1 in the LER model, this maximum is delayed until very high shear rates.

The elongational viscosity of the LEL and LER models reach an asymptotic limit of $\bar{\eta} = 4\eta_0$ at high elongation rates $\dot{\varepsilon}(\lambda_1-\lambda_2)$ whereas $\bar{\eta}$ for the SOF fluid increases linearly with increasing strain rate. For typical viscometric functions η and Ψ_1, the CEF model will predict $\bar{\eta}$ to decrease with increasing

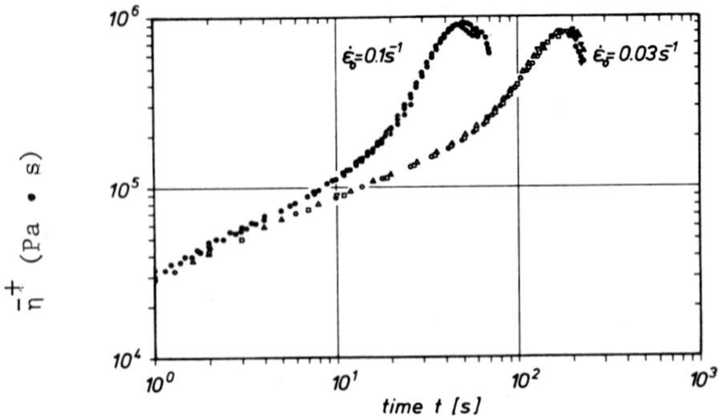

FIGURE 5. Elongational stress growth coefficient $\bar{\eta}^+$ vs. time for two elongation rates for a low-density polyethylene melt. [T. Raible, A. Demarmels, and J. Meissner, Polymer Bulletin, **1**, 397-402 (1979).

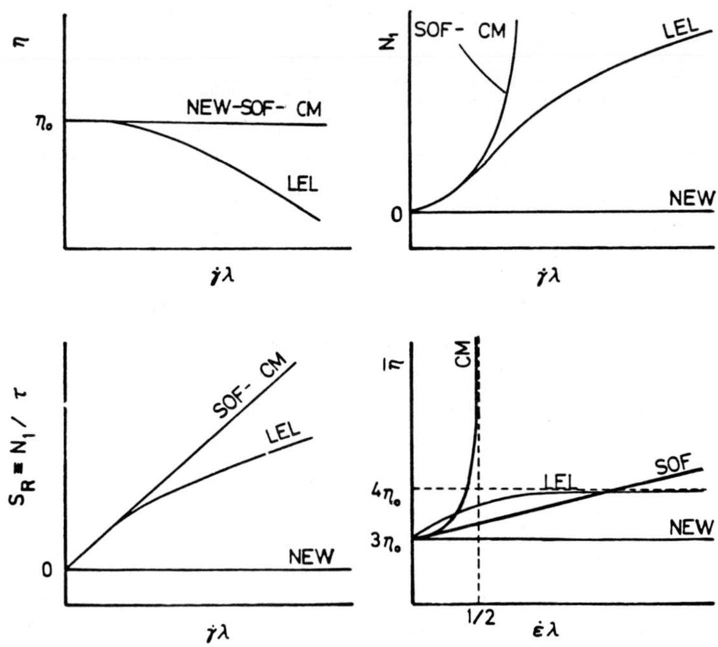

FIGURE 6. Material functions for limiting cases of the Giesekus model. The first normal stress difference is denoted by $N_1 = -(\tau_{xx} - \tau_{yy})$.

TABLE 3. Steady Material Functions for Several Differential Constitutive Equations

Model	η	Ψ_1	Ψ_2	$\bar{\eta}$ [a]
Second-Order Fluid	η_0	Ψ_{10}	Ψ_{20}	$3(\eta_0 + \Psi_{10}\dot{\varepsilon})$
Criminale-Ericksen-Filbey Equation	$\eta(\dot{\gamma})$	$\Psi_1(\dot{\gamma})$	$\Psi_2(\dot{\gamma})$	$3(\eta + \Psi_1\dot{\varepsilon})$
Convected Maxwell Model	η_0	$2\eta_0\lambda_1$	0	$3\eta_0 \dfrac{1}{(1 + \lambda_1\dot{\varepsilon})(1 - 2\lambda_1\dot{\varepsilon})}$
Oldroyd Fluid-B	η_0	$2\eta_0(\lambda_1-\lambda_2)$	0	$3\eta_0 \dfrac{1 + \lambda_2\dot{\varepsilon} - 2\lambda_1\lambda_2\dot{\varepsilon}^2}{(1 + \lambda_1\dot{\varepsilon})(1 - 2\lambda_1\dot{\varepsilon})}$
White-Metzner Model	$\eta(\dot{\gamma})$	$2\eta(\dot{\gamma})\lambda(\dot{\gamma})$	0	$3\eta(\dot{\gamma}) \dfrac{1}{(1 + \lambda(\dot{\gamma})\dot{\varepsilon})(1 - 2\lambda(\dot{\gamma})\dot{\varepsilon})}$
Giesekus[b] ($\lambda_2 = 0$) ($\alpha \neq 0,1$)	$\sqrt{(1-\alpha)/\alpha}\,\dfrac{\eta_0}{\lambda_1\dot{\gamma}}$	$\dfrac{\sqrt{2}}{\alpha}\left(\alpha(1-\alpha)\right)^{1/4}\dfrac{\eta_0\lambda_1}{(\lambda_1\dot{\gamma})^{3/2}}$	$-\dfrac{\eta_0\lambda_1}{(\lambda_1\dot{\gamma})^2}$	$\dfrac{2\eta_0}{\alpha}$

[a] In the CEF and White-Metzner model entries for $\bar{\eta}$, the magnitude of the rate of strain tensor $\dot{\gamma}$ is equal to $\sqrt{3}\,\dot{\varepsilon}$.

[b] The Giesekus results are asymptotically valid for large deformation rates.

$\dot{\varepsilon}$. The steady elongational viscosities of the CM and OLDB models are unbounded at a dimensionless extension rate $\lambda_1\dot{\varepsilon}$ of 1/2.

5. An Example Complex Flow: The Journal Bearing

As an example of how a complex flow can be used to gather information about a constitutive equation we consider the eccentric cylinder, or journal bearing, geometry. As shown in Fig. 7 the fluid is contained between inner and outer cylinders with radii R and R_0 respectively. The axes of the cylinders are displaced by an amount e, and the inner cylinder is rotated at an angular velocity Ω so that its linear velocity is $V = \Omega R$. When $e = 0$ the flow is viscometric, i.e., steady shear flow; by increasing e, the elongational character of the flow can be increased. This geometry is advantageous to study in the present context because in the small gap limit $\mu = (R - R_0)/R_0 << 1$ and $e = 0$ the flow is homogeneous. Also the periodicity of the flow and the absence of sharp corners facilitates analysis of the flow.

To describe the importance of 'elasticity' in the flow we introduce the Deborah number

$$De = \frac{\lambda_1 V}{R} \tag{5.1}$$

which gives the ratio of the characteristic time of the fluid, λ_1, to the characteristic time for the flow R/V. It is of primary interest to see how the six constitutive equations of Section 2 differ in the journal bearing at high De.

Perturbation solutions have been constructed for small gaps μ and small dimensionless eccentricities $\varepsilon = e/(R - R_0)$ for all of the models in Section 2 except the Giesekus model [9,10]. Finite element calculations [11] extend these results to high eccentricity for the convected Maxwell and Oldroyd-B fluids and to the Giesekus model with $\alpha = 0.5$ (with and without retardation times). The viscosity in the WM and CEF models is taken to be the power-law expression

$$\eta = m\dot{\gamma}^{(n-1)} \tag{5.2}$$

which is a good approximation for small gaps and small eccentricities. To facilitate comparisons between the models with memory (CM and WM) and those without (SOF and CEF), we take $\Psi_2 = 0$ in the SOF and CEF models, since this is the value of Ψ_2 given by the CM and WM models. In the Giesekus model calculations the retardation time is taken to be $0.001\lambda_1$ which is sufficient to prevent the formation of steep velocity gradients due to excessive shear thinning in the Giesekus model with $\lambda_2 = 0$. We now illustrate the sensitivity of these results to the constitutive equations.

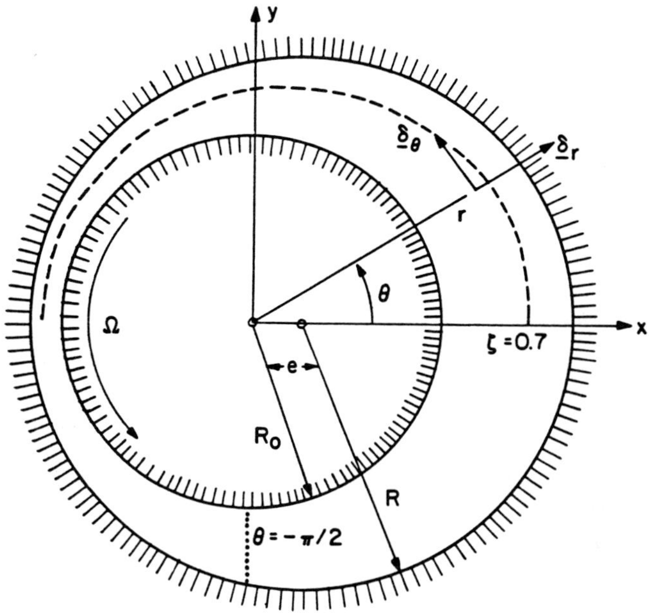

FIGURE 7. Journal bearing geometry. The dashed curve (– – –) indicates the position for which azimuthal profiles are given in subsequent figures; the dotted curve (·····) gives the location for radial profiles.

Figure 8 shows the dimensionless radial velocity as a function of dimensionless radial coordinate ζ defined as

$$v_r^* = v_r/V\mu\varepsilon \tag{5.3}$$

$$\zeta = (r-R_0)/[(R-R_0)(1 + \varepsilon\cos\theta)] \tag{5.4}$$

obtained from the perturbation calculations. Comparison of the results for the convected Maxwell, White-Metzner, and CEF models shows clearly that the influences of memory and shear thinning on the calculations are quite different. At De = 0 all models give the Newtonian radial velocity distribution. Increasing De for the two models with memory (CM and WM) is seen to cause the velocity to become more symmetric about the center of the gap. Less obvious from the figure is the development of a boundary layer at the inner cylinder ($\zeta = 0$) as De is increased. From continuity and no slip on the rotating cylinder we have $(\partial v_r/\partial\zeta)|_{\zeta=0} = 0$; clearly this derivative changes very rapidly in the wall region as De is increased. For the CEF fluid, opposite trends in v_r with increasing De are observed. The maximum value of v_r decreases and a boundary layer is formed at the outer cylinder rather than at the inner cylinder. The effect of a retardation time is shown in

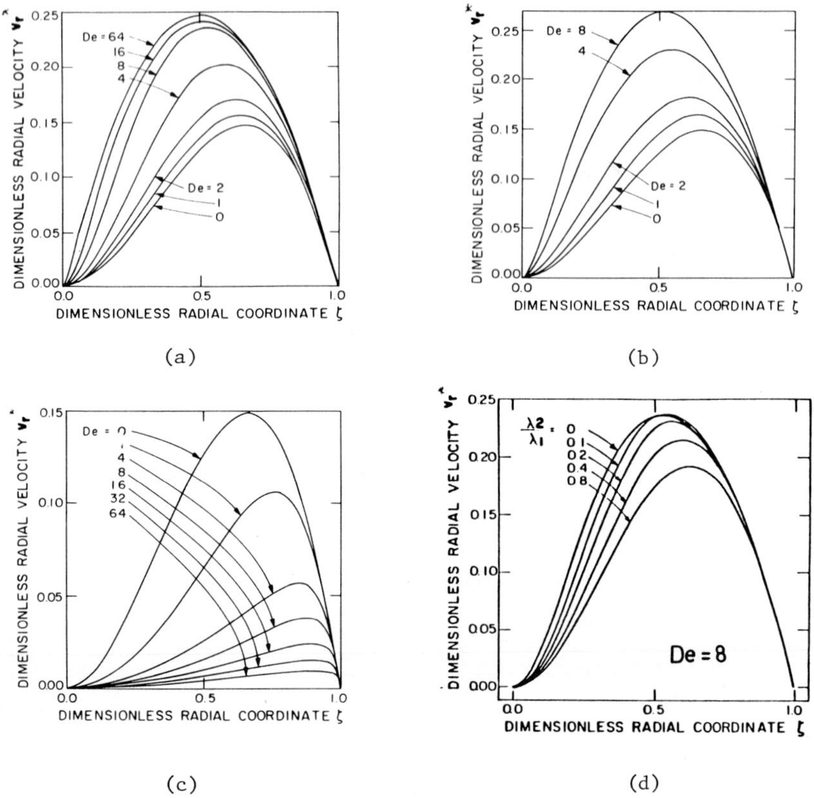

FIGURE 8. Radial profiles of the dimensionless radial velocity (Eq. 5.3) as a function of Deborah number for the (a) convected Maxwell model, (b) White-Metzner model, and (c) CEF model. In part (d), v_r* is shown for the Oldroyd-B fluid at De = 8 to demonstrate the effect of a retardation time.

---

the Oldroyd-B results to cause the radial velocity to tend to the Newtonian profile. The OLDB flow field is found to be the same as that for the convected Maxwell model at a lower 'effective' Deborah number given by

$$De_{eff} = De\left[1 - \frac{\lambda_2}{\lambda_1}\right] \tag{5.5}$$

Figure 9 shows the radial velocity profile for the LER model with $\lambda_2/\lambda_1 = 10^{-3}$. As De is increased the radial velocity first becomes more symmetric as in the CM and WM models; further increases in De from 1.48 to 15.5 cause the profile to shift back to Newtonian, although with a lower amplitude.

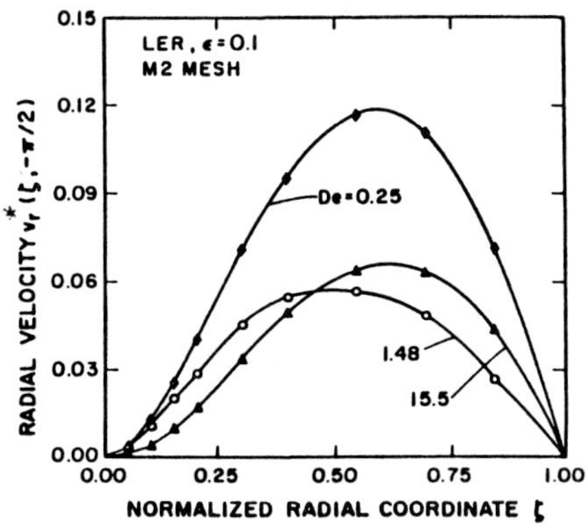

FIGURE 9. Dependence of the dimensionless radial velocity profile on radial position and De for the Giesekus model with α=0.5 and λ_2=0.001.

The tangential velocity profiles for the SOF, CEF, CM, and WM models are contrasted in Fig. 10 where the streamlines are compared for the different models at De = 10. In this figure the gap region has been cut at the widest point (θ = 0°) and unwrapped and stretched into a rectangular domain. At ε = 0.4 for which the results are shown the Newtonian fluid has a region of reversed flow next to the outer wall centered on θ = 0°. This zone is shaded in the figure. For the SOF model the flow field is the same as the Newtonian one as is known from the existence [12,13] and uniqueness [14] proofs for this model. Adding shear thinning without memory as in the CEF equation causes the secondary flow region to grow. The inclusion of memory with constant material functions (CM) causes the secondary flow region to shrink, whereas addition of both memory and shear thinning (WM) causes slight shrinkage and convection downstream. The influence of a retardation time in the Oldroyd-B model is again to give convected Maxwell model results with a lower effective De defined by Eq. 5.5.

For the LER model with λ_2/λ_1 = 0.001 the secondary flow region disappears at much smaller values of De than that shown for the preceding models. This is illustrated in Fig. 11 for the tangential velocity vs. ζ and in Fig. 12 in which the streamlines for the LER and CM models are contrasted. The recirculation is almost unchanged for the CM model from De = 0.0 up to about unity, whereas for the LER calculations no secondary flow is detected at De as low as 0.1. It is interesting to trace the critical De at which the recirculation disappears, De_{crit}, vs. the mobility factor α in the Giesekus model. This is shown in Table 4 for an eccentricity of ε = 0.4. Keep in mind that LER corresponds to α = 0.5 and CM to α = 0.

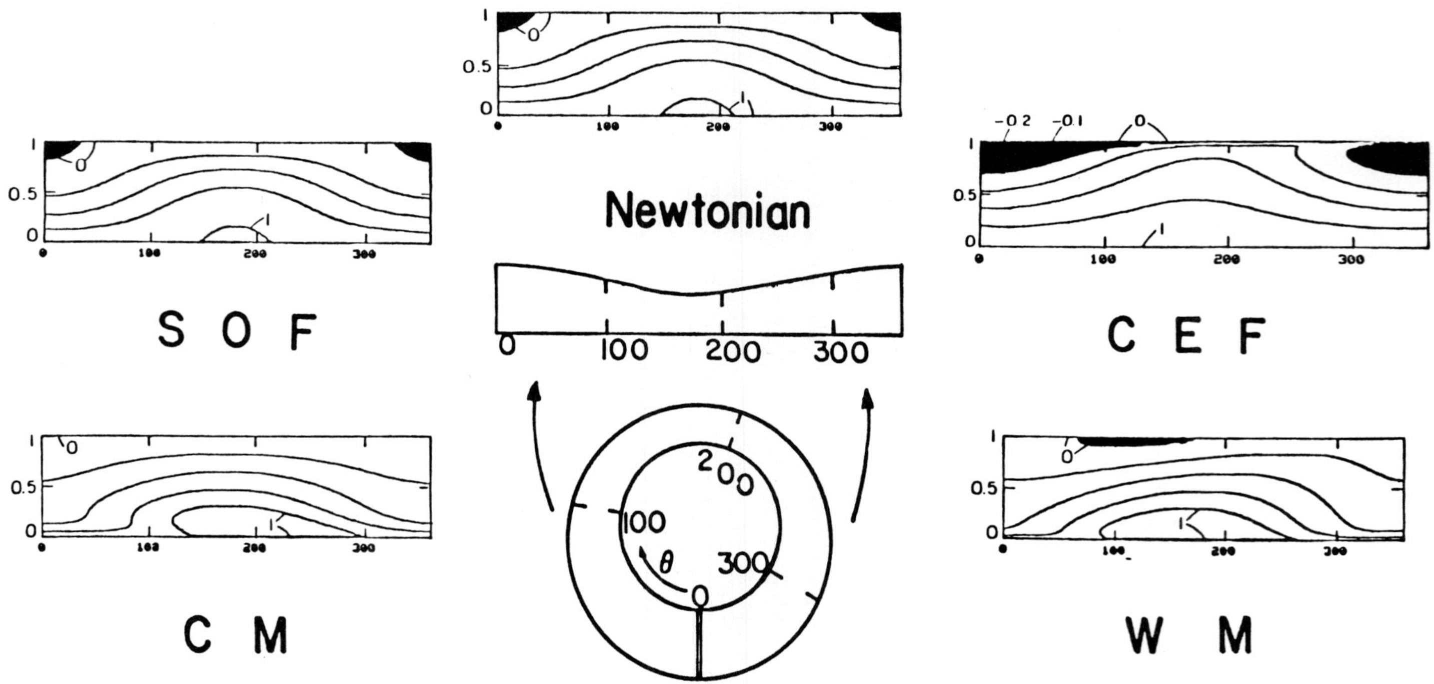

FIGURE 10. Contours of the tangential velocity for the Newtonian, second-order fluid, convected-Maxwell, CEF, and White-Metzner models. Results are shown for a Deborah number of 10 and for eccentricity ε = 0.4. The region of reversed flow near the outer cylinder is shaded. The unwrapping and stretching of the gap region is illustrated in the center of the figure.

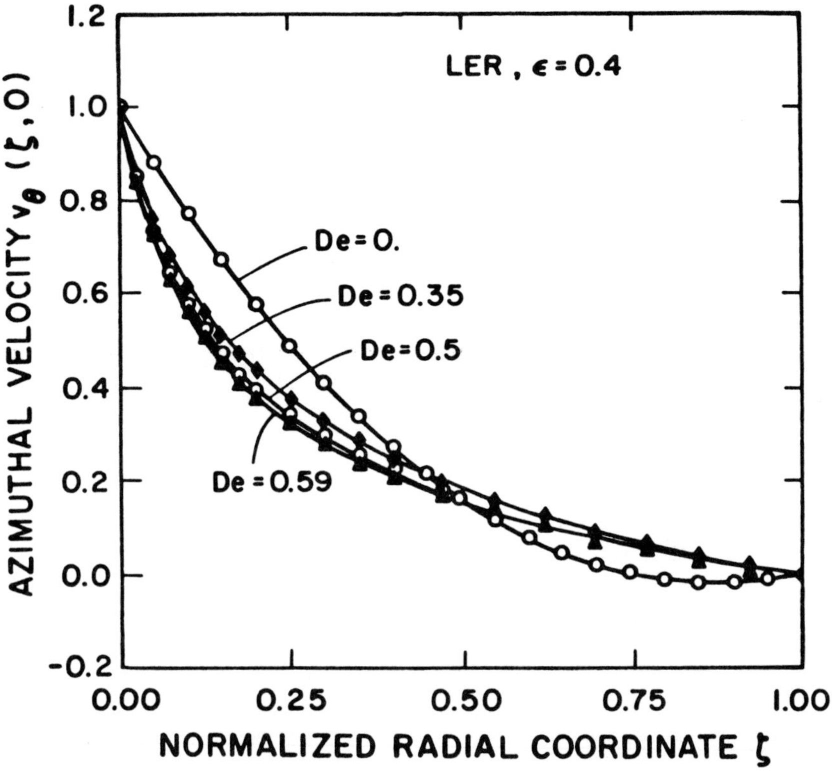

FIGURE 11. Dependence of the dimensionless tangential velocity on radial position and De for the Giesekus model with $\alpha=0.5$ and $\lambda_2=0.001$. The eccentricity is $\varepsilon = 0.4$ and the dimensionless gap is $\mu = 0.1$. For De = 0 the model reduces to Oldroyd-B which for the small value of λ_2 used is indistinguishable from the convected-Maxwell model. Note the small secondary flow region near $\zeta = 1.0$ for De = 0.

The sensitivity of the flow separation to the rheology can be seen more clearly in Figs. 13a and b which show, respectively, the variation in minimum eccentricity for flow separation with De and the angular position at which flow separation is observed to occur. Results are shown for different amounts of shear thinning as given by the power-law index n. Increasing amounts of shear thinning lower the minimum eccentricity required for separation for models with and without memory, at least as described by the WM and CEF equations. The CM and SOF models are obtained as the limiting cases of these with n = 1. Similarly, for both sets shear thinning seems to move the separated flow region downstream. The finite element calculations for the Giesekus model are qualitatively consistent with the WM predictions

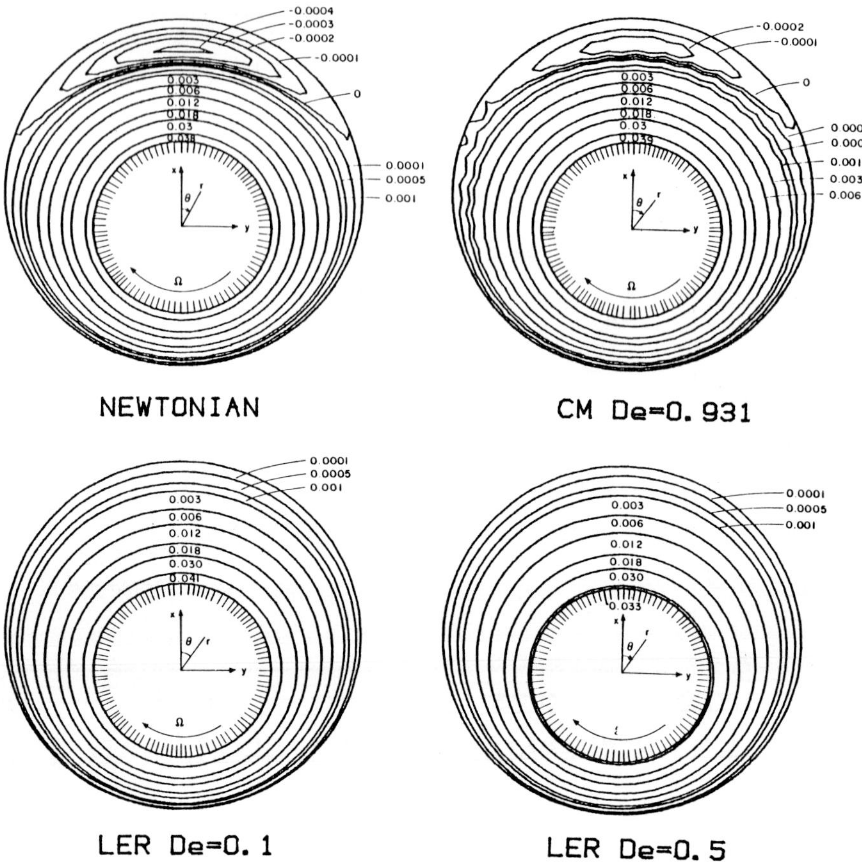

FIGURE 12. Streamlines for the Newtonian fluid, convected-Maxwell (CM) model, and Giesekus (LER) model. The dimensionless gap is $\mu = 0.1$, and the eccentricity is $\varepsilon = 0.4$. For the Giesekus model we take $\alpha = 0.5$ and $\lambda_2 = 0.001$. As De is increased from 0.1 to 0.5 in the Giesekus model, the only change in the streamlines is the development of a steep gradient in v_θ near the inner cylinder; this appears as a darkening near $\zeta = 0$ in the figure.

TABLE 4. Dependence of the critical Deborah number for flow separation on the Giesekus mobility parameter

α	De_{crit}
0.01	0.4
0.05	0.1
0.5	0.05

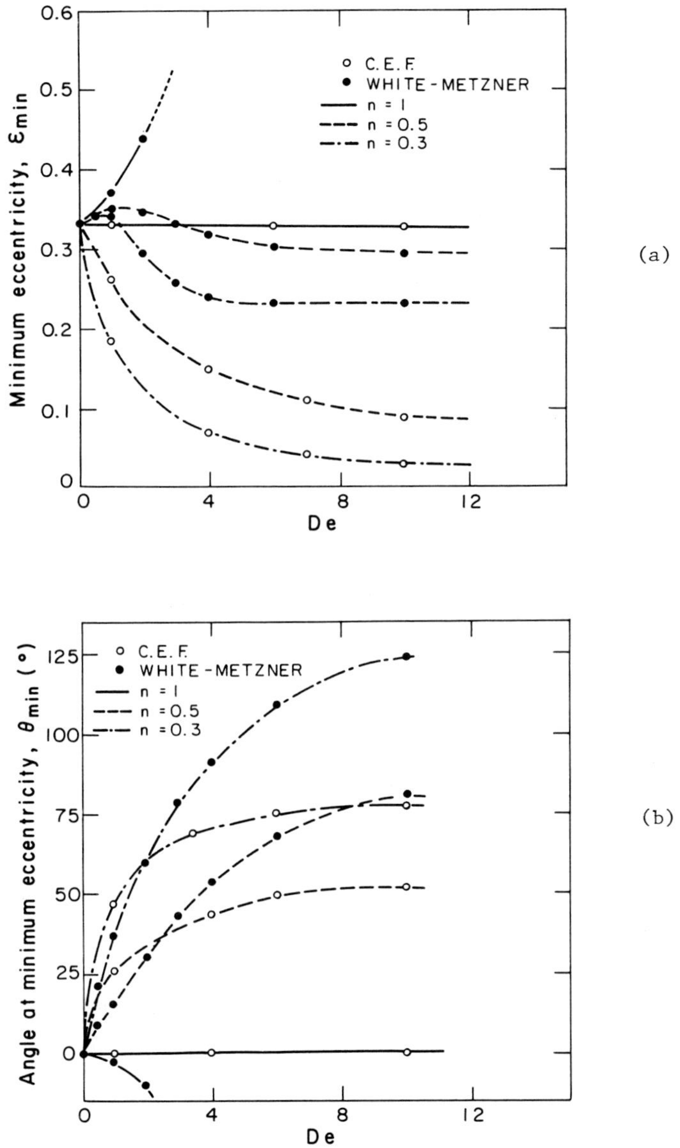

FIGURE 13. Dependence on De of (a) the minimum eccentricity ε_{min} for development of a separated flow region and (b) the azimuthal angle for the flow separation at the minimum eccentricity for the CEF and White-Metzner models with viscosity having a power-law index n (cf. Eq. 5.2). The second-order fluid and convected Maxwell models are the limiting cases of these, respectively, for n = 1.

at low De where ε_{min} increases over the Newtonian value. The importance of the retardation time is again a lowering of De_{eff}.

The constitutive equations we have been considering can also be easily distinguished in their predictions of the loads on the inner cylinder. This is shown in Fig. 14 for the SOF, CEF, CM, and WM models by means of the Deborah number dependence of the amplitude of the oscillatory pressure field and the phase angle of the pressure maximum. At large De, increasing the amount of shear thinning decreases the pressure amplitude for the fluids with no memory, but increases it for those with memory. The phase shift sets the direction of the load on the inner cylinder; for a Newtonian fluid the phase angle $\phi_p = -90°$ and the load is in the y-direction. The phase shift is less than $-90°$ and the load tends to center the inner cylinder for the four models shown, except the WM fluid with low n and moderate De.

The corresponding load results for the LER model are shown in Fig. 15. Here we present the magnitude of the force exerted by the fluid on the cylinder and the angle of the net force. We do this instead of giving a pressure amplitude and phase angle as before because in the large ε finite element results the pressure and stress fields are no longer sinusoidal. It is clear that the CM and LER models are quite different. The load bearing ability of the LER fluid drops <u>dramatically</u> with increasing De, and the direction of the force pushes the inner cylinder toward the <u>small</u> part of the gap.

It is interesting to compare, at least qualitatively, these constitutive predictions against experimental results on the journal bearing we have obtained in our laboratory. Experiments have been conducted on an aqueous polyacrylamide solution similar to the one shown in Figs. 1,2, and 4. Point velocity measurements have been made with a two-color laser doppler velocimeter [15,16]. Because of the highly elastic nature of the polyacrylamide solution (see Fig. 4), it is difficult to make measurements at low De. We show in Fig. 16 the variation in the tangential velocity with distance across the gap for an eccentricity $\varepsilon = 0.6$. In Fig. 16a the secondary flow regime for the Newtonian fluid is clearly evident; Fig. 16b shows that for the polyacrylamide solution the secondary flow is not present for $De \geqq 11$. This is consistent with the LER predictions and perhaps the CM perturbation results.

Although we cannot presently measure pressure distributions in the gap in the journal bearing, we can get some information about pressure variations from the shape of the free surface. In Fig. 17 we show the dimensionless

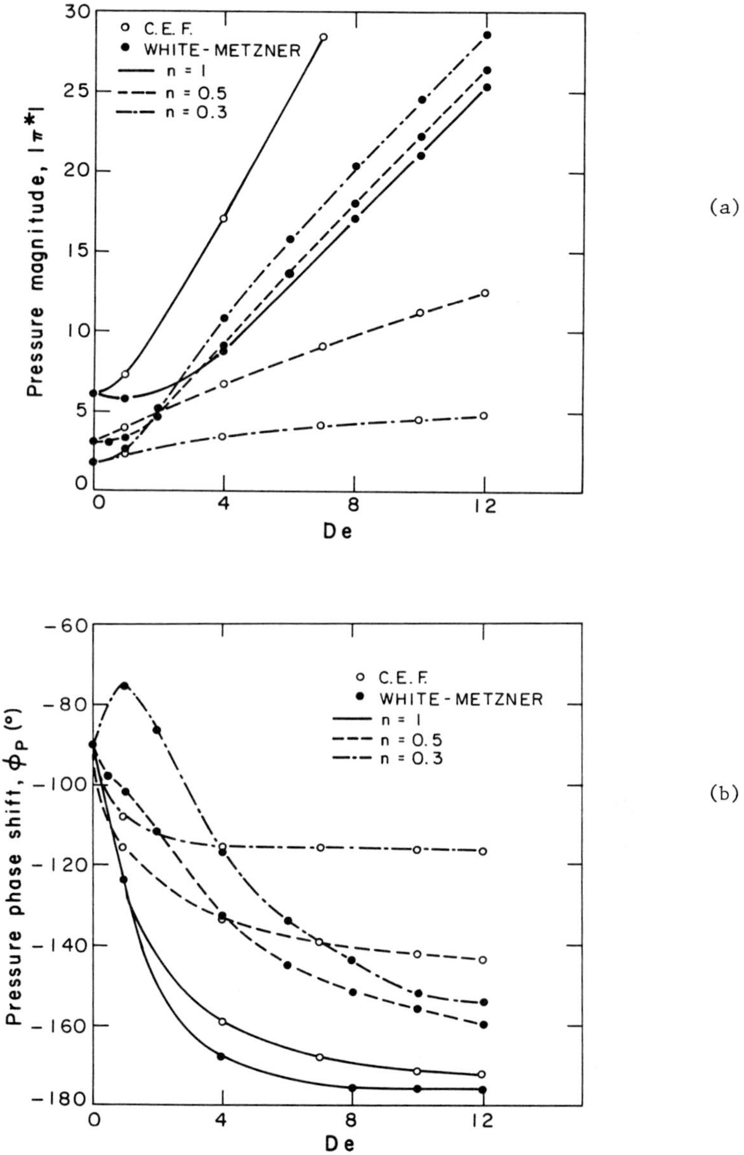

FIGURE 14. Dependence on De of (a) the amplitude of the pressure oscillation $|\pi^*|$ and (b) the phase shift ϕ_p of the pressure field for the CEF and White-Metzner models with viscosity having a power-law index n (cf. Eq. 5.2). The second-order fluid and convected Maxwell models are the limiting cases of these, respectively, for n = 1.

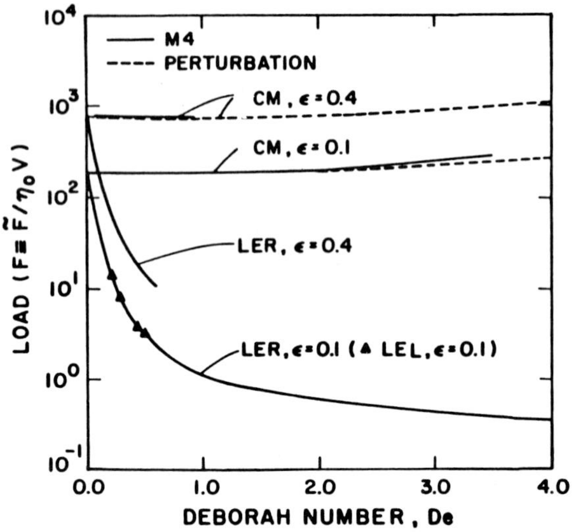

(a)

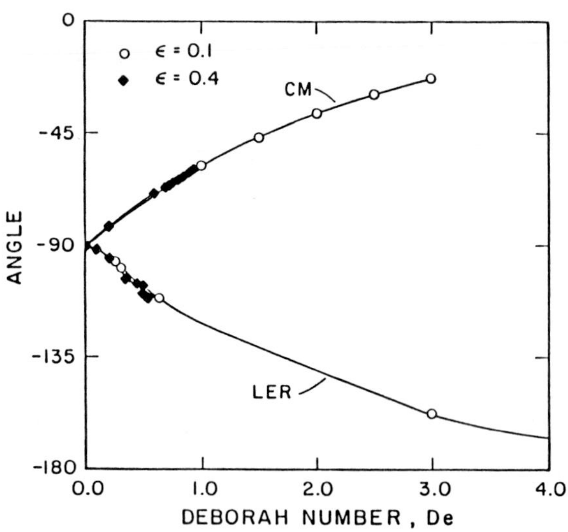

(b)

FIGURE 15. Dependence on De of (a) the magnitude of the dimensionless force exerted by the fluid on the inner cylinder and (b) the direction in which the force acts. Results are compared for the convected Maxwell model and the Giesekus model ($\alpha = 0.5$, $\lambda_2 = 0.001$); the dimensionless gap is $\mu = 0.1$. Note that the load scale in (a) is logarithmic so that the load for the Giesekus model decreases very rapidly with increasing De.

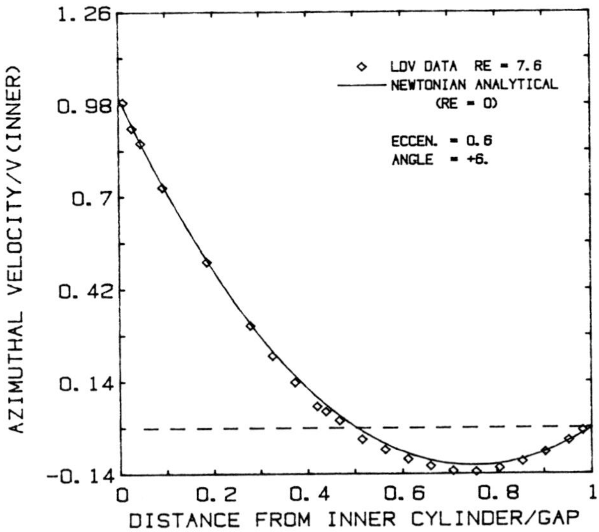

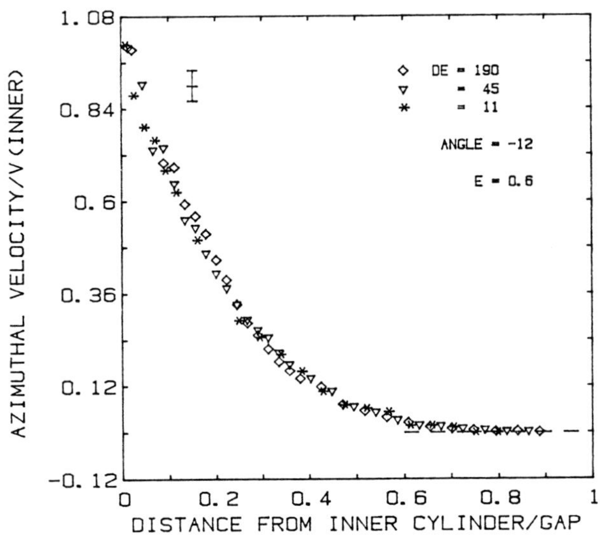

FIGURE 16. Measured values of the tangential velocity as a fucntion of radial position for (a) a Newtonian fluid (60% glycerin/40% water) and (b) a 1.5% aqueous polyacrylamide solution. Velocity is normalized by the inner cylinder speed; and radial position, by the gap. Measurements were made with a laser doppler velocimeter. The angles at which measurements were made are indicated and are measured from the center of the outer cylinder. The eccentricity was $\varepsilon = 0.6$, and the gap $\mu = 0.089$.

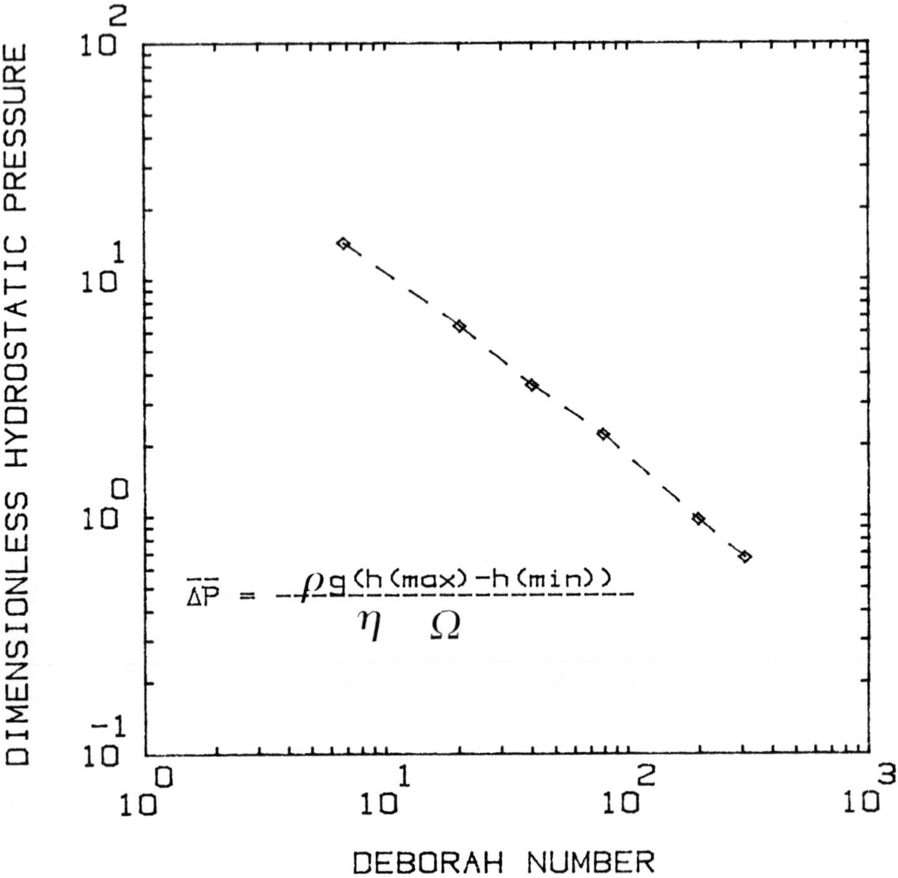

FIGURE 17. Dependence of twice the amplitude of the hydrostatic pressure oscillation on De for the same 1.5% polyacrylamide solution as in Fig. 16.

variation in the hydrostatic pressure defined by

$$\overline{\Delta p} = \frac{\rho g(h_{max} - h_{min})}{\eta\Omega} \tag{5.6}$$

where h_{max} and h_{min} are the maximum and minimum levels of the free surface above an arbitrary horizontal reference plane. Increasing De is seen to dramatically lower $\overline{\Delta p}$ which is consistent only with the LER calculations. In addition we find experimentally that the minimum pressure shifts upstream with increasing De, and this is predicted correctly by both the LER and WM models.

6. Conclusions

We have shown by the eccentric cylinder flow example how complex flows can complement the use of material functions in the evaluation of constitutive equations. Not only do we learn about the material behavior in complex, mixed kinematics and the ability of the constitutive equation to predict these, but also we have a useful method for quickly screening constitutive equations. This screening is illustrated in Table 5 which summarizes our computational and experimental comparisons for the journal bearing. Although the journal bearing is not primarily a shear-free flow, it nevertheless distinguishes clearly among models that differ more in steady elongational flow predictions than in steady shear flow.

We emphasize that this sort of comparison does not replace the need for extensive material function measurements but is an additional tool that should be used. As optical techniques for measuring velocity and stress fields become better developed and computational methods improve to allow use of more realistic constitutive equations, we can expect to see more quantitative constitutive equation evaluations and determination of material functions.

TABLE 5. COMPARISON OF OBSERVATIONS WITH MODEL CALCULATIONS

CONSTITUTIVE EQUATION	MATERIAL FUNCTIONS		CHANGE WITH INCREASING DEBORAH NUMBER		
	SHEAR FLOW	SHEAR-FREE FLOW	SIZE OF FLOW SEPARATION	PHASE SHIFT OF MINIMUM PRESSURE	MAXIMUM PRESSURE DIFFERENCE
CONVECTED MAXWELL	η, Ψ_1 CONSTANT	$\bar{\eta}$ UNBOUNDED	SHRINKS	DOWNSTREAM	INCREASES
CEF	$\eta,\ \Psi_1$ EXACT	$\bar{\eta} = 3(\eta + \Psi_1 \dot{\varepsilon})$	GROWS	DOWNSTREAM	INCREASES
WHITE-METZNER	η EXACTLY Ψ_1 APPROXIMATELY	$\bar{\eta}$ UNBOUNDED	DOWNSTREAM	UPSTREAM	INCREASES
GIESEKUS	$\eta,\ \Psi_1$ QUALITATIVELY	$\bar{\eta} \rightarrow 2\eta/\alpha$	DISAPPEARS	UPSTREAM	DECREASES
EXPERIMENTS	STEADY & SMALL-AMPLITUDE	NOT AVAILABLE	DISAPPEARS	UPSTREAM	DECREASES

REFERENCES:

1. H. Janeschitz-Kriegl, Polymer Melt Rheology and Flow Birefringence, Springer-Verlag, Berlin (1983).

2. H. Kramer and J. Meissner, in Rheology: Vol. 2 Fluids (G. Astarita, G. Marrucci, and L. Nicolais, Eds.), Plenum Press, New York (1980), pp. 463-468.

3. R. B. Bird, R.C. Armstrong, and O. Hassager, Dynamics of Polymeric Liquids Vol. 1: Fluid Mechanics, Wiley, New York (1977).

4. H. Giesekus, J. Non-Newtonian Fluid Mech., **11**, 69-109 (1982); **12**, 367-374 (1983); Rheol. Acta, **21**, 366-375 (1982).

5. A.I. Leonov, Rheol. Acta, **15**, 85-98 (1976); A.I. Leonov, E.H. Lipkina, E.D. Paskhin, and A.N. Prokunin, Rheol. Acta, **15**, 411-426 (1976).

6. T. Raible, Ph.D. Thesis No. 6751, ETH, Zurich (1981); Material und Technik, **10**, 159 (1982).

7. H.M. Laun, J. Rheol., "Prediction of Elastic Strains of Polymer Melts in Shear and Elongation", to appear (1985).

8. H. Munstedt, J. Rheol., **23**, 421-436 (1979); **24**, 847-867 (1980).

9. A. Beris, R.C. Armstrong, and R.A. Brown, J. Non-Newtonian Fluid Mech., **13**, 109-148 (1983).

10. A.N. Beris, Ph.D. Thesis, Massachusetts Institute of Technology, Cambridge, Massachesetts, 1985.

11. A.N. Beris, R.C. Armstrong, and R.A. Brown, J. Non-Newtonian Fluid Mech., **16**, 141-172 (1984).

12. H. Giesekus, Rheol. Acta, **3**, 59-71 (1963).

13. R.I. Tanner, Phys. Fluids, **9**, 1246-1247 (1966).

14. R.R. Huilgol, SIAM J. Appl. Math., **24**, 226-233 (1973).

15. J.V. Lawler, Sc.D. Thesis, Massachusetts Institute of Technology, Cambridge, Massachusetts 1985.

16. S.J. Muller, Sc.D. Thesis, Massachusetts Institute of Technology, Cambridge, Massachusetts 1985.

Acknowledgments

The authors gratefully acknowledge support of this work by the Office of Naval Research and the National Science Foundation.

FINITE ELEMENT METHODS FOR VISCOELASTIC FLOW

David S. Malkus

Mathematics Research Center and
Engineering Mechanics Department
University of Wisconsin-Madison
Madison, Wisconsin

ABSTRACT

Finite element methods to compute approximate solutions to flow problems involving the flows of viscoelastic fluids are discussed. The primary goals of such investigations are at least three: First, to evaluate the predictions of the many proposed constitutive theories for viscoelastic fluids. Second, to model measurement flows in various rheological measurement devices in order to quantify the deviation of the actual flow from the flow which must be presumed to interpret the measurement. Third, it is hoped that these methods will prove sufficiently robust to allow the simulation of idealized polymer processes with the aim of aiding in the design of such processes and the required apparatus. The focus of the current research of the author and a growing number of others is on two-dimensional, isothermal, steady flows of incompressible fluids. While these restrictions will be seen to be non-essential in theory, even the simplest calculations of non-viscometric flow solutions will be seen to require a high degree of computational complexity in practice. Nevertheless, the current finite element procedures seem to show promise in the continuing endeavor to understand this challenging class of problems.

INTRODUCTION

The idea to apply finite element methods to viscoelastic fluid flow problems, as opposed to finite difference methods, seems to have stemmed from the desire to model a variety of industrial polymer processes, which were often characterized by complex flow geometries. In the early work of Tanner, Nickel, and Bilger [1], one gets the clear idea that the hope was to develop a rather general-purpose code which could be used by the industrial rheologist to tackle geometrically complicated flow situations, particularly those flows like the extrusion of a jet of liquid material, in which the complex geometry itself was unknown a priori. Finite elements seem to be ideal tools for such modelling because of their geometric flexibility. Since ref. 1 appeared, it has been discovered that there were a number of unanswered questions about the modelling of viscoelastic fluids, which are even more basic than the choice of discretization scheme. Many of the problems currently being solved by this author and others are of an idealized nature and involve little geometric complexity, so that they could just as easily be attacked by finite differences, and the questions to which we seek answers could as well be asked about difference methods or element methods. There are a number of investigators who have, in good fluid-mechanical tradition, opted for finite difference methods [2-4]. Much of what is said here could equally apply to the difference approach, but in order to limit the present discussion to the manageable, only finite element methods will be discussed here.

The reader should be forewarned that the author's own predilection is for the development of general purpose methods which can be applied to broad classes of viscoelastic fluid flow problems in complex geometries without reprogramming. His hope is that once the basic difficulties involving choice of constitutive equation, convergence of nonlinear iteration methods, the effect of singularities of unkown order — and others still — are more firmly in hand, the vision on ref. 1 may be more completely realized. In the meantime, the author and other researchers have tended to solve problems which do not tax the geometric flexibility of the finite element formulation (and consequentially, do not illustrate its power). These are flows which might be classified as "measurement flows" in many cases — very well-characterized flows which can be established in laboratory situations for the purposes of making measurements of material properties. The lack of geometric complexity leads to flows which can be idealized as flows for which the equations of motion can be solved analytically; however, the real flow is some perturbation of the idealized flow: Cone-and-plate [5] flow actually has free surfaces, though it is observed that this does not lead to an unacceptable disturbance in the assumed form of the flow-field, at low enough shear-rate. The entrance into a slit-die [5] perturbs the flow-field in the entry region, but the velocity field and pressure gradient soon readjust themselves to those characteristic of channel flow, at least at low enough shear-rate. There are many other such flows, including plane flow over a transverse slot, discussed below. The common characteristic of such flows is that they are perturbations of known flows; we know rather well what qualitative features the numerical results must display. On the other hand, the effect the perturbation of the assumed flow may have on the measurements, particularly at the extreme range of shear-rate, is a question which seems to be appropriately addressed by numerical modelling. So such flows seem to be modest testing grounds for the methods under discussion here, yet also seem to raise questions to which numerical modelling may be able to supply useful answers.

We will see that investigations of such problems, humble though they be compared to what might be envisaged, raise thorny questions which are still only partially resolved. The questions seem to have at their heart the question of what constitutive equation should be used and why. The choice of constitutive equation will be seen to govern a basic decision about the kind of finite element method which must be used. It will be observed that different constitutive equations can lead to radically different behavior of the numerical method, yet there seems to be little to differentiate between them on physical grounds.

CONSTITUTIVE EQUATIONS

Perhaps the most basic choice in the design of a numerical method devolves from the fact that there are two seemingly different classes of constitutive equations which have been proposed by rheologists and which seem to be numerically tractable: the *differential* and the *integral* [6-9]. What is meant specifically by these terms will be defined shortly; the important thing to note is that these are intersecting classes of constitutive equations, yet neither class seems to entirely contain the other. At first sight, the two classes of constitutive equations seem to demand totally different numerical methods, forcing the numerical modeller to make an irrevocable choice of methodology at the outset. An important point to be observed here is that there is more in common between the two classes of equations than first meets the eye, and thus the potential exists for a more unified numerical approach than is the current practice.

General Differential Constitutive Equation

$$\begin{aligned} \boldsymbol{\sigma} &= -p\mathbf{I} + 2\mu(0)R\dot{\mathbf{e}} + (1-R)\boldsymbol{\Sigma} \\ \boldsymbol{\Sigma} &= \sum_{k=1}^{N} \boldsymbol{\tau}_k \end{aligned} \tag{1}$$

where $\mu(0)$ is the zero-shear viscosity, $\dot{\mathbf{e}}$ the strain-rate tensor, $R < 1$ is a retardation ratio defined by $R = \Lambda/T$ for a characteristic relaxation time, T, and retardation time, $\Lambda < T$, and p is the hydrostatic pressure function. The terms on the right-hand side of eq. (1) may be interpreted as an isotropic contribution to the stress, the Newtonian contribution to the stress — possibly arising from a Newtonian solvent — and the viscoelastic contribution to the stress, respectively.

The "extra stress" tensor, $\boldsymbol{\Sigma}$, determines the non-Newtonian contribution to the total stress tensor. In a differential constitutive equation, it is determined by one auxiliary equation for each of the "partial stresses," $\boldsymbol{\tau}_k$ [8],

$$\begin{aligned} \lambda_k \frac{\Delta \boldsymbol{\tau}_k}{\Delta t} + \mathbf{B_k} &= \mathbf{0} \\ \frac{\Delta \boldsymbol{\tau}}{\Delta t} \equiv \frac{\partial \boldsymbol{\tau}}{\partial t} &+ (\mathbf{v}\cdot\nabla)\boldsymbol{\tau} - \boldsymbol{\tau}(\nabla\mathbf{v})^T - (\nabla\mathbf{v})\boldsymbol{\tau} \end{aligned} \tag{2}$$

where λ_k is the kth relaxation time in a relaxation spectrum (which may be infinite). The T mentioned above can be taken to be a mean relaxation time in the spectrum. $\mathbf{B_k}$ is a tensor function which may depend on velocity gradients and stresses(in a possibly nonlinear fashion). Specific examples will be cited below. Note that the upper convected time derivative, $\frac{\Delta\tau}{\Delta t}$, has unknown stresses multiplying unknown velocity derivatives, and problems with such constitutive equations are inherently nonlinear, even for the simplest, linear choices of $\mathbf{B_k}$. With the obvious correspondence of notations, eqs. (1) and (2) are in the same form as the corresponding equations in ref. 6, generalized slightly in order to allow the possibility of more than one partial stress.

General Single-Integral Constitutive Equation

The single-integral designation is intended to make a distinction between the constitutive equations under consideration here and those involving multiple, iterated integrals. In fact, the theories described here may involve sums of more than one "single integral." While it makes perfect sense to consider transient flows with single-integral constitutive equations, the computational complexity implied by such flows is such that a new generation of bigger, faster computers would probably be required. Here we consider steady flows only, whereas no such restriction was imposed on our discussion of differential constitutive equations. The assumed general form of the stress field is given by

$$\boldsymbol{\sigma} = -p\mathbf{I} + 2\mu(0)R\dot{\mathbf{e}} + (1-R)\boldsymbol{\sigma}' \tag{3}$$

where the symbols in the first two terms are the same as those in eq. (1). The extra stress in eq. (3) is given by $\boldsymbol{\sigma}'$, which plays an analogous role to $\boldsymbol{\Sigma}$ in eq. (1). For the integral constitutive equations, however, the extra stress is given by an explicit calculating formula, rather than implicitly, as in eq. (1). It is assumed to have the form

$$\boldsymbol{\sigma}' = \sum_{l=1}^{M} \int_{-\infty}^{0} \mathbf{S_0}^{(l)}(\tau) m_l(\tau)\, d\tau \tag{4}$$

where m_l is a memory function, and $\mathbf{S_0}^{(l)}$ is a strain-measure. The memory function $m_l(\tau)$ has the following form in most examples of which we are aware:

$$m_l(\tau) = T^{-1} \sum_{k=1}^{N_l} G_k^{(l)} p_k^{(l)} exp(\frac{\tau p_k^{(l)}}{T}) \tag{5}$$

The constants $G_k^{(l)}$ and $p_k^{(l)}$ whether chosen experimentally or theoretically, must be chosen so that $\int_{-\infty}^{0} q(\tau) m_l(\tau)\, d\tau < \infty$, for all polynomials, q, of arbitrary degree. This is automatic unless $N_l = \infty$, which does occur in practice. In fact, one can see that by interchanging the order of integration in eq. (4) with the summation in eq. (5), the stress contribution of each of the single integrals in eq. (4) becomes a sum of partial stresses, each associated with a member of the relaxation spectrum, T/p_k, each given by integration of a strain-measure against a single exponential. Thus T/p_k in eq. (4) and λ_k in eq. (2) are analogous relaxtion times, and $\boldsymbol{\Sigma}$ and $\boldsymbol{\sigma}'$ are analogously expressed as sums of partial stresses.

For some differential constitutive equations, the relation between them and a rearrangement of eq. (4) into partial stress form is more than an analogy; the integral constitutive equation amounts to a direct solution of the hyperbolic equations (2) for a fixed velocity field along the streamlines, which turn out to be a double family of real characteristics [11]. Examples of this will be pointed out subsequently. It is important to note that this correspondence between integral and differential constitutive equations only holds in very special cases. There are constitutive equations derived on physical grounds, which have an integral form, yet cannot be identified as characteristic solutions of any differential constitutive equation given by eq. (2). Likewise, there are differential constitutive equations which do not seem to yield to closed-form solutions along the characteristics in the form of eq. (4).

The strain measures in eq. (4) are assumed to have the following dependencies:

$$\mathbf{S_0}^{(l)} = \mathbf{S_0}^{(l)}(\mathbf{E_0}(\tau), \mathbf{E_0}^{-1}(\tau), \nabla\mathbf{v}[\mathbf{x}(\tau)], \nabla\mathbf{v}[\mathbf{x}(0)]) \tag{6}$$

$\mathbf{x}(0)$ is the position at which $\boldsymbol{\sigma}'$ is evaluated in the current configuration of the fluid. $\mathbf{x}(\tau)$ is that particle's position at time τ in the past, and $\mathbf{v}$ is the velocity field carrying $\mathbf{x}(\tau)$ to $\mathbf{x}(0)$, inducing the stress field. $\mathbf{E_0}(\tau)$ is a deformation-gradient measuring the strain in the deformation at time τ relative to time 0, induced by $\mathbf{v}$. $\mathbf{E_0}(\tau)$ may be the usual $\frac{\partial\mathbf{x}(\tau)}{\partial\mathbf{x}(0)}$ or something similar but somewhat different. It can be chosen to correspond to various convected derivatives in differential models [8,9] and can physically represent the strain due to molecules which undergo motion relative to the motion of the continuum (non-affine motion) [9,12].

To compute the deformation-gradient, it is neccessary to construct the path $\mathbf{x}(\tau)$, which is a streamline; the gradient is a measure of strain at the historical time and position along that path and the present position. The following equations describe the path and strain:

$$\begin{aligned}
\dot{\mathbf{x}}(\tau) &= \mathbf{v}[\mathbf{x}(\tau)] \\
\mathbf{x}(0) &= \mathbf{x_0} \\
\dot{\mathbf{E}}_0(\tau) &= \mathbf{F}\left(\mathbf{x}(\tau), \nabla\mathbf{v}[\mathbf{x}(\tau)]\right)\mathbf{E_0}(\tau) \\
\mathbf{E_0}(0) &= \mathbf{I}
\end{aligned} \tag{7}$$

where $tr\mathbf{F} = 0$ for all $\tau < 0$. Note that this is a system of equations to be solved in reverse time. The first group is possibly nonlinear and determines the streamline; the second group is a linear system whose non-constant coefficient is determined by the streamline solution. It is also not hard to deduce that $\mathbf{E_0}(\tau)$ is the identity in the present configuration and maintains $det(\mathbf{E_0}(\tau)) = 1$ for all historical time [10].

Examples:

1. Differential only:

   (a) Phan-Tien/Tanner (PTT) [6]

   $$\begin{aligned}
   &N = 1,\ T = \lambda,\ \eta = (1 - R)\mu(0) \\
   &\mathbf{B} = \mathbf{B_1} = \boldsymbol{\tau} exp\{\frac{\lambda\epsilon}{\eta} tr\boldsymbol{\tau}\} - 2\eta\dot{\mathbf{e}} + \lambda\xi(\dot{\mathbf{e}}\boldsymbol{\tau} + \boldsymbol{\tau}\dot{\mathbf{e}})
   \end{aligned}$$

   ξ is a non-dimensional parameter of the model, as is ϵ; When the latter parameter is not zero, there is no known integral form of the PTT equation.

   (b) Leonov [6], White-Metzner, etc. [7].

2. Integral only:

   (a) Doi-Edwards [13]

   $$\begin{aligned}
   M = 1,\quad N &= \infty,\quad p_k = (2k-1)^2 \\
   G_k = C/Tp_k,\quad \mathbf{F} &= \nabla\mathbf{v},\ C = \frac{96\mu(0)}{\pi^4} \\
   \mathbf{S_0}(\tau) &= \phi_1(I, II)\mathbf{C_0^{-1}}(\tau) + \phi_2(I, II)\mathbf{C_0}(\tau) \\
   \mathbf{C_0}(\tau) &= \mathbf{E_0}^T(\tau)\mathbf{E_0}(\tau)
   \end{aligned}$$

The scalar functions ϕ_1 and ϕ_2 are functions of the two nontrivial principal invariants, I and II, of $\mathbf{C}_0^{-1}(\tau)$.

(b) Curtiss-Bird [14]

$$M = 2,\ N_1 = N_2 = \infty,\ p_k^{(1)} = p_k^{(2)} = p_k = (2k-1)^2$$
$$G_k^{(1)} = C_1/Tp_k,\ G_k^{(2)} = C_2/p_k^2,\ \mathbf{F} = \nabla\mathbf{v}$$
$$C_1 = \frac{96\mu(0)}{\pi^4(1+2\epsilon/3)},\ C_2 = \frac{\epsilon 192\mu(0)}{\pi^2(1+2\epsilon/3)}$$

$\mathbf{S_0}^{(1)}$ is the same as $\mathbf{S_0}$ of (2a), and $\mathbf{S_0}^{(2)}$ is a more complicated tensor function of $\mathbf{C}_0$, $\mathbf{C}_0^{-1}$, and $\nabla\mathbf{v}(0)$ [10,14]. ϵ is the "link-tension" coefficient, a non-dimensional parameter of the model; with $\epsilon = 0$, the Curtiss-Bird equation reduces to the Doi-Edwards.

3. Both differential and integral:

(a) Oldroyd/Jeffreys (taking $N = 1$ for simplicity, not necessity) [8,9]

$$T = \lambda,\ N = 1$$
$$\mathbf{B} = \boldsymbol{\tau} - 2(1-R)\mu(0)\dot{\mathbf{e}} + \lambda a(\dot{\mathbf{e}}\boldsymbol{\tau} + \boldsymbol{\tau}\dot{\mathbf{e}})$$

Note that taking $\xi = a$ and $\epsilon = 0$ in (1a) yields (3a).

(b) Johnson-Segalman (with a nonzero R) [9,12]

$$T = \lambda,\quad N = M = 1,\quad p_1 = 1$$
$$G_1 = \mu(0)/\lambda,\quad \mathbf{F} = a\dot{\mathbf{e}} - \dot{\boldsymbol{\omega}}$$
$$\mathbf{S_0}(\tau) = \mathbf{E}_0^{-1}(\tau)\mathbf{E}_0^{-\mathbf{T}}(\tau) - \mathbf{I}$$

(3b) is the integral form of (3a) (and thus of a special case of (1a)) [9]. $\dot{\boldsymbol{\omega}}$ is the vorticity tensor. The parameter a is a "slip parameter" in the derivation of (3a), which allows for non-affine motion of a strained molecular lattice with respect to the continuum [9,12].

CHARACTERISTICS AND NUMERICAL METHODS

Steady Plane Flow of a Johnson-Segalman Fluid

The author has performed computations with the fluid described by (3a) and (3b). Since his computations employ the integral form of the constitutive equation, he refers to the model as Johnson-Segalman. It turns out that, for the purposes of analyzing the system of equations to be solved, it is easier to deal with the differential form (3b). The analysis discussed here has only been carried out for this one differential constitutive equation (it does include several simple Maxwell - type equations for particular choices of a [9]). It is hoped that the conclusions hold for other differential models — and because of the analogy alluded to earlier — to integral models as well. In what follows, take $R = 0$ and $N = 1$.

The assumed form of the equations of steady motion is

$$\nabla \cdot \boldsymbol{\sigma} + \mathbf{f} = \rho(\mathbf{u} \cdot \nabla)\mathbf{u} \tag{8}$$

where $\mathbf{u}$ is the (two-dimensional) velocity field, $\boldsymbol{\sigma}$ is the stress field, $\mathbf{f}$ a body force, and ρ the fluid density. At present it is assumed that the flow is a plane flow, and that the material is exactly incompressible,

$$\nabla \cdot \mathbf{u} = 0 \tag{9}$$

For the Johnson-Segalman fluid with parameters as specified, the equations of stress, motion, and continuity take the form

$$\begin{aligned} &\lambda \frac{\Delta \boldsymbol{\tau}}{\Delta t} + \mathbf{B} = \mathbf{0} \\ &\rho(\mathbf{u} \cdot \nabla)\mathbf{u} + \nabla p - \nabla \cdot \boldsymbol{\tau} = \mathbf{0} \\ &\nabla \cdot \mathbf{u} = 0 \end{aligned} \tag{10}$$

The analysis of ref. 11 begins by observing that in this differential form, eqs. (10) are a quasilinear system, whose characteristics may be given explicitly in terms of a solution. The analysis of ref. 13 reveals that *below a critical stress*

I. For a fixed $\mathbf{u}$, the stress equations are *hyperbolic*.

II. Streamlines are a double family of real characteristics of the stress equations.

III. Solution for $\boldsymbol{\tau}$ by the method of characteristics yields a problem which is *elliptic* in $\mathbf{u}$. The stress is given by an integral stress calculator of the form of eq. (4).

It should be noted that the sense in which the equations of motion are elliptic must be carefully qualified: There are no real characteristics of the equations of motion alone, below the "sonic transition" (critical stress). But since there may be characteristics entering through boundaries at inflows (with corresponding outflows), stress data — or the equivalent — must be somehow specified there. So the whole problem is not elliptic in the usual sense, when characteristics cross domain boundaries.

There is a way to make the equations of motion elliptic in the usual sense when a given stress field is specified, and that is to rewrite eq. (10) in equivalent form,

$$\begin{aligned}
&\lambda \frac{\Delta \tau}{\Delta t} + \mathbf{B} = \mathbf{0} \\
&\rho(\mathbf{u} \cdot \nabla)\mathbf{u} + \nabla p - \mu \nabla^2 \mathbf{u} = \nabla \cdot \boldsymbol{\tau} - \mu \nabla^2 \mathbf{u} \qquad (10a) \\
&\nabla \cdot \mathbf{u} = 0
\end{aligned}$$

in which μ is a positive, viscosity-like constant. In an iterative solution method which uses the right-hand side of the middle equations of eqs. (10a) from a previous velocity iterate and the stresses calculated from the hyperbolic stress equations in that velocity field, solution for the next velocity iterate can be a truly elliptic problem. This is exploited in some of the methods to which we now turn our attention.

The Spectrum of Discrete Methods

The results of ref. 11 have interesting ramifications for numerical methods for viscoelastic fluids. In what follows, we assume that computations are to be carried out in a steady flow regime below the critical stress for type-change (numerical experiments by the author suggest that with the contitutive equations currently employed, the critical stress is so large as to be essentially inaccessible by current numerical methods). We assume that the characteristics of the equations to be solved for any constitutive equation of the forms considered in this paper are the same as they are in the Johnson-Segalman case. While the analysis has not been carried out in such generality, this assumption does not seem unwarrented in the light of numerical evidence. The methods currently employed by most researchers fall somewhere on the spectrum of methods given in Table 1.

METHOD	MIXED	STRESS ALT. $\mathbf{v},p$		PURE $\mathbf{v},p$	
KEY FEATURE	All at once	Variant 1 Total numerical	Variant 2 Meth. chars.	Numerical Streamline	Analytic Streamline
PDE TYPE	Mixed	Hyperbolic/Elliptic	Hyperbolic/Elliptic	Elliptic	Elliptic
UNKS.	$\mathbf{u},\tau,p$	$\tau \circlearrowright \mathbf{u},p$	$\tau \circlearrowright \mathbf{u},p$	$\mathbf{u},p$	$\mathbf{u}$
REFS.	[7],[15], [16],[17]	[18]	[18]	[7], [19], [20]	[7], [10]

Table 1

Spectrum of Finite Element Methods for Viscoelastic Flow.

At the extreme left end, there is a class of methods applicable, in principle, to any constitutive equation which has a differential form. These methods treat the quasilinear system as a system of mixed type and discretize velocity, pressure, and extra stress independently in a mixed/Galerkin method. At the extreme right end of the spectrum are methods which apply to those constitutive equations which have an integral form. They discretize only trial velocity fields in an iterative sequence of approximations to the solution of the equations of motion and continuity; even the pressure can be eliminated if a penalty approach is used. In these right-wing methods, stress is computed via the integral stress calculator of eq. (4) as a functional of the trial velocity fields. Each of the extreme right- and left-wing methods have their own distinct advantages and disadvantages. Left-wing, mixed methods need many stress unknowns — particularly if more than one relaxation time and corresponding partial stress must be discretized. Left-wing methods thus involve large matrices with large bandwidths, but transient problems can be solved by well-known time-stepping methods. Left-wing methods also require the specification of boundary data where characteristics emanate from boundaries. How to specify such data for partial stresses is a subtle matter [11], and the possiblity of over or under specifying boundary data is real one. On the other hand, left-wing methods give the Galerkin residual in terms of stress unknowns, which are primary variables; thus differentiation of the Galerkin residuals with respect to nodal variables as required to form Newton's method for solving the discrete nonlinear Galerkin equations is straight-forward.

The right-wing methods make some direct trades of advantage for disadvantage when compared to the left-wing methods. Only velocities need be discretized; computational work is not increased by adding more relaxation times. The specification of stress on incoming characteristics can be handled implicitly, in a way that seems to work rather well, if not theoretically justified at present. The incoming streamlines are assumed to be those of a known flow extending to infinity in a known geometry (such as channel flow), and thus the required contributions to the integrand of eqs. (4) and (11) are actually known analytically once the particle enters this "predecessor flow" [10,19]. On the other hand, transient methods seem nearly impossible, and Newton's method seems difficult, at best, because of the complicated functional relation between the stresses in the Galerkin residual and the velocity field. In most cases, it seems that the choice of left-wing and right-wing method is made a-priori by the form of the constitutive equation; integral models seem to demand right-wing methods and differential left-wing methods.

The middle-of-the-road methods in Table 1, at first sight, seem applicable only to differential models and thus seem to be left-wingers at heart, but on closer examination, they point the way to a unification of the two extremes. The middle-of-the-road methods are currently under development by Tanner and coworkers [18], and Variant 2 currently exists only as a boundary-integral method, but its obvious extension to a finite element method make it appropriate to this discussion. Both Variants 1 and 2 begin with an estimated velocity field and solve alternate hyperbolic and elliptic problems for stress and velocity, using ideas related to the obsevation of eqs. (10a). Variant 1 uses a discrete hyperbolic method to solve for the stress, and Variant 2 uses the fact that, in the current velocity iterate, the streamlines are characteristics, and thus the stress equations can be reduced to ordinary differential equations along them. These methods do not alleviate the drawbacks of the extreme methods, but rather represent a different balance of trade-offs. They require separate extra stress calculation, but fewer unkowns in each phase than the left-wing methods. Each added extra stress requires an extra pass through the stress-equation solver, not more simultaneously active unknowns. The middle-of-the-road methods require hyperbolic data, but Newton's method and transient analysis do not seem as forbidding as with right-wing methods.

However, the most important aspect of the middle-of-the-road methods is the bridge between the extreme methods they illuminate. Combination of Variant 2 and the analytic streamline version of the right-wing method seem to show real promise of allowing the development of a "supermethod," applicable to either differential or integral models, which shares as much code as possible in the computations with either kind of constitutive equation. The common portion of the code constructs the mesh and guides the nonlinear iterations (among other functions) and constructs streamlines by the procedures already in place in the analytic streamline version of the right-wing methods. Depending on whether the constitutive equation is integral or differential, a stress-integral evaluator or a stress ODE integrator is plugged in to evaluate the stress along the streamlines. The supermethod will take some time to develop, since solving hyperbolic equations numerically is not a trivial task. For example, one can forsee some difficulty associated with the specification of initial/end conditions for the stress ODEs on the streamlines. For incoming streamlines, stress can probably can be specified as data, but on closed streamlines, periodic end-conditions seem appropriate. How to do this without excessive "shooting" is not clear at present. But the possibility of the existence of a supermethod points out that the fundamental difference between integral and differential forms is only the question as to whether a closed-form quadrature is known for the stresses or integration of ODEs along the streamlines is required.

An alternative way to look at Table 1 is as an indicator of the degree of "Lagrangianess" of the various methods. From that point of view, the left and right wings of the table do not define extremes any longer. One may properly consider the left-wing methods to be totally Eulerian in that all quantities solved for are referred to fixed spatial locations and not to particles. as we move rightward on the table, streamlines — and thus particle paths — begin to play an increasingly important role. But the right-wing methods still retain the equations of motion in Eulerian form and are only partially Lagrangian in their need to track particles in order to compute stresses. At the far right wing of Lagrangianess is found Hassager's method [21], which uses a deforming mesh. This method has some drawbacks — particularly in flows with recirculations — but the method is worthy of note because it seems to be the only method currently employed which can do transient analysis with integral constitutive equations.

COMPUTATION WITH A SINGLE-INTEGRAL MODEL

It seems appropriate to present some numerical results in this paper, but in so doing we must of necessity limit our consideration to those results with which the author is most familiar. This means that we will look at the extreme right-wing method of Table 1, applied to a flow which is essentially a measurement flow. We first turn our attention to an overview of some of the problems associated with the implementation of a particle-tracking, pure $\mathbf{v}, p$ method. More details may be found in refs. 7 and 10.

Evaluating the History-Integral

The most basic problem to be dealt with in devising a numerical method for integral constitutive equations is the numerical approximation of the history-integrals in the stress calculator, eq. (4). Consider a typical memory function, $m(\tau)$; the most efficient way to compute the history-integral, presuming the components of the strain measure are known, is via a Gaussian formula,

$$\int_{-\infty}^{0} \mathbf{S}_{\mathrm{O}}(\tau) m(\tau)\, d\tau \approx \sum_{k=1}^{N_p} \omega_k \mathbf{S}_{\mathrm{O}}(\tau_k) \tag{11}$$

The formula is in the form of a weighted sum with weights, ω_k, of the integrand evaluated at discrete points in historical time, τ_k. For a single relaxation time, or $N = 1$ in eq. (5), the Gaussian formula is given by the well-known Laguerre family of formulas. For other memory functions (assumed to be sums of exponentials here), analogous formulas can be derived [10]. It should be pointed out that the number, N_p, of integration points is not explicitly related to the number, N, of relaxation times (which is often infinite). This is in distinct contrast to the situation with differential constitutive equations, where each relaxation time adds a new partial stress and a new set of stress equations to be solved. With the integral form, adding relaxation times need not increase N_p and the corresponding computational effort. Experience shows, however, that if adding relaxation times significantly spreads out the relaxation spectrum, an increased N_p may be required to maintain accuracy.

Evaluating the Strain

The author has chosen a basic approach to solving the tracking equations (7) which involves the choice of finite elements for which the solution for the particle pathline (streamline) is known exactly in piecewise fashion as it crosses each element. For this purpose, the quadrilateral element composed of four linear triangles defining the diagonals of the quadrilateral have been chosen; these are illustrated in Figure 1. The composite pattern is chosen to produce accurate satisfaction of the continuity equation with an element of sufficient over-all accuracy. Linear triangles in other arrangements will not do this. In order to compute the integrand of eq. (11), we observe that if a pathline is known, the strain measures of eq. (6) are readily computed from nodal variables and assumed interpolations if $\mathbf{E}_0(\tau)$ is known. $\mathbf{E}_0$ is the solution to the evolution equations for the deformation-gradient, which are part of eqs. (7). This solution is known piecewise analytically on each element of the type which is employed. We proceed as follows: We define $\mathbf{E}_0(\tau_1)$ to be the solution to the strain evolution equations back to historical time τ_1, at which the particle being tracked is located at the boundary of some triangle in the mesh. We define $\mathbf{E}_{\tau_1}(\tau)$ to be the solution to the following strain evolution equation, which is derived from eq. (7) by modifying the initial condition to an interface condition:

$$\begin{aligned} \dot{\mathbf{E}}_{\tau_1} &= \mathbf{F}\mathbf{E}_{\tau_1} \\ \mathbf{E}_{\tau_1}(\tau_1) &= \mathbf{I} \end{aligned} \tag{12}$$

It turns out that $\mathbf{E}_{\tau_1}$ can be easily determined analytically for the chosen elements [10]. The whole deformation-gradient back to $\tau < \tau_1$ can then be mutiplicatively accumulated by

$$\mathbf{E}_0(\tau) = \mathbf{E}_0(\tau_1)\mathbf{E}_{\tau_1}(\tau) \tag{13}$$

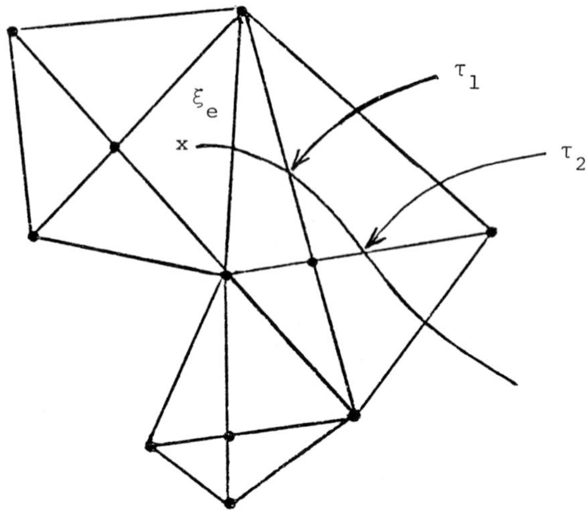

Figure 1
Composite Elements and Particle Paths for Right-Wing Method.

The computation of stress may thus proceed by piecing together streamlines and strain-measures along them analytically. This process is illustrated in Figure 1. Inspection of the discrete Galerkin equation

$$\sum_e \theta_e[\boldsymbol{\sigma}' \cdot \nabla \mathbf{v}^h + 2z(\nabla \cdot \mathbf{u}^h)(\nabla \cdot \mathbf{v}^h) + \rho(\mathbf{u} \cdot \nabla)\mathbf{u}^h \cdot \mathbf{v}^h - \mathbf{v}^h \cdot \mathbf{F}](\boldsymbol{\xi}_e) = 0 \quad (14)$$

shows that, with the computation of stress at a finite number of favored points in the domain, an approximation to the Galerkin residual of the equations of motion can be obtained. Eq. (14) comes from numerically integrating the usual Galerkin residual of the equations of motion (8); θ_e is the area of triangle e in the finite element mesh, $\boldsymbol{\xi}_e$ is the corresponding triangle centroid (spatial integration point). $\mathbf{v}^h$ is an arbitrary Galerkin test function, expanded in the same type of interpolations in which the desired discrete solution, $\mathbf{u}^h$, is expanded. The parameter, z, is the penalty parameter, which allows nearly exact satisfaction of the continuity equation (9) implicitly in the Galerkin formulation [10].

The Stream and Drift Functions

The construction of the particle paths of Figure 1 and the accumulation of the deformation-gradient via eq. (13) are the dominant computational cost of the method presented here. There are several interesting aspects of the chosen finite elements which contribute to the efficiency with which this can be done. Since the interpolating functions determining the test and trial functions on each triangle are linear, it follows that there is a quadratic stream function on each triangle, $\psi(\mathbf{x})$ (it actually turns out to be a quadratic spline on the whole mesh). Thus the boundary crossings illustrated in Figure 1 can be found in sequential triangles upstream of each $\boldsymbol{\xi}_e$ using the quadratic formula,

$$\begin{aligned} \psi(\mathbf{x_1}) &= \psi(\mathbf{x_2}) \\ l(x_2^1, x_2^2) &= 0 \end{aligned} \tag{15}$$

The relation involving $l(\cdot,\cdot)$ is the requirement that $\mathbf{x_2}$ be on the line defining one of the boundaries of the current triangle. Thus eq. (15) is a quadratic equation in one of the components of $\mathbf{x_2}$, either x_2^1 or x_2^2. To associate a historical time with $\mathbf{x_2}$, it turns out that for the chosen elements, the time of transit between two points on the same streamline can be determined from the *drift function*, which is known analytically [10].

$$\tau_1 - \tau_2 = w(\mathbf{x_1}) - w(\mathbf{x_2}) \tag{16}$$

The drift function is denoted by w; its analytic expression involves various cases depending on the characteristics of the pathline equations in eq. (7) [10]. Stress calculation can thus proceed by finding boundary crossing points only and accumulating strain at these points via eq. (13), using the times associated with the boundary crossings computed via eq. (16). The accumulation proceeds until the boundary crossing times in one element, τ_1 and τ_2, bracket one or more temporal integration points of eq. (11), at which time(s) the integrand of eq. (11) can be evaluated using the analytic expression for the strain [10] and eq. (13) with τ taken as the value of the temporal integration point(s). Since there are potentially six solutions to eqs. (15), when each boundary of the current triangle is considered as a possible candidate on which to find $\mathbf{x_2}$, the drift function is also indispensible in determining the actual boundary crossing of a particle as the one which gives the maximum value of τ_2 satisfying $\tau_2 < \tau_1$.

Solving the Nonlinear Equations

Here we briefly discuss the kind of numerical method which can be used to find the zero(s) of eq. (14). One may deduce that the functional dependence of the right-hand side of eq. (14) on the unknown velocity field is so complicated that Newton's method is difficult at best. On the other hand, it has been found that iterative methods based on the finite element approximation to the Stokes operator, which puts all nonlinearity on the right-hand side, do not have very favorable convergence properties, even at moderate Deborah numbers [2,7,19]. The method illustrated here is a Gauss-Newton type which begins from an estimate of the solution, $\mathbf{u_0}$, which represents nodal values of a finite element solution computed using the Stokes operator, but with inflow and outflow boundary conditions appropriate to the non-Newtonian solution (the "pseudo-Newtonian" solution). The iteration matrix is an approximation to the unavailable inverse Jacobian. The inverse Jacobian approximation begins using the Stokes approximation and continues, using rank-one updates, $\mathbf{r_i}$ and $\mathbf{v_i}$, to better approximate the inverse Jacobian as the iteration progresses. This method is known as the "inverse Broyden" method, and it is quite efficient, as it only requires a single matrix elimination proceedure to be carried out on $\mathbf{H_0}$.

$$
\begin{aligned}
&for\ k = 0\\
&\quad \mathbf{H_0} = \text{inverse Stokes' matrix},\\
&\quad \mathbf{u_0} = \text{pseudo-Newtonian solution},\\
&\quad \mathbf{R}(\mathbf{u_0}) = \text{residual of eq. (15) at } \mathbf{u_0}\\
&\quad \mathbf{s_0} = -\mathbf{H_0}\mathbf{R}(\mathbf{u_0})
\end{aligned}
\tag{17}
$$

$$
\begin{aligned}
&for\ k = 0, 1, 2, \ldots\\
&\quad \mathbf{u_{k+1}} = \mathbf{u_k} + \mathbf{s_k},\\
&\quad \mathbf{y_k} = \mathbf{R}(\mathbf{u_{k+1}}) - \mathbf{R}(\mathbf{u_k}),\\
&\quad \mathbf{v_k} = \mathbf{H_k}^T\mathbf{s_k}/(\mathbf{s_k}^T\mathbf{H_k}\mathbf{y_k}),\\
&\quad \mathbf{r_k} = \mathbf{s_k} - \mathbf{H_k}\mathbf{y_k},\\
&\quad \mathbf{H_{k+1}} = \mathbf{H_0} + \textstyle\sum_{i=0}^{k} \mathbf{r_i}\mathbf{v_i}^T,\\
&\quad \mathbf{s_{k+1}} = [1 - \mathbf{v_k}^T\mathbf{R}(\mathbf{u_{k+1}})]\mathbf{r_k},\\
&next\ k.
\end{aligned}
\tag{18}
$$

By rearrangement of eq. (18), it follows that

$$
\begin{aligned}
\mathbf{u_{k+1}} - \mathbf{u_k} &= \mathbf{s_k}\\
\mathbf{s_k} &= -\mathbf{H_k}\mathbf{R}(\mathbf{u_k}),
\end{aligned}
\tag{19}
$$

from which one may deduce that if $\mathbf{H_k}$ were the inverse Jacobian, eqs. (18) would be Newton's method. The inverse Broyden method given by eqs. (18) is a fairly robust method for computations with some constitutive equations, but it diverges at rather low Deborah numbers with others. The reason for this is not entirely clear at this point, but one avenue worth exploring fully is the possibility of finding more accurate approximations to the actual inverse Jacobian, and the author is currently involved in studying this matter.

A MODEL PROBLEM

Plane Flow over a Transverse Slot

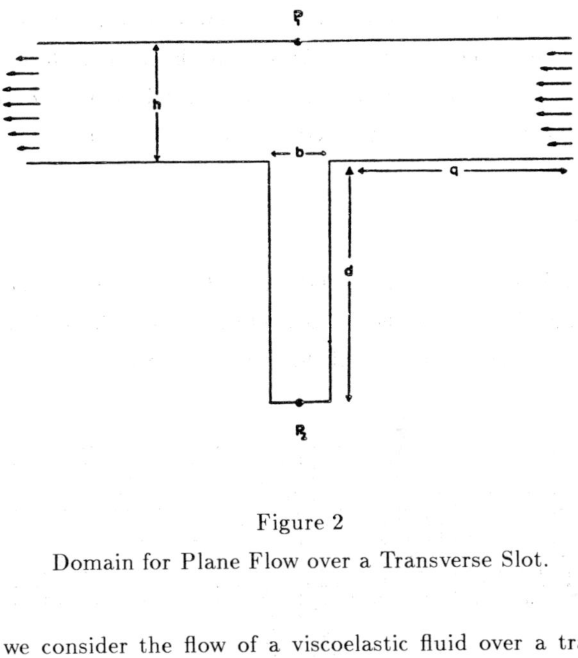

Figure 2

Domain for Plane Flow over a Transverse Slot.

Here we consider the flow of a viscoelastic fluid over a transverse slot [3,4,10,17,20,22-25]. The author [10,26] has been investigating the prediction that at negligible Reynolds number (based on slot width), the difference between the thrust on the wall at the top and bottom of the slot, pictured in Figure 2

$$P_e = P_1 - P_2 \tag{20}$$

is related to the first normal stress difference, N_1, in a simple fashion. This problem has a long history and is discussed in some detail in refs. 3,4,10,17,20, and 22-25. It is of practical importance, because it seems feasible to design measurement devices based on the relation between P_e and N_1, which predict the latter based on continuous measurement of the former and the wall shear-stress in undisturbed flow, σ [10,23,24]. It has been found experimentally that the $P_e - N_1$ relation accurately describes the behavior of various viscoelastic fluids in flows over slots [23,24]. This appears to happen in spite of the fact that the assumptions behind the derivation [22] of eq. (20) are violated [10,23]. The purpose of the investigation undertaken by the author is to explore this apparent anomoly in detail, using his numerical techniques. Significant progress has been made in these investigations since the appearence of ref. 10; however, these results are best presented in a future publication. Here we will focus on some interesting qualitative features of plane flow over a transverse slot and the apparent relation of these qualitative features to the numerically observed hole-pressure. The results will also illustrate the dramatic and somewhat puzzling differences resulting from using different constitutive equations to study the same flow.

Numerical Results

Figure 3 illustrates a mesh of 1008 crossed-triangle macroelements, which has been designed by the author to discretize the domain of Figure 2 and to investigate hole-pressure problem of the previous subsection.

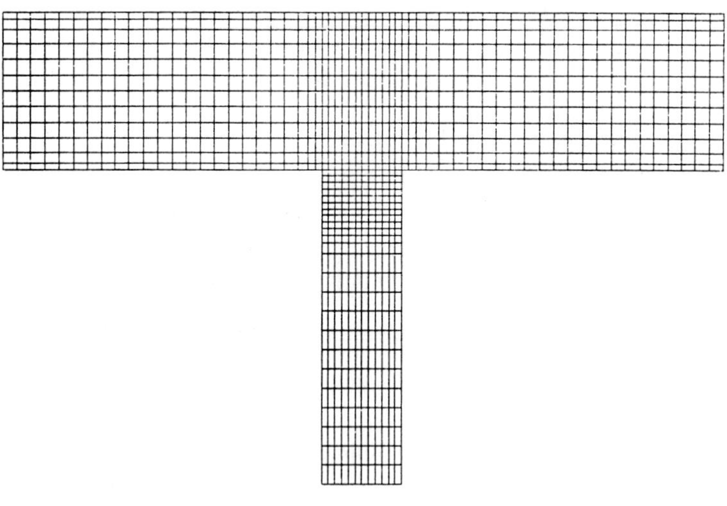

Figure 3
Mesh Discretizing the Domain of Figure 2 with 1008 Macroelements.

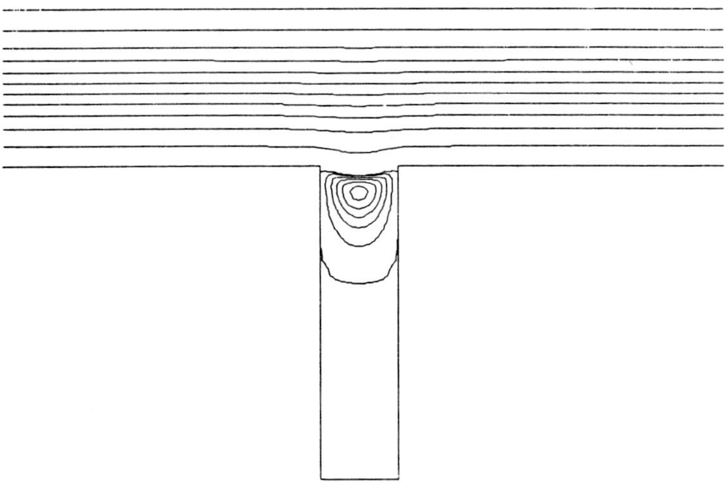

Figure 4

Stokes-flow in the Domain of Figure 2, Using the Mesh of Figure 3.

The distribution of elements was determined from extensive numerical experimentation, and the author believes it produces more accurate results than do any of those of ref. 10. The domain parameters of Figure 2 are $h = 1$, $b = 0.5$, and $q = 2$. Figure 4 shows a plot of the streamlines in Stokes-flow (Newtonian, zero Reynolds number) obtained using this mesh; this will serve as a useful baseline in what follows.

We note that the symmmetry of the streamlines is perfect to numerical accuracy, and any very slight deviations from symmetry in Figure 4 arise from the streamline integration used for graphical purposes [25]. This integration does not make use of the element-wise conic sections on triangles, but an approximate method. In all streamline plots shown here, the streamlines are not equally spaced contours throughout because the vortex in the slot is so faint in comparison to the flow outside. Instead, there are ten equally spaced contours outside the slot, the streamline at the wall in which the slot is cut, and five equally spaced contours in the slot. These latter contours have values which evenly divide the range of values between the streamline at the wall and the maximum value of the stream function.

We turn our attention to the non-Newtonian flow illustrated in Figure 5. Among other things, Figure 5 illustrates the fact that there are some constitutive equations with which rather high shear-rates can be obtained. Two non-dimensional numbers are useful in quantifying the degree of nonlinearity due to non-Newtonian effects,

$$\begin{aligned} D_e &\equiv \dot{\gamma}(T - \Lambda) \\ L_o &\equiv N_1/\sigma \end{aligned} \tag{21}$$

D_e is the "Deborah number" and L_o the "stress-ratio." $\dot{\gamma}$ is the shear-rate at the wall of the die, far from the slot, and N_1 and σ are taken at the same location. For the flow pictured in Figure 5, $D_e = 17.9$, and $L_o = 4.1$; the computation used the Curtiss-Bird equation [10,14], with the same parameters given in ref. 10 for the "EFN Fluid." This is intended to model a high-viscosity, concentrated polymer system without significant branching of the molecular chains. The resulting viscosity is so high ($\approx 10^4$ poise at zero shear) that the Reynolds number is negligible, even when shear-thinning is accounted for.

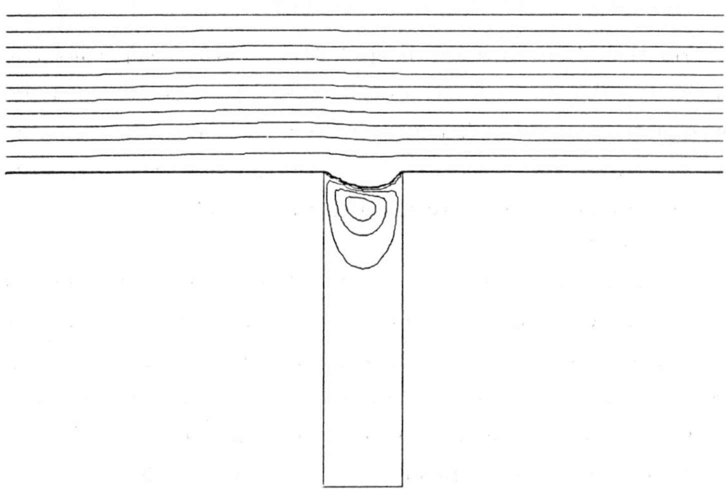

Figure 5

Flow of a Curtiss-Bird Fluid at $D_e = 17.9$.

The computations of ref. 10 were carried out with Gaussian memory formulas (eq. (11)) with ten points; computations were reported there up to $D_e = 7.5$. It has since been determined that the ten-point Gaussian formula loses its accuracy in simple shearing flow (and thus in the undisturbed die-flow) for D_e much larger than that. Furthermore, one observes that the ten-point formula is inaccurate enough to lead to an artificial shear-stress maximum [10,20] for very high D_e. For that reason, the computations leading to Figure 5 were carried out with 18-point formulas for each memory-integral. With the 18-point formula, no difficulty in convergence of the Broyden iterations was observed — no "barrier" or "high Weissenberg number problem" which has been observed for other constitutive equations [2,7,16,20]. The run was terminated when the residual measure of ref. 10 had been reduced by a factor of 84 (a similar reduction factor to those reported in ref. 10). The residual appeared to be continuing to converge, but the computation was terminated because the experience of ref. 10 shows that most quantities of interest have stabilized to a couple of decimal digits with a comparable residual reduction. This residual reduction was accomplished in 18 inverse Broyden iterations. The computation took 15 hours of VAX 11/780 CPU time on a VAX with a standard configuration, using double precision hardware.

There are several interesting features of Figure 5. The flow is from right to left, and we observe substantial vortex distortion of the type observed in ref. 4. We also note that there is little disturbance in the streamlines near the mouth of the slot when compared to those in the Stokes-flow of Figure 4 or the streamlines in the flow visualizations of ref. 4. Also, of the five equally-spaced streamlines in the slot, three are crowded together near the dividing streamline; this indicates that the velocity of fluid in the vortex is lower compared to the velocity at the dividing streamline in the non-Newtonian case than it is in Stokes-flow. This appears to be due to shear-thinning; the viscosity in undisturbed flow at the wall in the case of Figure 5 is only 9% of its zero-shear value. The very high driving stresses in the die are not transmitted to the slot, and the fluid slides by without feeling much effect of the slot.

We turn our attention to the flow of another fluid in the same geometry, using the same mesh as before. The author really would have liked to have shown results involving a Johnson-Segalman fluid at significant D_e, but these are not available. Johnson and Segalman found that a value of their slip-parameter of $a = 0.87$ gave good agreement of viscometric functions with published data [12]. With this value of a, a very high zero shear viscosity (implying negligible Reynolds number), and $R = 0$ in eq. (3), the inverse Broyden iterations of eq. (18) diverge dramatically at about $D_e = 1.0$. Adding a substantial Newtonian viscosity with $R = 1/3$ helps some, as one might expect, but divergence ensues at a disappointing $D_e = 1.5$. In order to obtain an integral constitutive equation which shear thins less than Curtiss-Bird, behaves better than Maxwell, does not imply *very* peculiar lattice motion which would be the result of taking a very small slip-parameter [9], and which allows computation with at least a moderate D_e, the author just tinkered with the Johnson-Segalman equation. Favorable results were obtained by adding a Wagner-type damping function

$$\phi(II) = exp(-c\sqrt{II - 3}) \tag{22}$$

which just multiplies the strain-measure, $\mathbf{S_O}$, of the Johnson-Segalman model. II is the second principal invariant of the *non-affine* strain measure, $\mathbf{E_O^{-1}E_O^{-T}}$ No physical reasoning lies behind this; damping functions are intended to account for the effect of breaking temporary junctions in the lattice under large strain. To what extent this would happen if non-affine motion of the lattice is allowed, or, if so, whether the damping function for a non-affine strain-measure should take the form of eq. (24) is unknown. At any rate, the flow of Figures 6 and 7 use this concocted model, and take $c = 0.1$; the other parameters were as just described for the Johnson-Segalman with a retardation time. The flow pictured in Figure 6 has $D_e = 2.5$, $L_o = 1.8$, and viscosity reduced to 46% of its zero-shear value. The flow of Figure 7 is at $D_e = 3.7$, but $L_O = 1.6$, which is lower than the stress-ratio of Figure 6; in Figure 7, the viscosity is 40% of its zero-shear value. The stress-ratio of the concocted fluid model goes through a maximum, due to the retardation time; the flow of Figure 6 has shear-flow at the wall which is just about at that maximium. 16 Laguerre integration points were used for the quadrature of eq. (11) when computing the flows of Figures 6 and 7.

In both the flows of Figure 6 and Figure 7, the Broyden iteration scheme had a tougher time converging than it did with the Curtiss-Bird fluid at a

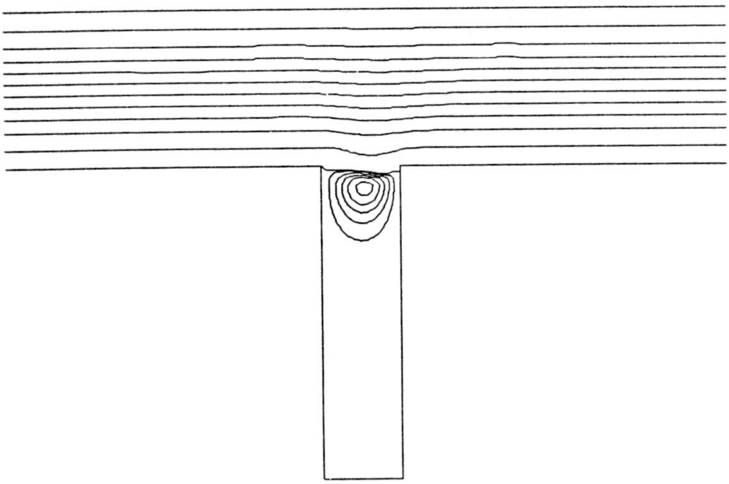

Figure 6

Flow with the Concocted Constitutive Equation at $D_e = 2.5$.

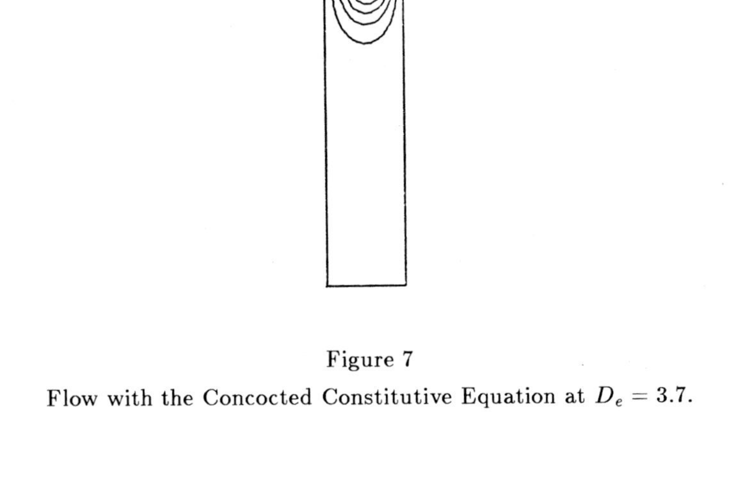

Figure 7

Flow with the Concocted Constitutive Equation at $D_e = 3.7$.

much higher D_e and L_o. The $D_e = 3.7$ case took 40 iterations to get only a factor of 25 reduction in the residual. Attempts to converge at higher D_e failed. The $D_e = 2.5$ case took 20 iterations to get a factor of 20 reduction in the residual. Thereare some qualitative differences between the flows of Figures 6 and 7 and the flow of Figure 5. These may be related to the greater difficulty in convergence, though this cannot be stated with any certainty. We note that the streamlines just outside the slot are, if anything, more distorted by the presence of the slot than they are in Stokes-flow. Careful comparison of Figures 4, 6, and 7 show that the streamline nearest the slot is in virtually the same position in both Stokes-flow and the non-Newtonian flow, but that in both Figures 6 and 7, the streamline dips measureably futher down. Also, in distinct contrast to the flow of Figure 5, numerical results reported elsewhere [4,10], and flow visualizations [4], the vortices in Figures 6 and 7 are higher up and nearer the slot mouth than they are in Stokes-flow. It seems that with the concocted constitutive equation, the flow in the slot interacts with and disturbs the channel-flow to a much greater extent than it does in the Curtiss-Bird case.

It would be easy to dismiss the differences between the two constitutive equations discussed here as an unphysical consequence of the artificially concocted constitutive equation; however, behavior like that observed with the concocted equation seems to be characteristic of *all* the integral constitutive equations the author has tested, *except* the Curtiss-Bird. And the behavior seems to be worse for other equations; the tinkering applied to the Johnson-Segalman model has produced the second best constitutive equation to the Cirtiss-Bird, measured by the ability to get to high D_e and/or L_o. The difference between the two constitutive equations seems to devolve from the amount of shear-thinning at a given elasticity; this can be quantified by observing L_o as a function of σ in simple shear. The Curtiss-Bird equation leads to an L_o vs. σ curve with positive curvature, while the concocted equation has negative curvature. This leads to a prediction [22], which is observed numerically, that P_e/N_1 should be larger at fixed σ for the concocted equation than it is for Curtiss-Bird. Referring to Figures 5 and 7, we can see that the component of cross-stream normal force acting along the streamline should be less for the Curtiss-Bird fluid than it is for the concocted equation, since in the Curtiss-Bird case, the streamlines are much more nearly parallel to the wall. Thus we are brought back to the old arguement of Tanner and Pipkin [26], in which the non-zero values of P_e observed experimentally are attributed to elastic restoring forces acting on a control volume over the slot. This kind of explanation of viscoelastic hole-pressure seems to be borne out in these numerical results, at least on a qualitative level. It is also interesting to observe that the same interaction between slot and channel-flow which causes relatively higher hole-pressure seems to cause numerical difficulty. It is tempting to speculate that the numerical difficulty may be due to an impending breakdown of steady channel-flow near the slot mouth, due to this interaction.

CONCLUSIONS

The author believes that the numerical methods decribed here represent an auspicious beginning to the task of numerically modelling non-Newtonian flows; however, they represent only a beginning. There needs to be a better theoretical understanding of the nature of the equations and the resulting approximation schemes before definitive progress can be made. Recent work illuminating the nature of the characteristics of the equations moves in that direction and serves to illustrate common ground between the different classes of constitutive theories proposed by rheologists. It is hoped that this can lead to a more unified numerical approach to this challenging and demanding class of problems. In the meantime, current numerical techniques, though only partially understood, seem to be capable of uncovering physically important consequences of non-Newtonian fluid behavior.

REFERENCES

1. R. I. TANNER, R. E. NICKEL, and R. W. BILGER, Finite element methods for the solution of some incompressible non-Newtonian fluid mechanics problems with free surfaces, *Comp. Meth. Appl. Mech. Eng. 16*, 155 (1975).
2. H. COURT, K. WALTERS, and R. DAVIES, Long-range memory effects in flows involving abrupt changes in geometry. Part 4: Numerical simulation using integral rheological models, *J. Non-Newtonian Fluid Mechs. 8*, 95 (1981).
3. G. D. RICHARDS and P. TOWNSEND, A finite element computer model of model of the hole-pressure problem, *Rheologica Acta 20*, 261 (1981).

4. T. COCHRANE, K. WALTERS, and M. F. WEBSTER, On Newtonian and non-Newtonian flow in complex geometries, *Phil. Trans. Roy. Soc. London, 301*, 163 (1982).
5. K. WALTERS, "Rheometry," Chapman and Hall, London (1975).
6. R. I. TANNER, Constitutive equations for the computing person, This volume (1985).
7. M. J. CROCHET, A. R. DAVIES, and K. WALTERS, "Numerical Simulation of Non-Newtonian Flow," Elsevier, Amsterdam (1984).
8. C. J. S. PETRIE, "Elongational Flows," Pittman, London (1979).
9. C. J. S. PETRIE, Measures of deformation and convected derivatives, *J. Non-Newtonian Fluid Mechs. 5*, 53 (1975).
10. D. S. MALKUS and B. BERNSTEIN, Flow of a Curtiss-Bird fluid over a transverse slot using the finite element drift-function method, *J. Non--Newtonian Fluid Mechs. 16*, 77(1984).
11. D. D. JOSEPH, M. RENARDY, and J. C. SAUT, Hyperbolicity and change of type in the flow of viscoelastic fluids, *Arch. Rat. Mech. Anal.*, to appear.
12. M. W. JOHNSON and D. SEGALMAN, A model for viscoelastic fluid behavior which allows non-affine deformation, *J. Non-Newtonian Fluid Mechs. 2*, 255 (1977).
13. M. DOI and S. F. EDWARDS, Dynamics of concentrated polymer systems. Parts I-IV, *J. C. S. Faraday 74 and 75*, 1789 (1978), 1802 (1978), 1818 (1978), 38 (1979).
14. C. F. CURTISS and R. B. BIRD, Kinetic theory for polymer melts. Parts I and II, *J. Chem. Phys. 74*, 2016 (1981).
15. M. CROCHET and R. KEUNINGS, Die swell of a Maxwell fluid: Numerical prediction, *J. Non-Newtonian Fluid Mechs. 7*, 199 (1980).
16. M. A. MENDELSON, P.-W. YEH, R. A. BROWN, and R. C. ARMSTRONG, Approximation error in finite element calculation of viscoelastic fluid flows, *J. Non-Newtonian Fluid Mechs. 10*, 31 (1982).
17. N. R. JACKSON and B. FINLAYSON, Calculation of hole pressure: II Viscoelastic fluids, *J. Non-Newtonian Fluid Mechs. 10*, 71 (1982).
18. M. B. BUSH, R. I. TANNER, and N. PHAN-TIEN, A boundary element investigation of extrudate swell, *J. Non-Newtonian Fluid Mech.*, to appear.
19. M. VIRIYAYUTHAKORN and B. CASWELL, Finite element simulation of viscoelastic flows, *J. Non-Newtonian Fluid Mechs. 8*, 245 (1981).

20. S. DUPONT, J. M. MARCHAL, and M. CROCHET, Finite element simulation of viscoelastic fluids of the integral type, Preprint, 1984.

21. O. HASSAGER and C. BISGAARD, A Lagrangian finite element method for the simulation of non-Newtonian liquids, *J. Non-Newtonian Fluid Mech.* *12*, 155 (1983).

22. K. HIGASHITANI and W. G. PRITCHARD, A kinematic calculation of intrinsic errors in measurements made with holes, *Trans. Soc. Rheology 16*, 687 (1972).

23. A. S. LODGE and L. de VARGAS, Positive hole-pressures and negative exit pressure generated by molten low-density polyethelyne flowing through a slit die, *Rheologica Acta 22*, 151 (1983).

24. R. D. PIKE and D. G. BAIRD, Evaluation of the Higashitani and Pritchard analysis of the hole pressure using flow birefringence, *J. Non-Newtonian Fluid Mech. 16*, 211 (1984).

25. D. S. MALKUS, Finite element simulation of flows with integral constitutive equations, *Comm. Appl. Num. Meths.*, to appear.

26. R. I. TANNER and A. C. PIPKIN, *Trans. Soc. Rheol. 13*, 471 (1969).

Acknowledgements: The numerical techniques for integral constitutive equations described here were developed jointly by the author and B. Bernstein (Dept. Mathematics, I. I. T.). The author acknowledges the support of the National Science Foundation (Grants MCS 79-03542, 81-02089, and 83-01433, and the Supercomputer Initiative, which provided access to the University of Minnesota Cray - 1A). Further development of the computational method for single-integral models is sponsored by the United States Air Force, under Grant No. 84-NM-399. At the Mathematics Research Center, University of Wisconsin - Madison, the author's research is sponsored by the United States Army under Contract No. DAAG29-80-C-0041.

CONSTITUTIVE EQUATIONS FOR THE COMPUTING PERSON

Roger I. Tanner

Mathematics Research Center
University of Wisconsin-Madison
Madison, Wisconsin and
Department of Mechanical Engineering
University of Sydney
Sydney, Australia

1. INTRODUCTION

Aside from the simple viscometric and elongational flow cases, the solution of viscoelastic fluid flow problems usually entails the selection of a constitutive equation connecting stress and kinematics; broad statements to the effect that the stress depends on the history of the deformation are not generally sufficiently explicit to enable progress to be made. After selecting a constitutive equation, and except in the flows mentioned above and in regular perturbation calculations, the use of the computer is almost mandatory in order to find solutions. For example, one may consider the problem of finding the flow field in a Newtonian, creeping flow in an abrupt transition from one pipe diameter to another. In this simple case, there is no useful alternative to computation known to the author. The aim of this article is to point out some precautions in choosing a constitutive equation when setting out to do computing; some comments on computing methods are also given.

The equations we have to solve are

$$\underset{\sim}{\nabla}\,\underset{\sim}{v} = 0 \qquad \text{(incompressibility condition)} \qquad (1.1)$$

$$\underset{\sim}{\nabla}\,\underset{\sim}{\sigma} = \rho(\underset{\sim}{a} - \underset{\sim}{f}) \qquad \text{(equation of motion)} \qquad (1.2)$$

$$\underset{\sim}{\sigma} = -p\underset{\sim}{I} + \underset{\sim}{\tau} \qquad \text{(definition of stress)} \qquad (1.3)$$

Here $\underset{\sim}{v}$ is the velocity vector in a fixed (Eulerian) system, $\underset{\sim}{\sigma}$ is the stress tensor, $\underset{\sim}{a}$ and $\underset{\sim}{f}$ are the acceleration and body force (per unit mass) vectors respectively, p is the pressure, $\underset{\sim}{I}$ is the unit tensor, and $\underset{\sim}{\tau}$ is the part of the stress tensor which is determined by the motion. We shall call $\underset{\sim}{\tau}$ the extra stress tensor and write constitutive equations connecting $\underset{\sim}{\tau}$ and the kinematic quantities. In the examples to be presented, we shall neglect the acceleration and body forces. This is often a good approximation.

Incompressibility is assumed, so (1.1) holds. We shall not discuss the energy equation but in many practical cases the effect of temperature changes on the flow may be just as important or even more important than the rheological effects[1]. When seeking a computed solution, the variation of temperature is not difficult to include, but it obscures issues, and hence its omission is felt to be justified.

A large menu of constitutive equations is now available[1,2]. Not all of them are good representations in all flow types and it is useful to consider the sort of kinematic fields that may be encountered. Table 1 gives a 'rheological heptathlon' which classifies flow fields.

Table 1. Flow Types

Small Strain Motions	(weak)
Viscometric Flows	(weak)
Elongational Flows	(strong)
Time-dependent Shearing	(weak)
Unsteady Elongation/Recoil	(strong)
Single Step of Shear	(weak)
Multiple Shear Steps	(weak)

The adjectives 'weak' and 'strong' refer to the separation of two particles joined by a spring in a flow; in strong flows a large separation of the particles eventually occurs; in weak flows the particles always remain close to one another[1].

Viscometric flows are, locally, steady shearing motions (e.g. Poiseuille flow, simple shearing); the other categories are self-explanatory.

It is more difficult to simulate the later members of Table 1 but it may not always be necessary. For example, except for the Lagrangian code of Hassager[3], most general computer programs do not allow jumps of strain, since they are based on an Eulerian representation of the velocity field; hence these codes do not need to simulate jump-strain behaviour accurately. Thus here we advocate what is normal in applied mechanics: depending on the kinematic problem in hand one may choose one of several mathematical representations of the material. It is not always necessary to choose the most complex rheological model and in many cases simplicity will be an advantage.

2. FAMILIES OF CONSTITUTIVE EQUATIONS

We shall briefly consider the behaviour of (a) Rate-type equations, (b) Green-Rivlin or integral type expansions, (c) Differential constitutive equations.

(a) Rate-type Equations.

In this category we may write down the Criminale-Ericksen-Filbey[4] (CEF) equation which is an exact representation of isotropic fluid behaviour in viscometric flows:

$$\underset{\sim}{\tau} = \eta \underset{\sim}{A}_1 + (\nu_1 + \nu_2)\underset{\sim}{A}_1^2 - \frac{1}{2}\nu_1 \underset{\sim}{A}_2 \tag{2.1}$$

If we denote the velocity gradient tensor by $\underset{\sim}{L}$, so that

$$\underset{\sim}{L}^T = \underset{\sim}{\nabla}\underset{\sim}{v} \quad \left(L_{ij} = \frac{\partial v_i}{\partial x_j} \text{ in Cartesian tensor notation}\right) \tag{2.2}$$

where $\underset{\sim}{L}^T$ is the transpose of $\underset{\sim}{L}$, then

$$\underset{\sim}{A}_1 = \underset{\sim}{L} + \underset{\sim}{L}^T \tag{2.3}$$

and

$$\underset{\sim}{A}_2 = \frac{\partial \underset{\sim}{A}_1}{\partial t} + (\underset{\sim}{v}\cdot\underset{\sim}{\nabla})\underset{\sim}{A}_1 + \underset{\sim}{L}^T\underset{\sim}{A}_1 + \underset{\sim}{A}_1\underset{\sim}{L} \tag{2.4}$$

$\underset{\sim}{A}_1$ is called the first Rivlin-Ericksen tensor, and $\underset{\sim}{A}_2$ is the second Rivlin-Ericksen tensor, t is time. Thus $\underset{\sim}{A}_2$

is the "time derivative" or rate-of-change of $\underset{\sim}{A}_1$ in the sense specified. By inserting $\underset{\sim}{A}_2$ in place of $\underset{\sim}{A}_1$ on the right-hand-side of (2.4) one may similarly define the rate-of-change of $\underset{\sim}{A}_2$ thus generating $\underset{\sim}{A}_3$ and hence an infinite sequence of Rivlin-Ericksen tensors $(\underset{\sim}{A}_n)$.

The functions η, ν_1 and ν_2 are functions of $\mathrm{tr}\,\underset{\sim}{A}_1^2$; they are material functions which may be found from viscometric experiments. When η, ν_1 and ν_2 are constant, (2.1) is part of the sequence of order or Rivlin-Ericksen equations, and we call (2.1) in this special case a <u>second-order</u> equation. (The first-order equation, in which $\nu_1 = \nu_2 = 0$, is the Newtonian fluid). Similarly, higher-order fluid equations containing $\underset{\sim}{A}_3$, $\underset{\sim}{A}_4$, etc. can be written down. Aside from the problems of defining the many coefficients in the higher-order fluids, it is also difficult to compute with these representations because of the repeated spatial differentiation of the velocity field involved in the $(\underset{\sim}{v}\,\underset{\sim}{\nabla})\underset{\sim}{A}_n$ terms. Experience[5] shows that even the second-order fluid is difficult to attain convergence with when the ratio $\nu_1\sqrt{\mathrm{tr}\,\underset{\sim}{A}_1^2}\,/\eta$ is much greater than unity. Generally, the conclusion is that this class of models should be avoided for computing. The exceptions are the Newtonian fluid and the generalized Newtonian fluid. In the latter, $\nu_1 = \nu_2 = 0$ and the viscosity is a function of the shear rate $\dot{\gamma}\left(\equiv \sqrt{\tfrac{1}{2}\,\mathrm{tr}\,\underset{\sim}{A}_1^2}\right)$. Neither of these is useful in describing viscoelastic effects on flow; they are, like all other members of the rate family, inelastic, recoil-less fluid models. None of these models performs strongly in any of the categories of Table 1 except the viscometric flows, for which (2.1) is exact.

(b) <u>Integral-type Expansions</u>.

Green and Rivlin[6] introduced the integral expansion form

$$\underset{\sim}{\tau} = \int_{-\infty}^{t} \mu_1(t - t')\underset{\sim}{G}(t')dt' \tag{2.5}$$

$$+ \iint_{-\infty}^{t} \mu_2(t - t', t - t'')\underset{\sim}{G}(t')\underset{\sim}{G}(t'')dt'dt'' + \iiint_{-\infty}^{t} \cdots .$$

where μ_1 and μ_2 are kernel functions characteristic of the material, and $\underset{\sim}{G}$ is a strain measure. Suppose a particle occupies a place $\underset{\sim}{r}$ at time t', and place $\underset{\sim}{x}$ at time t; we regard $\underset{\sim}{x}$ as a fixed Eulerian reference frame. Then if we define

$$\underset{\sim}{F}(t') = \frac{\partial \underset{\sim}{r}}{\partial \underset{\sim}{x}} \left(F_{ij} = \frac{\partial r_i}{\partial x_j}\right) \tag{2.6}$$

then

$$\underset{\sim}{G}(t') \equiv \underset{\sim}{F}^T \underset{\sim}{F} - \underset{\sim}{I} \ . \tag{2.7}$$

Thus (2.5) is recognized as a nonlinear generalization of Boltzmann's linear theory of viscoelasticity[1,2]. It is difficult to obtain experimental information about the double and higher integrals, and most computing has been restricted to single-integral models. Of the available single-integral models the K-BKZ[7,8] form seems to be among the most successful approximate representations available for polymer melts and solutions. In this case

$$\underset{\sim}{\tau} = \int_{-\infty}^{t} m(t - t')\left\{\frac{\partial U}{\partial I_{c^{-1}}} \underset{\sim}{C}^{-1} - \frac{\partial U}{\partial I_c} \underset{\sim}{C}\right\} dt' \tag{2.8}$$

Here the strain $\underset{\sim}{C}(t')$ is just $\underset{\sim}{G} + \underset{\sim}{I}$, and I_c is the first invariant of $\underset{\sim}{C}$. The function U depends on I_c and $I_{c^{-1}}$ only. Problems of utilizing this and related forms are discussed by Malkus[9]. The sense in which (2.8) is an optimal choice among single-integral theories is discussed elsewhere[1]; one of the advantages of the K-BKZ model is that it gives at least reasonable behaviour over the whole heptathlon (Table 1).

(c) Differential Constitutive Equations.

We will consider here several constitutive relations of the general form

$$\lambda \frac{\Delta \underset{\sim}{\tau}}{\Delta t} + \underset{\sim}{B} = \underset{\sim}{0} \tag{2.9}$$

where λ is the (single) relaxation time. The convected time derivative of the extra-stress tensor $\underset{\sim}{\tau}$, $\Delta\underset{\sim}{\tau}/\Delta t$ is defined by

$$\frac{\Delta\tau_{ij}}{\Delta t} \equiv \frac{\partial\tau_{ij}}{\partial t} + v_k \frac{\partial\tau_{ij}}{\partial x_k} - \tau_{ik}L_{jk} - L_{ik}\tau_{kj} \tag{2.10}$$

The extra-stress τ_{ij} is related to the total stress σ_{ij} here by

$$\tau_{ij} = \sigma_{ij} + p\delta_{ij} - \eta_s(L_{ij} + L_{ji}) \tag{2.11}$$

where p is a pressure, δ_{ij} is the unit tensor, and η_s is a constant ('solvent') viscosity. The tensor $\underset{\sim}{B}$ in (2.9) defines the specific constitutive model under discussion. Here we treat the Maxwell, the Oldroyd-B, the Leonov and the Phan-Thien-Tanner (PTT) models[1]. The values of $\underset{\sim}{B}$ and η_s for these models are given in Equations (2.12-2.14).

Oldroyd-B and Maxwell: $\underset{\sim}{B} = \underset{\sim}{\tau} - \eta\underset{\sim}{A}_1$ (2.12)

where η is a (constant) viscosity. If we set η_s in (2.11) equal to zero we have the classical convected Maxwell model, and if it is not zero then we have the Oldroyd-B model; essentially the latter contains the former plus the "base" or "solvent" viscosity term $\eta_s\underset{\sim}{A}_1$.

For the PTT model we set

$$\underset{\sim}{B} = \underset{\sim}{\tau}\exp\{\frac{\lambda\varepsilon}{\eta}\,\mathrm{tr}\,\underset{\sim}{\tau}\} - \eta\underset{\sim}{A}_1 + \frac{\lambda\xi}{2}(\underset{\sim}{A}_1\underset{\sim}{\tau} + \underset{\sim}{\tau}\underset{\sim}{A}_1) \tag{2.13}$$

where ε and ξ are dimensionless constants and the rigidity is η/λ. Crochet and Keunings[10] discovered that it was essential to have a non-zero η_s with this model to obtain a monotonically increasing shear stress-shearing rate curve in simple shearing. In the PTT model the value of ε controls the elongational viscosity and ξ controls the shear viscosity; when both are zero one has the Maxwell-Oldroyd-B case.

For the Leonov model we have

$$\underset{\sim}{B} \equiv \frac{1}{2}\underset{\sim}{\tau}\{\frac{\underset{\sim}{\tau}}{2\mu} - 2\mu\underset{\sim}{\tau}^{-1} + \frac{1}{3}\underset{\sim}{I}[2\mu\,\mathrm{tr}(\underset{\sim}{\tau}^{-1}) - \frac{\mathrm{tr}\underset{\sim}{\tau}}{2\mu}]\} \tag{2.14}$$

where μ is a modulus and $\underset{\sim}{I}$ is the unit tensor. It is also possible to have a viscous response with this model; i.e. η_s not zero in Equation (2.11). For other models in this family see Giesekus[11] and Bird et al.[2].

There are of course many other useful consitutive models which do not follow the pattern (2.9). The main advantage of the Maxwell-type models is that when one is dealing with steady flows, and if the velocities are given, then the streamlines are the characteristic curves of the partial differential equations (2.9), and by integrating along the streamlines the problem is reduced to solving three or four <u>ordinary</u> differential equations at each iteration step. This is straightforward and accurate in cases where no recirculating regions occur. Boundary conditions on the stresses τ_{ij} need to be supplied at inlet planes, in addition to the usual mechanical boundary conditions on traction and/or velocity. We thus have an iterative boundary or finite-element scheme for solution. Step (1) finds the Stokes solution for velocity and tractions; Step (2) solves for the streamlines; Step (3) finds the stresses by solving Equation (2.9) along the streamlines; Step (4) finds the pseudo-body forces from the nonlinear terms; Step (5) repeats Step (1) until convergence (or divergence!) is obtained.

This procedure, as well as many other finite element and finite difference procedures, quickly becomes unstable when dealing with the Maxwell fluid model. When the Weissenberg number (Wi), defined as

$$Wi = \lambda\dot{\gamma}_w \tag{2.15}$$

where $\dot{\gamma}_w$ is a characteristic shear rate, exceeds order unity, instability sets in, apparently irrespective of computing method. Hence one wonders whether one is dealing with a computational instability or one built into the problem by the Maxwell equation. Therefore we now consider the stability of the rheological problem.

3. STABILITY CONSIDERATIONS

The problems set out above are generally nonlinear and need iteration in order to produce a solution. Unless a stable, attainable solution exists, computing exercises are likely to be fruitless. Sometimes there may be a genuine flow instability, observable experimentally. An example of this is given by Keunings[12]. Another cause of problems is lack of convergence due to inadequate numerical schemes[13]. We shall suppose that these problems have been avoided. There still remains the possibility that the description of the material (constitutive equation) chosen is the cause of lack of convergence. Several known problems are:

(a) Non-monotonicity of the shear-rate ($\dot{\gamma}$) - shear stress (τ) curve in simple shearing. When $d\tau/d\dot{\gamma} < 0$ it is easily shown that the material is unstable and lack of convergence will result. This fault can be rectified by adding on a 'solvent' viscosity term as was done by Crochet and Keunings[10] for the PTT model (2.13).

(b) The second-order fluid (and higher-order fluids) are unstable in unsteady shearing flows[14,15] and should not be used in a description of such flows.

(c) A potential instability exists in the tendency of some writers to ignore the second term $\frac{\partial U}{\partial I_c}$ in (2.8) and write $\frac{\partial U}{\partial I_{c^{-1}}}$ as $\phi(I_c, I_{c^{-1}})$. In a very rapid sequence of strains, (2.8) behaves as a rubberlike body and the 'potential' U acts in a way similar to the strain-energy function in rubber elasticity; if there is no such 'potential' U, we may be in danger of inventing a function ϕ that enables us to do work on the surroundings in a cyclic sequence of rapid strains.

(d) In a certain differential equation model[16] constraints on the parameter values were needed to ensure stability in unsteady shearing flows.

We now consider the differential family of constitutive equations, and in particular we shall consider the Maxwell-Oldroyd models (2.12). Ho and Denn[17] have investigated the simple shearing flow of a Maxwell fluid and have found no instability (including the case when inertia was considered). It is clearly important to discover if inertialess stabilities exist in flows of the Maxwell and Oldroyd-B fluids. As a step in this direction Phan-Thien[18] has analyzed the flow between coaxial discs (torsional flow) of infinite extent, separated by a distance d along the z-axis.

We now consider this solution. The Newtonian flow was solved by von Kármán in 1921 and has been widely discussed.

Phan-Thien[18], using an exact solution in the von Kármán form for the Maxwell fluid, studied the stability of the flow. We follow his discussion here, omitting inertia. In this case the (steady-state) solution is exact; particles move in circles around the axis of the discs and the speed increases linearly from one disc to another.

Suppose the plane $z = 0$ rotates at an angular speed Ω_0 and the plane $z = d$ rotates at Ω_d. Both plates are set in motion at $t = 0^+$, and the flow is transient. The constitutive equation used is the Oldroyd-B fluid:

$$\underset{\sim}{\tau}' = \underset{\sim}{\sigma} + p\underset{\sim}{I} = \eta_s \underset{\sim}{A}_1 + \underset{\sim}{S} \tag{3.1}$$

where $\underset{\sim}{S}$ is a Maxwell stress (Equations (2.9) and (2.12)) and η_s is the "solvent" viscosity; when $\eta_s = 0$ we return to the Maxwell case. We let the viscosity associated with the Maxwell element be η and the associated time constant be λ; let $\eta_0 = \eta + \eta_s$. The retardation time is then

$$\lambda_2 = \lambda\eta_s/\eta_0 = (1 - \beta)\lambda \tag{3.2}$$

and β is the retardation parameter (η/η_0) ranging from 0 (Newtonian) to 1 (Maxwell).

The steady-state solution in general (including inertia) has a velocity field of the form

$$\underline{v} = \left(r \frac{\partial H}{\partial z}, rG, -2H\right) \tag{3.3}$$

where G, H are functions of z and t; in the inertialess case $H = 0$ and G is linear in z (torsional flow). For the inertialess case we assume a form (3.3) and a development of the solution in series

$$\left.\begin{aligned} g &= \xi + g_1 + \text{higher order terms} \\ h &= h_1 + \ldots \\ \tau'_{\theta\theta} &= 2\beta\, Wi + \widehat{tt} + \ldots \\ \tau'_{\theta z} &= \beta + \widehat{tz} + \ldots \end{aligned}\right\} \tag{3.4}$$

where $\xi = z/d$, $t' = (\Omega_d - \Omega_0)t$, $G = (\Omega_d - \Omega_0)g + \Omega_0$, and $H = d(\Omega_d - \Omega_0)h$; h and g are functions of ξ and t'. The Weissenberg number Wi defined here is $\lambda(\Omega_d - \Omega_0)$ and the stresses $\tau'_{\theta\theta}$ and $\tau'_{\theta z}$ are dimensionless; they are formed by dividing the relevant components by $\eta(\Omega_d - \Omega_0)$. Substitution of the forms (3.4) in the equations of motion and the constitutive equations and linearizing about the steady solution, one finds a set of eight linear ordinary differential equations. A normal mode Fourier analysis of the form

$$h_1 = \hat{h}[\exp(2\pi i\, k\xi) - 1] \tag{3.5}$$

shows that the kth mode has a critical Weissenberg number $Wi_{c,k}$ of

$$Wi_{c,k} = \frac{\pi k}{\sqrt{\beta(2\beta + 3)}} \tag{3.6}$$

The most unstable mode, Wi_c, is when $k = 1$, hence if

$$Wi > \frac{\pi}{\sqrt{\beta(2\beta + 3)}} \tag{3.7}$$

the flow is unstable to infinitesimal disturbances of the form assumed; it is possible that there are other, more unstable, disturbances. Note that $Wi_c \to \infty$ as $\beta \to 0$ (Newtonian case) which is consistent with expectations. When $\beta = 1$ (Maxwell case) the critical Weissenberg number is 1.405.

This result is highly significant as it shows that there are cases where a fundamental instability occurs in a creeping viscoelastic flow. Hence no numerical scheme can be expected to find a stable solution since none exists. Physically, it appears that the large $\sigma_{\theta\theta}$ is the cause of the instability; there is no such instability in the corresponding plane shearing flow where $\sigma_{\theta\theta}$ is absent. More cases of instabilities like this example have since been discovered and hence it appears likely that the Oldroyd-B fluid model imposes definite Weissenberg number limits. It is therefore thought that these models are difficult to compute with[1] and we turn to other, more tractable (and more realistic) models.

4. SOME RESULTS FOR LEONOV AND PTT MODELS

The results obtained for the Oldroyd-B model by Crochet and Keunings[10] show more stability than the corresponding Maxwell results[1]. However, a limiting Wi still exists according to the reasoning in the previous section and means of dealing with these problems are needed. Figure 1 shows the elongational responses for the four models under consideration, for a single time constant. Experiments[1] show a rise of 2-10 times in the elongational viscosity for low density polyethylene melts. The PTT model gives the most realistic elongational response; when ε is set to zero it is a special case of the general Oldroyd model and the elongational viscosity becomes unbounded at a finite Weissenberg number.

The plane swelling problem is shown in Figure 2. The flow exits across plane CF and we wish to predict the swelling ratio χ. In computations neither the PTT nor the Leonov models exhibited any instability up to the Weissenberg numbers shown in Figure 3 and convergence was rapid (~4 or 5 iterations) at each Weissenberg number. Calculations were performed with a boundary element code. The shear-thinning effect in the Leonov model is clearly dominant and swelling is small. This is in accordance with previous experience with shear-thinning inelastic materials[19]. The material is not "elastic" enough to swell much; alternatively one can see from Figure 1 that the elongational stress in unrealistically small. On the other hand the PTT model swells more quickly; Phan-Thien[20] has shown that it is capable of modelling real polymer extensional flows quite accurately. This deficiency in the Leonov model has been confirmed in work by the author and

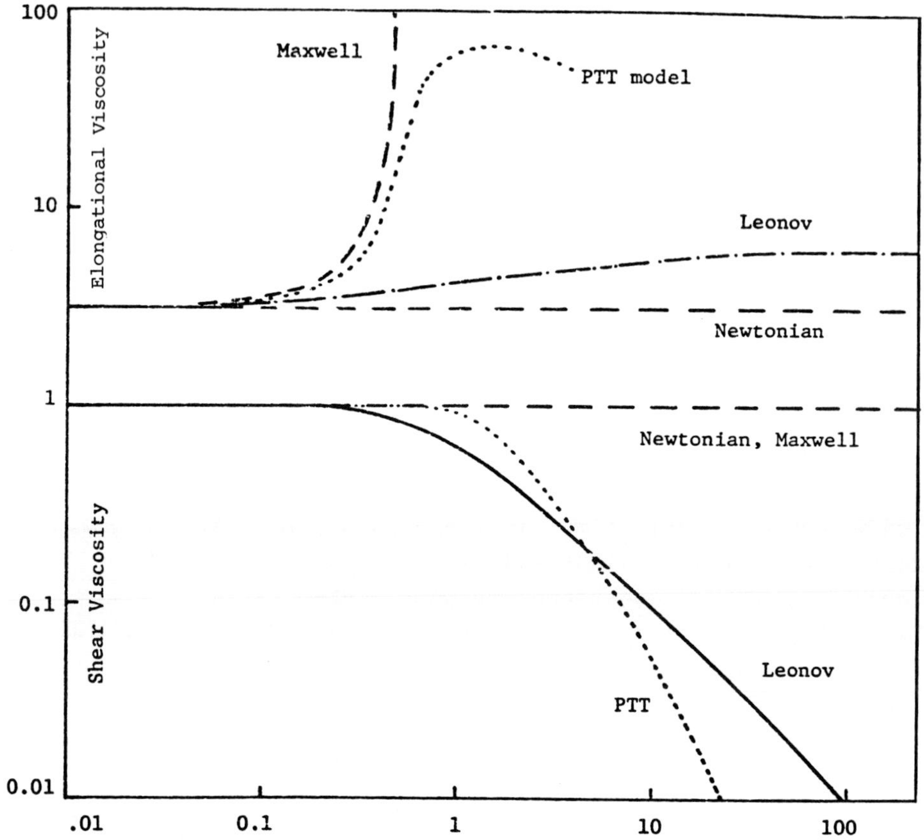

Dimensionless Shearing Rate $(\lambda\dot{\gamma})$ or Elongational Rate $(\lambda\dot{\varepsilon})$

Figure 1. Comparison of shearing (below) and elongational (above) steady viscosities for various models. In the PTT-model $\varepsilon = 0.01$, $\xi = 0.1$. η_s is zero for all models shown and η and λ are 1.0. Note the rapid increase for the Maxwell model when $\lambda\dot{\varepsilon} = 0.5$; the Oldroyd-B model response is similar. The Leonov model elongation response is nearly Newtonian.

(The shearing velocity field $\underline{v} = (\dot{\gamma}y, 0, 0)$; the shear viscosity is $\tau_{xy}/\dot{\gamma}$; the elongational field is $\underline{v} = \left\{\dot{\varepsilon}x, -\frac{\dot{\varepsilon}}{2}y, -\frac{\dot{\varepsilon}}{2}z\right\}$ and the elongational viscosity is $(\tau_{xx} - \tau_{yy})/\dot{\varepsilon}$).

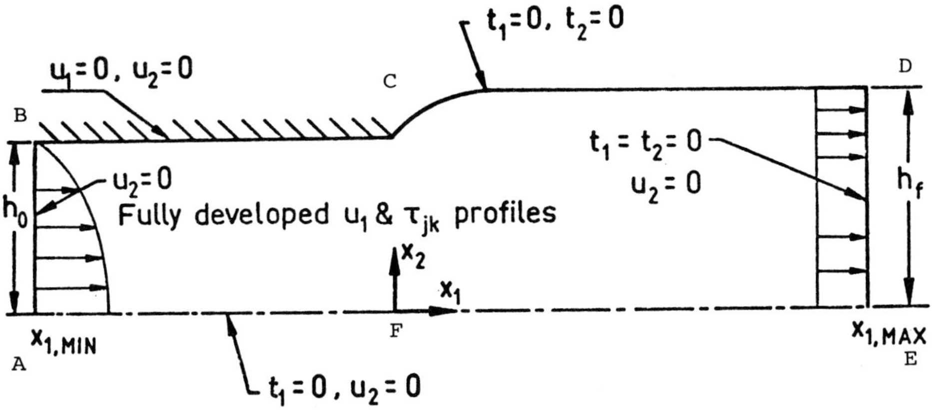

Figure 2. Die-swell problem. BC is solid wall, AFE is the centerline, and CD is a free surface. The swelling ratio χ is h_f/h_0. The upstream die is 'long', so the flow is fully developed upstream. Both plane flow and axisymmetric versions of the problem may be solved. $\dot{\gamma}_w$ is the value of $\frac{\partial u}{\partial y}$ at the wall in the fully developed region upstream. For a creeping flow with a single time constant (λ), χ is a function of $\lambda\dot{\gamma}_w$ for fixed $\eta_s, \eta, \xi, \varepsilon$ and μ in the models.

Mr. X.-L. Luo (to be published) on film-blowing. The Leonov[21] model is not "stiff" enough in an elongational flow of this type to model real materials well. (It seems likely that one can alter the model to rectify this problem, but we have not done this; if it were done, then the Leonov model would be less attractive as it would contain more parameters).

The boundary element method is able to reach reasonable Weissenberg numbers with a PTT model as is shown in Figures 4 and 5; in Figure 5 Wi = 4.0; more recent results up to Wi = 7 are now available. The isothermal results in Figure 6 (found using a finite element code that also can deal with realistic temperature variations; it is not necessary to use a 'mixed method' in these computations[10,22]) plot the swelling ratio against $\lambda\dot{\gamma}_w$ and show the effect of parameter changes on swelling with the PTT model. The parameter ε (broadly) controls the maximum in the elongational viscosity η_T; the higher is ε, the lower is the maximum in η_T. This has a dramatic effect on swelling

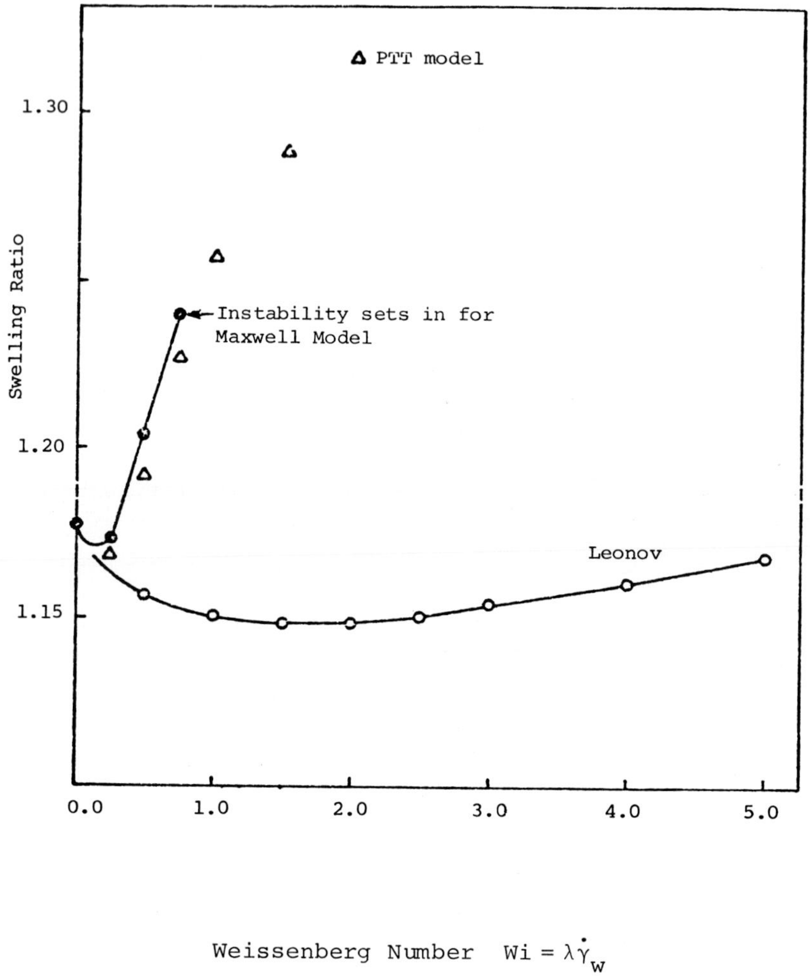

Figure 3. Swelling of Plane Jet. In the PTT and Leonov models no instability occurred up to the Wi given here. Parameters are:

PTT model $\lambda = 1$, $\eta = 0.889$, $\eta_s = 0.111$, $\xi = 0.1$, $\varepsilon = 0.01$

Maxwell model $\lambda = 1$, $\eta = 1.0$, $\eta_s = 0$

Leonov model $\lambda = 1$, $\eta = 1.0$, $\eta_s = 0$, $\mu = 0.5$.

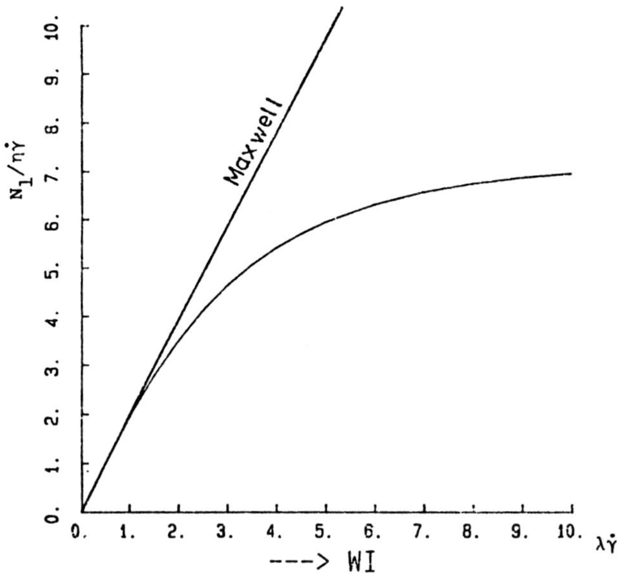

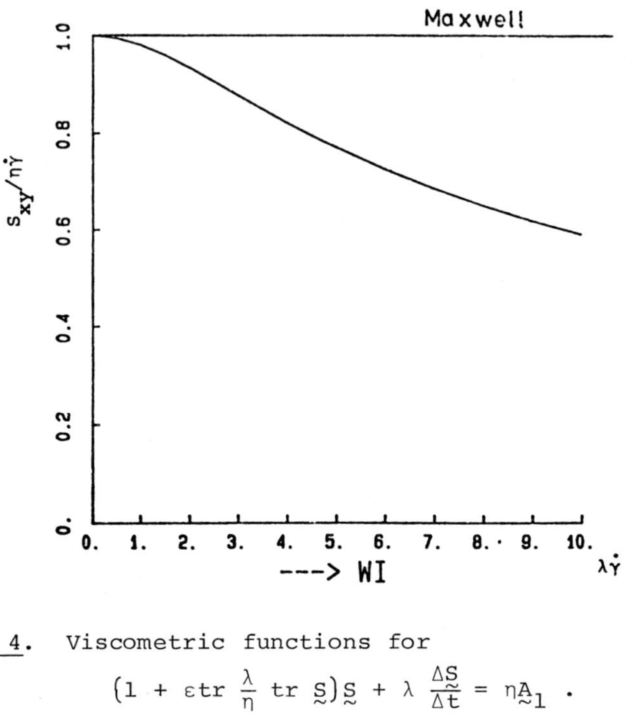

Figure 4. Viscometric functions for

$$\left(1 + \varepsilon \mathrm{tr}\,\frac{\lambda}{\eta}\,\mathrm{tr}\,\underset{\sim}{S}\right)\underset{\sim}{S} + \lambda\,\frac{\Delta \underset{\sim}{S}}{\Delta t} = \eta \underset{\sim}{A}_1 \ .$$

(For ε small this equation behaves almost identically to Equation 2.13; here $\varepsilon = 0.01$).

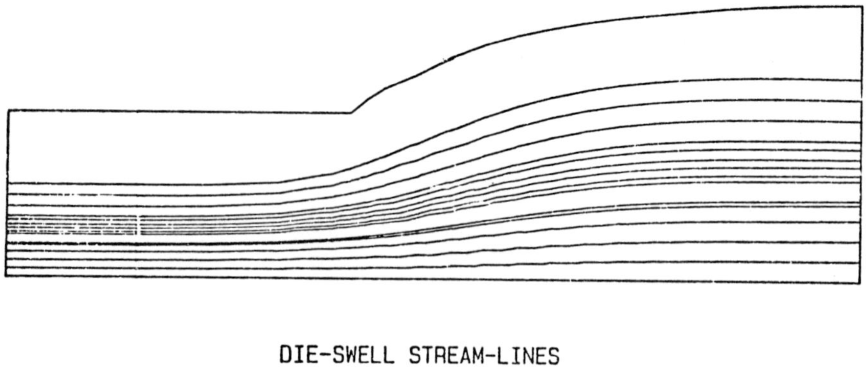

Figure 5. Die-Swell for PTT model in Figure 4; $\varepsilon = 0.01$, $\lambda\dot{\gamma}_w = 4.0$.

as may be seen by comparing the curves for $\xi = 0.0$ with $\varepsilon = 0.015$ (near-Maxwell) and $\varepsilon = 0.30$ (Leonov-like). Similarly, the parameter ξ governs shear-thinning; as ξ increases, the fluid shear-thins more. Comparing the two curves for $\varepsilon = 0.015$ we see the rapid reduction in swelling due to shear-thinning.

6. CONCLUSION

We have shown that the choice of constitutive equation may be crucial to the success of a computation. In particular, the Oldroyd-B and Maxwell fluids tend to be unstable and difficult to obtain convergence with. Given a suitable constitutive equation, then it is possible to reach useful Weissenberg numbers with a variety of computational schemes; it is curious that the more realistic constitutive models are more stable in computations. More investigation of the stability of flows is clearly urgently needed in order to reduce wasted effort in fruitless computation. Finally, the increased stability enables one to adopt simpler, more flexible computational schemes and thus avoid the expensive 'mixed methods'[22] which may become impractical if used with differential models containing multiple relaxation times. Aspects of computations using integral models are discussed by Malkus[9].

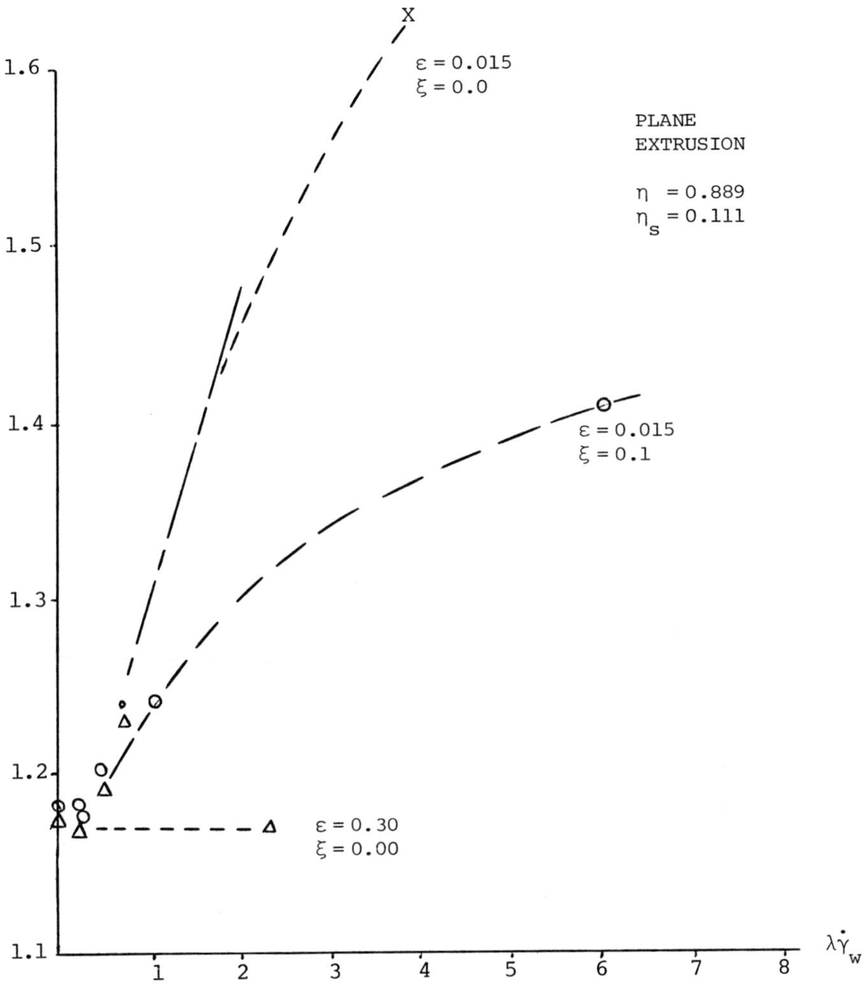

Figure 6. Plane swelling for PTT models using various parameters in Equation 2.13. Results were found using a finite-element code.

REFERENCES

1. Tanner, R. I., 1985, Engineering Rheology, Oxford Univ. Press.
2. Bird, R. B., R. C. Armstrong and O.Hassager, 1977, Dynamics of Polymeric Liquids, Vol. 1, Wiley, New York.
3. Hassager, O., 1985. This volume.
4. Criminale, W. O., J. L. Ericksen and G. L. Filbey, 1958, Arch. Ratl. Mech. Analysis 1, 410.
5. Reddy, K. R. and R. I. Tanner, 1978, J. Rheol. 22, 661.
6. Green, A. E. and R. S. Rivlin, 1957, Arch. Ratl. Mech. Anal. 1, 1.
7. Kaye, A., 1962, College of Aeronautics, Cranfield, Note No. 134.
8. Bernstein, B., E. A. Kearsley and L. J. Zapas, 1963, Trans. Soc. Rheol. 7, 391.
9. Malkus, D., 1985. This volume.
10. Crochet, M. J. and R. Keunings, 1982, J. Non-Newt. Fluid Mech. 10, 85.
11. Giesekus, H., 1984, Proc. IX Intl. Congr. on Rheology, Mexico 1, 39.
12. Keunings, R., 1984, Proc. IX Intl. Congr. on Rheology, Mexico, 1, 699.
13. Tanner, R. I., 1982, J. Non-Newt. Fluid Mech. 10, 169.
14. Coleman, B. D., R. J. Duffin and V. J. Mizel, 1965, Arch. Ratl. Mech. Analysis 19, 100.
15. Joseph, D. D., 1981, Arch. Ratl. Mech. Analysis 75, 251.
16. Tanner, R. I. and J. M. Simmons, 1967, Chem. Eng. Sci. 22, 1079.
17. Ho, T.-C. and M. M. Denn, 1978, J. Non-Newt. Fluid Mech. 3, 179.
18. Phan-Thien, N., 1983, J. Non-Newt. Fluid. Mech. 13, 325.
19. Pearson,J. R. A. and S. M. Richardson (eds.), 1983, Computational Analysis of Polymer Processing, Applied Sci. Publishers, London.
20. Phan-Thien, N., 1978, J. Rheology, 22, 259.
21. Leonov, A. I., 1976, Rheol. Acta. 15, 85.

22. Crochet, M. J., A. R. Davies and K. Walters, 1984, Numerical Simulation of Non-Newtonian Flow, Elsevier, Amsterdam.

I thank Dr. N. Phan-Thien and Dr. M. B. Bush for help with this research, which was supported by the Australian Research Grants Scheme. I also thank the Mathematics Research Center for hospitality during 1984.

Index

R

S

T